Tropical Forests and Global Atmospheric Change

Tropical Forests & Global Atmospheric Change

EDITED BY

Yadvinder Malhi

Oxford University Centre for the Environment, Oxford, UK

AND

Oliver L. Phillips

Earth and Biosphere Institute, School of Geography, University of Leeds, UK

Originating from a Royal Society Discussion Meeting first published in Philosophical Transactions of the Royal Society B: Biological Sciences

OXFORD
UNIVERSITY PRESS

OXFORD

UNIVERSITY PRESS

Great Clarendon Street, Oxford OX2 6DP

Oxford University Press is a department of the University of Oxford.
It furthers the University's objective of excellence in research, scholarship,
and education by publishing worldwide in

Oxford New York

Auckland Cape Town Dar es Salaam Hong Kong Karachi
Kuala Lumpur Madrid Melbourne Mexico City Nairobi
New Delhi Shanghai Taipei Toronto

With offices in

Argentina Austria Brazil Chile Czech Republic France Greece
Guatemala Hungary Italy Japan Poland Portugal Singapore
South Korea Switzerland Thailand Turkey Ukraine Vietnam

Oxford is a registered trade mark of Oxford University Press
in the UK and in certain other countries

Published in the United States
by Oxford University Press Inc., New York

British Library Cataloguing in Publication Data

Data available

Library of Congress Cataloging in Publication Data

Data available

Typeset by Newgen Imaging Systems (P) Ltd., Chennai, India
Printed in Great Britain
on acid-free paper by
Antony Rowe Ltd., Chippenham

ISBN 0 19 856705 7 (Hbk.)   978 0 19 856705 9
ISBN 0 19 856706 5 (Pbk.)   978 0 19 856706 6

10 9 8 7 6 5 4 3 2 1

Preface

We live in an age of rapid changes in the global environment, changes caused by prolific human activity and unprecedented in human history or prehistory. Amongst the most prominent of the changes is the alteration of the composition of the atmosphere, with associated consequences for the Earth's climate. Examples of this alteration include rising concentrations of CO_2 and other greenhouse gases, rising global temperatures, changing precipitation patterns, pollution hazes causing changes in sunshine, and acceleration of global nitrogen cycling. All terrestrial ecosystems are bathed in this altering atmosphere, and interact with it continuously through physiological processes such as photosynthesis, respiration and transpiration. It seems likely that these ecosystems are also shifting in structure, function and composition.

Tropical forests are the richest, and perhaps the least understood, of Earth's ecosystems. They store more biomass carbon and harbour more biodiversity than any other living system on earth, but in the early twenty-first century they are now under unprecedented threat, not only from deforestation and fragmentation, but also the more insidious threats of accelerating atmospheric and climate change. How they respond to these pressures over the next few decades has enormous implications for the kind of planet we leave to future generations. It therefore matters to all of us.

This book is aimed at all interested researchers, students, and policy makers. Most readers will have a professional or academic involvement in understanding how tropical forests work, the threats they face, and how their responses could in turn affect the global environment.

Our aim is to make a coherent attempt to tackle a very complex question: how the world's most biodiverse ecosystems might fare in an era of rapid atmospheric change. To achieve this we needed to ask original thinkers from a wide range of disciplines to focus on this common concern, using new field data or new syntheses of existing data. The editors actively solicited contributions to the book from among the world's leading authorities, and asked our contributors to produce focussed, concise contributions to provide state-of-the-art understanding. Many of these contributions are based partly on a symposium held at the Association for Tropical Biology annual meeting in Panama City, Panama, in July 2002, or on a set of more detailed, discursive articles published in March 2004 in the Philosophical Transactions of the Royal Society of London, while other chapters are entirely new to this volume. Overall this book builds on contemporary advances in several distinct scientific disciplines, all key to our larger question.

Among *atmospheric scientists*, the rapid rate of change was first highlighted by the long-term observations of CO_2 concentrations at Mauna Loa, Hawaii, starting in the late 1950s. These observations were a harbinger of the realization that human industrial and economic activities are now of sufficient scale to be altering fundamental properties of the atmosphere at regional and global scales. Anthropogenic atmospheric change will certainly become more significant through the century. Atmospheric CO_2 concentrations are likely to reach levels unprecedented for at least 20 million years, and nitrogen-deposition rates and climates are predicted to move far beyond those experienced for millions of years. In **Part I (Contemporary atmospheric change in the tropics)** Malhi and Wright show how the climate of tropical forest regions has changed over recent decades, while Cramer *et al.* use a comparison of sophisticated vegetation and climate models to assess the potential impacts of future CO_2 increase, climate change, and deforestation on tropical forests. Laurance reminds us that climate change impacts will be played out against a background of ongoing tropical deforestation, and explores forestclimate interactions in fragmented tropical landscapes.

Anticipated climatic and atmospheric change led to a suite of *physiological studies* on tree species in the 1980s and 1990s aimed at examining the response of trees to rising CO_2 and/or temperatures. These graduated from laboratory studies on saplings to free air CO_2 enrichment of forest stands or soil warming experiments in some biomes, although ecophysiological studies of tropical forests so far have been limited, and there have been no free air experiments in mature tropical forests. In **Part II (Atmospheric change and ecosystem processes)** evidence from the ecophysiological literature is reviewed and distilled (with sometimes differing viewpoints) by Chambers & Silver, Körner, and Lewis *et al.* Changes in water availability have potentially the most drastic impacts on tropical forest viability, and many climate models anticipate some increase in moisture stress for tropical forests. Meir and Grace tackle the important issue of how humid tropical ecosystems respond to drought stress.

Understanding of tropical ecosystems tends to lag behind that of temperate and boreal systems for two reasons: (i) most research funding is focused on temperate latitudes; and (ii) the high biodiversity of most tropical ecosystems limits the ability to generalize from experimental studies of a few species. Given these constraints, particularly useful insights can come from *in-field monitoring* of tropical ecosystems. In the 1990s, synthesis of such monitoring of old-growth forests has yielded disturbing evidence of rapid contemporary change, including acceleration in tree death and growth, increases in biomass, and shifts in ecosystem structure. Large lianas are becoming more dominant in western Amazonia, and in a central Amazonian landscape canopy tree species have

gained at the expense of shade-tolerant subcanopy trees. In **Part III (Observations of contemporary change in tropical forests)** overviews of these results, and potential explanations, are presented by Baker *et al.*, Phillips *et al.*, Lewis *et al.*, and Laurance. Sampling issues are explored by Chave *et al.* Barlow and Peres describe how Amazon forests respond to fire, an increasing threat due to land-use changes and likely increases in moisture stress.

We may gain further insights into the resilience or vulnerability of tropical forests to climate change by examining the paleoecology of tropical forests under past climates. In **Part IV (The past and future of tropical forests)** interactions between Amazonian forest and climate are examined over a time scale ranging back from the origin of flowering plants to the end of the 21st century. Maslin takes the really long view, and reviews the extent and composition of tropical broadleaved forests in the Americas over more than 65 million years. On a more recent timescale, Mayle and Bush present a comprehensive review of 20,000 years of Amazonian climate and vegetation extent. There is now agreement that tropical regions were substantially cooler and generally drier as recently as the last glacial maximum. This led to the suggestion that large areas of tropical forest retreated into refugia surrounded by savanna. While evidence for this retreat is stronger for Africa, for Amazonia there has been vigorous debate as to the extent of this forest retreat. Resolving the debate is crucial to understanding the sensitivity and resilience of tropical forests to climate change. Computer models can provide a process-level understanding of how forests change in response to atmospheric change, and can be tested against paleoecological evidence before being applied to future climates. Cowling *et al.* employ a physiological vegetation model to explore the nature of Amazonian vegetation at specific periods in recent Earth history and then apply this model to a projected late 21st century climate. An alternative perspective on this debate is developed by Pennington *et al.*, who use genetic approaches to explore the times of speciation of tree species in seasonally dry tropical forests.

The chapters in this book tackle the subject of atmospheric change in tropical forests from a wide variety of approaches. In the final chapter, we draw these stands together and present a personal interpretation of the prospects for tropical forests in the 21st century atmosphere, and what the priorities should be for both research and policy in coming decades.

We sincerely thank those without whom this book would not have been possible: William Laurance for originally inviting us to prepare a symposium, Joe Wright and his colleagues who organized the ATB meeting in Panama, the Royal Society for helping us to develop a special theme issue, and Oxford University Press staff, especially Ian Sherman, Abbie Headon and Anita Petrie for their encouragement in seeing this project through. In particular, we thank the authors for their original contributions, and the numerous referees who ensured that the manuscripts were of the highest standards.

We hope the final product is worthy of our collective effort.

Oliver L. Phillips
Yadvinder Malhi
March 2005

Contents

Contributors

Salomon Aguilar, Center for Tropical Forest Science, Smithsonian Tropical Research Institute, Unit 0948, APO AA 34002–0948, USA.

Miguel Alexiades, Department of Anthropology, University of Kent, UK.

Samuel Almeida, Museu Paraense Emilio Goeldi, 66077–530 Belém, Brazil.

Ana Andrade, Biological Dynamics of Forest Fragments Project, National Institute for Amazonian Research (INPA), C.P. 478, Manaus, AM 69011–970, Brazil.

Luzmila Arroyo, Museo Noel Kempff Mercado, Santa Cruz, Bolivia; Missouri Botanical Garden, St. Louis, USA.

Timothy R. Baker, Max-Planck-Institut für Biogeochemie, Postfach 100164, 07701 Jena, Germany; Earth and Biosphere Institute, School of Geography, University of Leeds, Leeds LS2 9JT, UK.

Jos Barlow, Centre for Ecology, Evolution and Conservation, School of Environmental Sciences, University of East Anglia, Norwich NR4 7TJ, UK.

Richard A. Betts, Meteorological Office, Hadley Centre for Climate Prediction and Research, Exeter, UK.

Alberte Bondeau, Potsdam Institute for Climate Impact Research, Department of Global Change and Natural Systems, PO Box 60 12 03, D-14412 Potsdam, Germany.

Sandra Brown, Winrock International, 1621 North Kent Street, Suite 1200, Arlington, VA 22209, USA.

Mark B. Bush, Department of Biological Sciences, Florida Institute of Technology, 150 W. University Blvd, Melbourne, Florida 32901, USA.

Jeffrey Q. Chambers, Tulane University, Ecology and Evolutionary Biology, New Orleans, LA 70118, USA.

Jerome Chave, Laboratoire Evolution et Diversité Biologique UMR 5174 CNRS/UPS, bâtiment 4R3, 118 route de Narbonne F-31062 Toulouse, France.

Guillem Chust, Laboratoire Evolution et Diversité Biologique UMR 5174 CNRS/UPS, bâtiment 4R3, 118 route de Narbonne F-31062 Toulouse, France.

James A. Comiskey, Smithsonian Institution, Washington DC, USA.

Richard Condit, Center for Tropical Forest Science, Smithsonian Tropical Research Institute, Unit 0948, APO AA 34002–0948, USA; Smithsonian Tropical Research Institute, Apartado 2072, Balboa, Republic of Panamá.

Sharon A. Cowling, Department of Geography, University of Toronto, 100 St. George Street, Toronto, Ontario, M5S 3G3, Canada.

Peter M. Cox, Meteorological Office, Hadley Centre for Climate Prediction and Research, London Road, Bracknell, Berkshire, RG12 2SY, UK.

Wolfgang Cramer, Potsdam Institute for Climate Impact Research, Department of Global Change and Natural Systems, PO Box 60 12 03, D-14412 Potsdam, Germany; Institute of Geoecology, Potsdam University, PO Box 60 15 53, D-14415 Potsdam, Germany.

Claudia I. Czimczik, Max-Planck-Institut für Biogeochemie, Jena, Germany; University of California Davis, USA.

Anthony Di Fiore, Department of Anthropology, New York University, New York, USA.

Christopher W. Dick, Smithsonian Tropical Research Institute, Apartado 2072, Balboa, Republic of Panamá.

Terry Erwin, Smithsonian Institution, Washington DC, USA.

Virginia J. Ettwein, Department of Geography, University College London, 26 Bedford Way, London, WC1H 0AP, UK.

John Grace, School of GeoSciences, University of Edinburgh, Edinburgh EH9 3JU, UK.

Andres Hernandez, Center for Tropical Forest Science, Smithsonian Tropical Research Institute, Unit 0948, APO AA 34002–0948, USA.

Niro Higuchi, Instituto Nacional de Pesquisas Amazônicas, 69011–970 Manaus, Brazil.

Chris D. Jones, Meteorological Office, Hadley Centre for Climate Prediction and Research, Exeter, UK.

Timothy Killeen, Museo Noel Kempff Mercado, Santa Cruz, Bolivia; Center for Applied Biodiversity Science, Conservation International, Washington DC, USA.

Christian Körner, Institute of Botany, University of Basel, Schonbeinstrasse 6, CH-4056 Basel, Switzerland.

Caroline Kuebler, Center for Applied Biodiversity Science, Conservation International, 1919 M Street, NW, Suite 600, Washington, DC 20036, USA.

Suzanne Lao, Center for Tropical Forest Science, Smithsonian Tropical Research Institute, Unit 0948, APO AA 34002–0948, USA.

Susan G. Laurance, Smithsonian Tropical Research Institute, Apartado 2072, Balboa, Republic of Panamá.

William F. Laurance, Smithsonian Tropical Research Institute, Apartado 2072, Balboa, Republic of Panamá; Biological Dynamics of Forest Fragments Project, National Institute for Amazonian Research (INPA), C.P. 478, Manaus, AM 69011–970, Brazil.

Matt Lavin, Department of Plant Sciences, Montana State University, Bozeman, MT 59717, USA.

Simon L. Lewis, Earth and Biosphere Institute, School of Geography, University of Leeds, Leeds LS2 9JT, UK.

Jon Lloyd, Max-Planck-Institut für Biogeochemie, Postfach 100164, 07701 Jena, Germany; Earth and Biosphere Institute, School of Geography, University of Leeds, Leeds LS2 9JT, UK.

Thomas E. Lovejoy, Biological Dynamics of Forest Fragments Project, National Institute for Amazonian Research (INPA), C.P. 478, Manaus, AM 69011–970, Brazil.

Wolfgang Lucht, Potsdam Institute for Climate Impact Research, Department of Global Change and Natural Systems, PO Box 60 12 03, D-14412 Potsdam, Germany.

Yadvinder Malhi, School of Geography and the Environment, University of Oxford, Oxford OX1 3TB, UK; School of GeoSciences, University of Edinburgh, Edinburgh EH9 3JU, UK.

Mark Maslin, Environmental Change Research Centre, Department of Geography, University College London, UK.

Francis E. Mayle, Institute of Geography, School of GeoSciences, University of Edinburgh, Drummond Street, Edinburgh EH8 9XP, UK.

Patrick Meir, School of GeoSciences, University of Edinburgh, Edinburgh EH9 3JU, UK.

Abel Monteagudo, Herbario Vargas, Universidad Nacional San Antonio Abad del Cusco, Cusco, Peru; Proyecto Flora del Perú, Jardin Botanico de Missouri, Oxapampa, Perú.

Henrique E. M. Nascimento, Biological Dynamics of Forest Fragments Project, National Institute for

Amazonian Research (INPA), C.P. 478, Manaus, AM 69011–970, Brazil.

David A. Neill, Missouri Botanical Garden, c/o Herbario Nacional del Ecuador, Quito, Ecuador.

Percy Núñez Vargas, Herbario Vargas, Universidad Nacional San Antonio Abad del Cusco, Cusco, Peru.

Alexandre A. Oliveira, Department of Biology, University of São Paulo, Avenida Bandeirantes 3900, Ribeirão Preto-São Paulo, SP 14040–901, Brazil.

Jean Olivier, Laboratoire Evolution et Diversité Biologique, CNRS/UPS Toulouse, France.

Walter Palacios, Fundacion Jatun Sacha, Quito, Ecuador.

Sandra Patiño, Max-Planck-Institut für Biogeochemie, Jena, Germany; Alexander von Humboldt Biological Research Institute, Bogotá, Colombia.

Susan K. Pell, Lewis B. and Dorothy Cullman Program for Molecular Systematics Studies, New York Botanical Garden, Bronx, New York, 10458 USA.

Colin A. Pendry, Royal Botanic Garden Edinburgh, 20a Inverleith Row, Edinburgh EH3 5LR, UK.

R. Toby Pennington, Royal Botanic Garden Edinburgh, 20a Inverleith Row, Edinburgh EH3 5LR, UK.

Carlos A. Peres, Centre for Ecology, Evolution and Conservation, School of Environmental Sciences, University of East Anglia, Norwich NR4 7TJ, UK.

Rolando Perez, Center for Tropical Forest Science, Smithsonian Tropical Research Institute, Unit 0948, APO AA 34002-0948, USA.

Oliver L. Phillips, Earth and Biosphere Institute, School of Geography, University of Leeds, Leeds LS2 9JT, UK.

Nigel Pitman, Center for Tropical Conservation, Duke University, Durham, USA.

Darién Prado, Cátedra de Botánica, Facultad de Ciencias Agrarias, Universidad Nacional de Rosario, S2125ZAA, Zavalla, Argentina.

Carlos A. Quesada, Earth and Biosphere Institute, Geography, University of Leeds, UK; Departamento de Ecología, Universidade de Brasilia, Brazil.

José E. L. S. Ribeiro, Department of Botany, National Institute for Amazonian Research (INPA), C.P. 478, Manaus, AM 69011–970, Brazil.

Mario Saldias, Museo Noel Kempff Mercado, Santa Cruz, Bolivia.

Ana C. Sanchez-Thorin, Smithsonian Tropical Research Institute, Apartado 2072, Balboa, Republic of Panamá.

Sibyll Schaphoff, Potsdam Institute for Climate Impact Research, Department of Global Change and

Natural Systems, PO Box 60 12 03, D-14412 Potsdam, Germany.

J. Natalino M. Silva, CIFOR, Tapajos, Brazil; EMBRAPA Amazonia Oriental, Belém, Brazil.

Whendee L. Silver, Division of Ecosystem Science, University of California, Berkeley, CA 94720.

Stephen Sitch, Joint Centre for Hydro-Meteorological Research, Centre for Ecology and Hydrology, Maclean Building, Crowmarsh Gifford, Wallingford OX10 8BB, UK.

Ben Smith, Department of Physical Geography and Ecosystems Analysis, University of Lund, Sölvegatan 13, S-22362 Lund, Sweden.

Steven A. Spall, Meteorological Office, Hadley Centre for Climate Prediction and Research, London Road, Bracknell, Berkshire, RG12 2SY, UK.

Armando Torres Lezama, INDEFOR, Universidad de Los Andes, Mérida, Venezuela.

John Terborgh, Center for Tropical Conservation, Duke University, Box 90381, Durham, NC 27708, USA.

Barbara Vinceti, School of GeoSciences, University of Edinburgh, Edinburgh EH9 3JU, UK; International Plant Genetic Resources Institute, Via dei Tre Denari 472/a, 00057 Maccarese (Fiumicino), Rome, Italy.

Rodolfo Vásquez Martínez, Proyecto Flora del Perú, Jardin Botanico de Missouri, Oxapampa, Perú.

James Wright, Department of Geography, University of Southampton, UK.

PART I

Contemporary atmospheric change in the tropics

Late twentieth-century patterns and trends in the climate of tropical forest regions

Yadvinder Malhi and James Wright

We present an analysis of the mean climate and climatic trends of tropical rainforest regions over the period 1960–98, with the aid of climatological databases. Up until the mid-1970s most regions showed little trend in temperature, and the western Amazon experienced a net cooling probably associated with an interdecadal oscillation. Since the mid-1970s all tropical rainforest regions have experienced a strong warming at a mean rate of $0.26 \pm 0.05°C$ per decade, in synchrony with a global rise in temperature that has been attributed to the anthropogenic greenhouse effect. Over the study period precipitation appears to have declined sharply in northern tropical Africa (at 3–4% per decade), declined marginally in tropical Asia, and showed no significant trend in Amazonia. There is no evidence to date of a decline in precipitation in eastern Amazonia, a region thought vulnerable to climate-change induced drying. The strong drying trend in Africa suggests that this should be a priority study region for understanding the impact of drought on tropical rainforests. Only African and Indian tropical rainforests appear to have seen a significant increase in dry season intensity. The El Niño-Southern Oscillation is the primary driver of interannual temperature variations across the tropics, and of precipitation fluctuations for large areas of the Americas and Southeast Asia.

Introduction

The global atmosphere is undergoing a period of rapid human-driven change, with no historical precedent in either its rate of change, or its potential absolute magnitude (IPCC 2001). There is considerable concern at how this change may affect Earth's ecosystems, and in turn how these system responses may feed back to accelerate or decelerate global change. This threat is spawning considerable laboratory and field research into ecosystem responses to climate. This research gains its justification from the global change agenda. However, there is a gap between global context and local research that can sometimes feel unsatisfactory.

On one hand, global warming trends over the twentieth century have been examined compre-

hensively at global or regional levels (e.g. Jones et al. 1999; New et al. 2001; Giorgi 2002), particularly through the work of the IPCC (e.g. Folland et al. 2002). Among other issues, this has prompted concern at how ecosystems may respond to such trends. Yet there have been few attempts to examine climate trends at the level of specific biomes, and in particular the tropical rainforest biome.

On the other hand, there has recently been a surge of research aimed at understanding the interactions between climate change and tropical rainforest ecology and function, which is reflected in several of the chapters in this book. However, in their justification of their research agenda, tropical forest researchers generally have had to rely on broad or global climatic trends reported by the

IPCC, or generalities about warming or drying climates. But what are the recent trends in the climate of the region or biome being studied, and how do these trends compare with general trends in tropical rainforest regions, and with projections from global climate models?

In Malhi and Wright (2004), we presented a detailed analysis of spatial patterns and recent trends in the world's tropical forest regions since the 1960s. In this chapter, we present a summary of the results presented there, with updates where new studies and literature have appeared. For colour maps of the patterns described the reader is referred to the original paper.

We use recent global climatic datasets to examine the spatial variability of tropical rainforest climates. In public imagination, the climate of the tropical rainforest is frequently characterized as 'ever-warm and ever-wet'. In reality, however, although the 'ever-warm' is generally true (but not always), tropical rainforest regions can span a wide range of intensity and seasonality of rainfall. Moreover, many regions can experience significant interannual, interdecadal, and perhaps intercentennial variability in rainfall, and the regularity and intensity of drought can have a major impact on forest structure and adaptation. In the face of this spatial and temporal heterogeneity, there can be a danger of extrapolating too much from intensive studies of tropical rainforest in one locale; thus a global synthesis of tropical rainforest climate helps to put the climate of any particular site into context. The new global observational climatologies can be a powerful tool with which to address questions in tropical rainforest ecology.

We focus on three climatic parameters: temperature, precipitation, and dry season intensity. There are other atmospheric parameters that are varying and may have a direct effect on tropical forests, the principal examples being carbon dioxide concentrations, direct and diffuse solar radiation, and nitrogen deposition. Their possible influences are reviewed by Lewis, Malhi, and Phillips in Chapter 4, this volume. We also examine in particular detail the role of the El Niño-Southern Oscillation, which is the primary driver of interannual climate variability in the tropics.

In this chapter we address the following questions:

1. What is the spatial heterogeneity of the contemporary climate of tropical rainforest regions, and the representativeness of particular study regions?

2. How is interannual variability in temperature, precipitation, and drought stress in different regions related to the El-Niño-Southern Oscillation, the primary driver of present-day climatic variability in the tropics?

3. What are the trends in temperature, precipitation, and drought stress in the last 40 years, and how do we interpret these trends in terms of possible artefacts, longer-term oscillations, and net anthropogenic trends?

Methods

Before embarking on this analysis, we emphasize one important caveat: the ocean–atmosphere system may exhibit a number of multi-decadal internal oscillations, and trends observed over a 39-year period may not reflect longer-term variations. Hence caution must be applied in attributing these short-term changes to a longer-term trend caused by anthropogenic influences. Malhi and Wright (2004) illustrated this feature by demonstrating a century-long record of precipitation from Manaus in central Amazonia. In this record there was an evidence of a multidecadal oscillation in precipitation, which resulted in the precipitation trend in the last 40 years ($-0.5 \pm 5.5\%$ per decade) being different from the century-long trend ($2.6 \pm 1.2\%$ per decade).

The datasets and analysis procedures used in this chapter have been described in Malhi and Wright (2004) and this description is not repeated here. However, we do review briefly our definition of a dry season index. In most tropical terrestrial ecosystems, the most evident climatic factor is the duration and intensity of the dry season. The mean evapotranspiration rate (E) of a fully wet tropical rainforest is about 100 mm per month (Shuttleworth 1989; Malhi et al. 2002a); hence a common definition of dry season is when the precipitation (P) is less than 100 mm per month (i.e. when the forest is in net water deficit), and a common parameter for dry season length is the number of months per year with $P < 100$ mm. However, this does not capture the importance of degrees of intensity: a dry season where P drops to 10 mm per month is more severe than one where it hovers at 90 mm per month. Similarly, many regions (particularly in Africa) experience a 'double-dip' pattern in rainfall, with two short dry seasons, which are less severe than one long dry season of equivalent total duration. Similarly, the dry season length does not capture the importance of soil 'memory'— that is, whether a soil is fully hydrated or only partially hydrated in the months preceding a dry month.

We developed a dry season index (DSI) in Malhi and Wright (2004) to represent the strength and duration of the dry season, calculated from precipitation data using a simple water balance model that incorporates seasonal variations of precipitation and field-calibrated evapotranspiration. The DSI as defined here does not take into account seasonal variations and long-term trends in temperature and solar radiation, except as implicitly correlated terms for the particular central Amazonian calibration site. Any net warming trend would be expected to increase actual water stress. Hence our DSI should be thought of as a descriptor

of dry season length and intensity, rather than a measure of water stress.

To maintain simplicity in the summary charts we have divided the tropical rainforest area into sub-regions (not always contiguous), which are shown in Fig. 1.1. The names ascribed to each region are convenient geographical terms, rather than referring to political entities. The climate results were overlaid with masks of tropical forests cover which were derived from FAO Global Forest Resources Assessment 2000 (FAO Forestry Paper 140, available online at www.fao.org/forestry/fo/fra) and coarsened to 0.5° resolution to match the climate data.

The mean climate of tropical rainforest regions

Globally, the mean annual precipitation of the tropical rainforest region is 2180 mm, the mean temperature is 25.4°C, and the mean insolation is 16.5 GJ yr^{-1} (Table 1.1). There is considerable variation in precipitation around this mean, with east Malesia and north-west Amazonia the wettest tropical rainforest regions with about 3000 mm of rain per year, and almost all of Africa much drier than the mean. There is less variation in mean temperature and solar radiation, although the

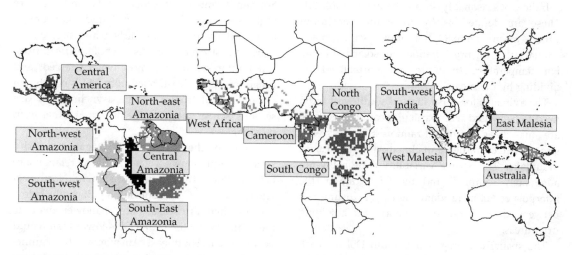

Figure 1.1 A map of the tropical rainforest sub-regions allocated for this analysis. Names of sub-regions are convenient geographical descriptors rather than reflecting political identities. (See Plate 1(a–c)).

Table 1.1 The mean climate of tropical rainforest regions over the period 1960–98 (1960–95 for solar radiation)

Region	Mean			Seasonal Variation			Dry Season		SD Peak DSI (mm)
	P (mm)	S (GJ yr^{-1})	T (°C)	P Fractional	S Fractional	T (°C)	Length (months)	Peak DSI (mm)	
Central America	2206	17.3	24.0	2.4	0.42	4.4	4.7	126	15.2
North-west Amazonia	2962	15.6	25.9	1.2	0.13	2.3	0.6	15	13.9
South-west Amazonia	2194	16.3	25.7	1.8	0.17	3.2	3.3	89	25.1
Central Amazonia	2420	15.6	26.2	1.6	0.23	2.4	2.3	56	15.2
North-east Amazonia	2260	16.6	26.0	2.0	0.26	2.2	3.7	82	26.7
South-east Amazonia	2103	16.1	25.5	2.3	0.31	2.9	4.5	149	21.8
West Africa	1601	16.6	26.3	2.7	0.28	4.6	5.8	165	14.4
Cameroon	1782	15.2	24.6	2.1	0.18	3.0	4.3	82	13.5
North Congo	1689	17.6	24.5	1.7	0.14	2.5	3.6	104	21.4
South Congo	1530	16.5	23.8	1.8	0.26	2.7	4.3	118	12.1
South-west India	1993	18.3	25.9	3.2	0.32	4.0	5.9	148	22.7
West Malesia	2584	17.1	25.3	2.1	0.31	3.2	2.9	72	9.2
East Malesia	3094	16.7	25.4	1.4	0.18	1.6	1.2	23	15.3
Australia	1702	20.1	24.2	3.5	0.22	6.7	7.0	193	24.1
Pantropical mean	2178	16.5	25.4	2.0	0.24	3.2	3.7	—	—

Notes: P = precipitation, S = solar radiation, T = temperature. Seasonal variation in precipitation and temperature is normalized by dividing by annual mean values. SD Peak is the standard deviation of the annual peak DSI values.

equatorial African forests are generally cooler by virtue of being at high elevation, and outer tropical rainforests tend to have slightly cooler annual means. For a given dry season regime, these cooler forests suffer less water stress.

Indices of seasonality are also shown in Table 1.1. These are defined as (mean annual maximum value – mean annual minimum value)/mean annual value for precipitation and solar radiation. For temperature they are not normalized by dividing by the mean.

The estimates for Australia and south-west India need to be treated with particular caution, as these are small areas of tropical rainforest in areas with strong spatial gradients in climate associated with mountains. This spatial variability is probably not adequately sampled, and the 0.5° scale of the interpolated climate data may not distinguish between tropical rainforest areas and adjoining drier areas.

The spatially averaged maximum DSI for each region is also shown in Table 1.1. As would be expected, the outlying tropical rainforest regions show the greatest seasonality and dry season intensity (Australia, west Africa, south-west India, Central America), and the 'core' tropical regions the weakest dry seasons (north-west and central Amazonia, all of Malesia). However, there are significant longitudinal trends, and the core regions of tropical Africa (north and south Congo and Cameroon) are still fairly water-stressed, as is north-east Amazonia. As a whole, African tropical rainforests are the most water-stressed, and Asian forests the least. Maximum DSI is not shown for the global tropics, as the asynchrony of dry seasons across such a large area make this term meaningless.

A plot of the maximum DSI against the dry season length shows a general linear relationship between these two terms (not shown). North-east Amazonia and Cameroon fall significantly below this line; here dry seasons are 'shallower' than the mean and hence the peak DSI is lower than would be expected. South-east Amazonia falls significantly above this line, and experiences more intense dry seasons than the mean.

Variation in temperature 1960–98

Interannual variability in temperature

The pantropical temperature anomaly for the tropical rainforest regions for the period 1960–98 is shown in Fig. 1.2. The multivariate El Niño index is also shown in the same figure. Over this time-scale, the interannual variability of temperature in tropical rainforest regions is clearly greater than any net trend. This variation clearly shows a strong correlation with the ENSO, with mean temperatures about 1°C higher during El Niño events

(positive ENSO index) than during La Niñas. The Pearson correlation between mean temperature and the multivariate ENSO index for various lag times is shown in Fig. 1.3. Globally, the mean temperature of tropical rainforests lags behind the ENSO index with a mean lag time of two months, and a correlation coefficient $r = 0.64$. The correlation is greatest in Southeast Asia ($r = 0.64$, lag = 4 months), intermediate in Africa ($r = 0.57$, lag = 4 months), and least in the Americas ($r = 0.52$, lag = 1 month). Considering sub-regions, the greatest correlations are in western Malaysia ($r = 0.62$, lag = 3 months),

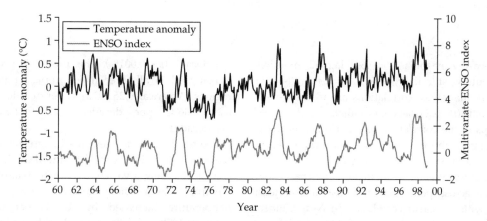

Figure 1.2 Time series of pantropical temperature anomaly relative to the period 1960–98 and the multivariate ENSO index.

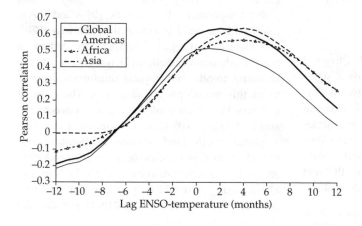

Figure 1.3 The cross-correlation function between mean temperature in each tropical rainforest continent and ENSO.

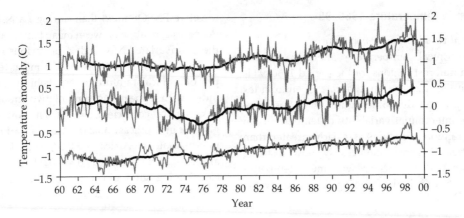

Figure 1.4 Time series of the tropical rainforest temperature anomaly (centre line), compared with the global (land and sea) temperature anomalies for the same period for the Northern (top line) and Southern (bottom line) Hemispheres. The thicker lines are 4-year moving averages.

north-east Amazonia ($r = 0.60$, lag = 2 months), and south-west India ($r = 0.60$, lag = 2 months) (not shown); the lowest correlations are in south-west Amazonia ($r = 0.24$, lag = 0 months).

Maps of the correlation between deseasonalized temperature and the ENSO index are presented in Malhi and Wright (2004). In Africa the strongest temperature influence is felt in the Congo basin, and in Asia across peninsular and insular Southeast Asia, with a weaker correlation in New Guinea. The Americas show the most striking spatial gradients, with a strong ENSO influence on temperatures in northern and eastern Amazonia, and in Central America, and almost no ENSO influence in south-west Amazonia. No tropical rainforest region is cooler during El Niño events.

Pantropical trends in temperature

A simple linear trend through the global tropical rainforest temperature time series yields a significant ($p < 1\%$) temperature trend of $+0.08 \pm 0.03°C$ per decade (i.e. a net increase of $0.31°C$ between 1960 and 1998).[1] The pantropical mean time series shows a net cooling during the period from 1960 to 1974 ($-0.08 \pm 0.11°C$ per decade) and a subsequent strong warming; since 1976 the net warming has been at a rate of $0.26 \pm 0.05°C$

per decade ($p < 0.001\%$). This post-1976 warming is observed in all three major tropical rainforest regions (Americas $0.26 \pm 0.07°C$ per decade, Africa $0.29 \pm 0.06°C$ per decade, Asia $0.22 \pm 0.04°C$ per decade).

The results reported here are in agreement with the general trends observed by the IPCC (Folland *et al.* 2002) using a variety of climate datasets. For the tropics as a whole (land and seas), surface temperature increased by $0.08°C$ per decade between 1958 and 2000, a trend with can be divided into a cooling period from 1958 to 1978 ($-0.09 \pm 0.12°C$ per decade), and a warming period since 1978 ($+0.10 \pm 0.10°C$ per decade). The pattern of relatively invariant or slightly cooling global temperatures between the 1940s and 1970s, followed by a rapid warming since the mid-1970s, is a feature of the global temperature pattern in both Northern and Southern Hemispheres, and the warming trend over tropical rainforests appears to fit this global pattern (Fig. 1.4). The rates of warming in tropical rainforest regions between 1976 and 1998 ($0.26 \pm 0.05°C$ per decade) compares with a global mean land surface temperature rise of $0.22 \pm 0.08°C$ per decade between 1976 and 2000, and lie between the rates observed for the Northern Hemisphere ($0.31 \pm 11°C$ per decade) and the Southern Hemisphere ($0.13 \pm 0.08°C$ per decade) as a whole (Jones *et al.* 2001).

---

[1] All significant values calculated in this manuscript are estimated from the two-tailed Mann–Kendall rank statistic.

Regional trends in temperature

The mean temperature trends in each region are summarized in Fig. 1.5, for the overall period (Fig. 1.5(a)), and the period 1976–98 (Fig. 1.5(b)). All regions showed a net warming over the overall period except central and western Amazonia, and all regions showed strong warming since 1976 of between 0.15 and 0.4°C per decade. The cooling in western Amazonia occurred in the early 1970s and was followed by steady warming, and it is this cooling that causes the dip in global mean tropical rainforest temperatures over the same period. Close examination of individual station time series shows that this cooling–warming in western Amazon is real and not an artefact of station switching (Malhi and Wright 2004). The overall positive trend is partially influenced by positive temperature anomalies of 1.0–1.5°C during the 1983 and 1998 El Niño events, but the trend is only reduced by 0.03°C per decade when these peaks are removed.

Victoria *et al.* (1998) conducted an analysis of 17 stations in the Brazilian Amazon from 1913

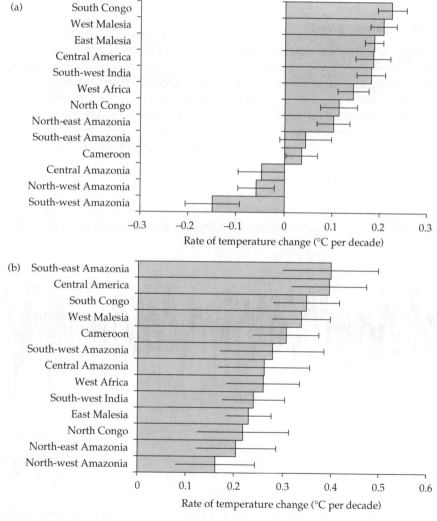

Figure 1.5 Temperature trends in each tropical rainforest sub-region (a) for the period 1960–98; (b) for the period 1976–98.

to 1995, taking care to eliminate stations with discontinuities in methodologies. They report little variation in the period 1910–40, a decade-long dip by 0.3°C in the 1950s, and a period of rapid warming since the mid-1970s at a rate of about 0.25°C per decade. New *et al.* (2001) analysed the temperature trends for the whole African continent over the twentieth century, and observed a warming up to the1940s, a slight cooling up to the mid-1970s, and a rise of about 0.8°C since then. Hence the overall pattern of rapid and pantropical recent warming is confirmed, but in the longer-term trends there can be significant regional differences.

Variations in precipitation 1960–98

The mean monthly precipitation anomaly over tropical rainforest regions for the period 1960–98 is shown in Fig. 1.6, calculated using the same procedure as for temperature. Compared to temperature, the precipitation record is much more variable, both spatially and temporally.

Interannual variability in precipitation
As with the temperature time series, the interannual variability of precipitation in tropical rainforest regions is greater than any net trend. Globally, this variation shows a reasonably strong

inverse correlation with the ENSO, with less precipitation during El Niño events. This reduction in precipitation is principally caused by the warming of the eastern Pacific causing enhanced convection in this region and compensatory zones of air subsidence in northern South America and Southeast Asia, which suppress rainfall in these normally highly convective regions. The correlation between the precipitation anomaly and the ENSO index for various lag times is shown in Fig. 1.7. Globally, the mean precipitation of tropical rainforests does not lag behind the ENSO index (lag time 0 or −1 months) and the correlation is weaker than that for temperature ($r = -0.44$, compared to $r = 0.64$ for temperature). The correlation is greatest in Southeast Asia ($r = -0.48$, lag $= -3$ months), intermediate in the Americas ($r = -0.35$, lag $= 0$ months), and very weak for Africa ($r = -0.12$, no distinctive lag). The comparison with temperature in Africa is particularly noteworthy: although African forests are consistently warmer during El Niño events, they are not generally much drier; hence the higher temperatures are not induced by local drought, but perhaps by more general atmospheric and oceanic changes, such as increases in surface temperatures in the Indian and Atlantic Oceans. In almost every tropical rainforest region, the peak changes in precipitation precede

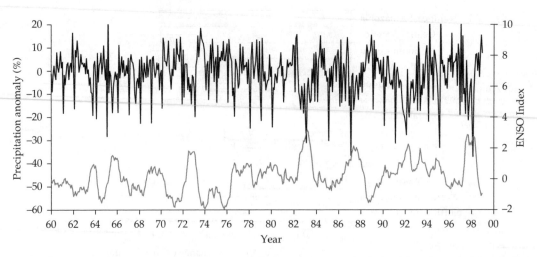

Figure 1.6 Monthly time series of pantropical precipitation anomaly for the period 1960–98 (top line) and the multivariate ENSO index (bottom line). The percentage is calculated relative to the pantropical mean precipitation for tropical rainforests (181.5 mm per month, or 2177 mm yr^{-1}).

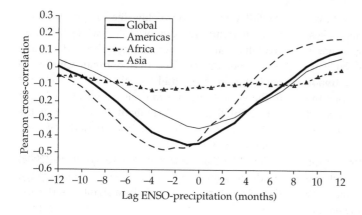

Figure 1.7 Cross-correlation function of the precipitation anomaly and the ENSO index for the three tropical rainforest continents.

changes in temperature, and often also precede peaks in the ENSO index.

Maps of the ENSO correlation with deseasonalized precipitation have been shown in Plate 1. Throughout Africa the correlation is negative but weak. It appears that some El Niños do significantly affect tropical Africa, but the teleconnection from the tropical Pacific is variable, and appears to be dependent on the exact seasonal timing of the El Niño, and the extent to which it influences Indian and Atlantic Ocean temperatures (Nicholson *et al.* 2000). In contrast, in both the Americas and Asia many tropical rainforest regions show a strong response to ENSO. The largest inverse correlations (reduced precipitation during El Niños) are seen in north-eastern Amazonia and in a band running from Borneo to New Guinea. In contrast, many regions at the dry fringe of the tropical rainforest belt (southern and northern tips of Amazonia, northern Central America, northern fringe of Congo, parts of Southeast Asia) show a weak positive correlation, with increased precipitation during El Niño events.

Trends in precipitation

A simple linear trend through the pantropical anomaly time series yields a decline in annual precipitation rates of -22 ± 17 mm per decade, or $-1.0 \pm 0.8\%$ per decade ($p < 0.05$). This is a net decrease of about 86 mm between 1960 and 1998. There is little pantropical trend between 1960 and the mid-1970s, and a more marked decline since then. Because of the high interannual variability, the

trends are generally of low significance. It must be emphasized that the trend over four decades may not necessarily be indicative of a longer-term trend.

In the three tropical rainforest continents, there is no overall significant trend in precipitation in the American tropical rainforests ($-0.6 \pm 1.1\%$ per decade), a moderately significant drying trend in Asia ($-1.0 \pm 1.1\%$ per decade, $p < 0.05$), and a strong drying trend in Africa ($-2.4 \pm 1.3\%$ per decade, $p < 0.0001$). It is this strong drying trend in Africa that drives the overall pantropical decline.

A map of spatial variations in the linear trends is presented in Malhi and Wright (2004). Most tropical regions show a drying trend over this period, with the exception of north-east Amazonia and Central America. However, at a regional level the interannual variation is non-significant in many regions. The only significant trends are in eastern Malesia ($-1.0 \pm 1.4\%$ per decade; $p < 0.05$), much of Africa (West Africa $-4.2 \pm 1.2\%$ per decade ($p < 10^{-5}$); north Congo $-3.2 \pm 2.2\%$ per decade ($p < 0.05$), south Congo $-2.2 \pm 1.8\%$ per decade ($p < 0.1$)), and south-west India $-3.5 \pm 2.9\%$ per decade ($p < 0.05$). One noteworthy feature is that the drying trends tend to be strongest in regions which are least directly affected by El Niño events, whereas the trends are small or even reversed in some of the El Niño susceptible regions (Malesia, north-east Amazonia). A highly significant drying trend in north-west Amazonia is most likely a station-switching artefact (Malhi and Wright 2004) and is not shown here.

In Asia, a modest drying trend is observed over the entire period, with few regional anomalies, again suggesting that station switching is not a major problem. Because of the interannual variability, in most areas of Africa and Asia the trends are not significant at sub-regional level, but the broad spatial consistency makes the overall trend marginally significant for Asia ($p < 0.05$), and highly significant for Africa ($p < 10^{-4}$).

The most noteworthy feature in the precipitation results is the strong drying trend in northern African tropics. This pattern is confirmed by Nicholson *et al.* (2000), who examined meteorological data from a much larger dataset in Africa (1400 weather stations, including many in the Congo Basin). They demonstrated a strong drying trend in northern sub-Saharan Africa, centred on major droughts in the Sahel but extending into the tropical rainforest belt of west Africa and north Congo. The drying peaked in the 1980s, when most of west Africa and the Congo basin where anomalously dry (the non-rainforest regions of east Africa, on the other hand, were anomalously wet). In the 1990s there are hints of some recovery, particularly in the eastern most sectors (east Nigeria, Cameroon, Gabon) but in most regions

rainfall was still well below the long-term mean. Rainfall trends in equatorial Africa appear more spatially variable, and do not show the same strong net drying. The drying trend appears to have started in the mid-twentieth century: overall in the west Africa/north Congo tropical rainforest belt, rainfall levels were about 10% lower in the period 1968–97 than in the period 1931–60 (Nicholson *et al.* 2000).

For Brazilian Amazonia, Marengo *et al.* (2004) conducted an analysis of extensive rain gauge data. They reported regionally a significant decline in rainfall in northern Amazonia since the mid-1970s. River data from northern Amazonia indicate wetter periods in the mid-1970s, and drier periods in the 1980s. Multi-decadal variations in precipitation in Amazonia seem to be dominating over any long-term trend. This was also demonstrated by an analysis of a long term rainfall record from Central Amazonia by Malhi and Wright (2004).

Strength and intensity of dry season

The mean annual maximum dry season index (DSI) over the three tropical rainforest continents is shown in Fig. 1.9. Globally, there is no significant

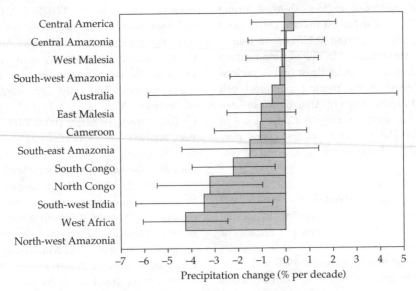

Figure 1.8 Bar chart of precipitation trends in each tropical rainforest sub-region for the period 1960–98. The trend for north-west Amazonia seems to be an artefact generated by the interpolation method and is not shown.

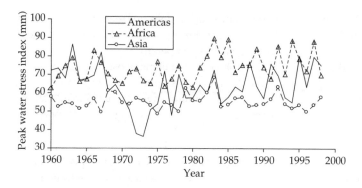

Figure 1.9 Time series of the dry season index over the three tropical rainforest continents.

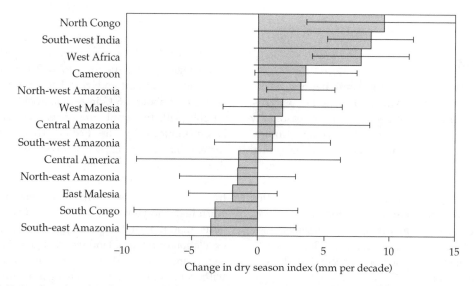

Figure 1.10 Bar chart of trends in dry season index for each tropical rainforest sub-region for the period 1960–98.

trend in maximum DSI (-0.5 ± 1.7 mm per decade). However, for DSI it is particularly important to consider local trends rather than pantropical averages, because of the asynchrony of the time of peak DSI in different regions. On a continental level, there is no significant trend in Asia (0.2 ± 1.2 mm per decade) or the Americas ($+0.4 \pm 3.2$ mm per decade), but there has been a moderately significant increase in dry season intensity in Africa ($+2.8 \pm 2.0$ mm per decade, $p < 0.05$).

A bar chart of trends in maximum dry season index in the various sub-regions is shown in Fig. 1.10. The drying trend in Africa is driven by a

steady increase in dry season intensity throughout the northern African tropics (west Africa $+7.8 \pm 3.3$ mm per decade, $p < 0.0001$; north Congo $+9.5 \pm 5.3$ mm per decade, $p < 0.01$; Cameroon $+3.6 \pm 3.7$ mm per decade, $p < 0.10$), perhaps slightly offset by a modest weakening of dry season in the southern Congo (-3.2 ± 3.3 mm per decade, $p < 0.05$). A similar drying trend is seen in south-west India ($+8.5 \pm 5.9$ mm per decade, $p < 0.01$). In contrast, a significant reduction in dry season index in this period is indicated for the Australian tropical rainforest belt (-8.4 ± 6.4 mm per decade, $p < 0.01$). Once again, on the basis of the limited dataset used here, the trend for much of the DRC needs to be treated with caution. However, this pattern of general drying in northern sub-Saharan Africa is clearly observed in the more comprehensive analysis by Nicholson *et al.* (2000), as is the peak in drying throughout the African tropical rainforest belt in the 1980s.

The contrast between, for example, the hydrological regimes of north Congo (steadily dry, seasonal, and getting drier), Malaysian (generally very wet but occasionally very drought-stressed), and north-west Amazonian (almost always wet) tropical rainforests is marked, and suggests that successful trees in these regions need to adopt very different survival strategies.

Discussion

The climate of tropical rainforest regions clearly exhibits considerable variation in rainfall patterns, from no seasonality to high degrees of seasonality and interannual variability. The 'average' tropical rainforest (Table 1.1) can be said to have an annual rainfall of 2180 mm, a dry season of 3–4 months, a mean annual temperature of 25.2°C with a seasonal range of 3.2°C, and a mean insolation of 16.5 GJ, but there is considerable variation, particularly in the hydrological parameters. As a whole, African tropical rainforests are generally drier, at higher elevation, and cooler than those of other continents.

On an interannual timescale, it is clear that the ENSO is the primary driver of temperature variations through most of the tropics, and of precipitation in the Americas and Southeast Asia. The correlation between tropical rainforest temperature and the ENSO index is particularly remarkable. The teleconnections between ENSO and rainfall in the tropical rainforest regions of Africa are less clear, and appear to depend on the seasonal timing of events. For example, the strong El Niño of 1998 appeared to have little influence on rainfall in African tropical rainforests (Nicholson *et al.* 2000). It is noteworthy that temperature in African forests still appears to respond to the ENSO signal even when there is no precipitation response; hence, the

rise in temperature does not seem to be driven by local drought, but by a general increase in mean tropical surface temperatures.

It is also clear that climatic oscillations on multi-decadal timescales are important in many tropical climates, and caution should be used in attributing many of the trends observed here to anthropogenic climate change. A particular example is the apparent multi-decadal oscillation in temperature in western Amazonia, a feature that was also noted by Botta *et al.* (2002). The strong drying trend in the northern African tropics may also be driven by an oscillation in Atlantic Ocean temperatures, but global warming may also be driving a secular shift in Atlantic temperatures.

A substantial and rapid warming of about 0.26°C per decade is evident in all tropical rainforest regions since the mid-1970s. The recent rise is highly significant and global in extent: the net increase of about 0.6°C in recent decades is comparable with the amplitude of temperature oscillation induced by the ENSO. This feature is remarkable and perhaps not noted sufficiently in discussions of century-long trends: on the physiologically relevant timescale of years to decades, tropical rainforests have experienced rapid warming. The pantropical extent of this rise suggests that it is not induced by a local climatic oscillation or by local land use change effects. This trend since the 1970s seems to be synchronous with, and consistent with, changes in the global climate over the same period that have been ascribed to the anthropogenic greenhouse effect (IPCC 2001). It thus seems likely that this recent trend is indeed a signal of the anthropogenic greenhouse effect. Climate models suggest that a warming by 2–5°C can be expected in tropical rainforest regions over this century (Cramer *et al.*, Chapter 2, this volume), a change that seems likely to have a substantial impact on tropical rainforest physiology (e.g. Cowling *et al.*, Chapter 16, this volume).

There is much greater uncertainty in how tropical precipitation regimes will respond to changes in the global atmosphere. Globally, evaporation rates are expected to increase, the atmosphere is expected to become more humid, and rainfall rates to increase. Indeed, satellite observations of upper-tropospheric humidity from 1980 to 1997 show statistically positive trends of 0.1% yr^{-1} for the zone 10°N to 10°S. However, any pantropical trends in precipitation are expected to be eclipsed by strong regional variations as atmospheric circulation patterns shift. There is little agreement among climate models on the future pattern of rainfall in the tropics.

The results presented here confirm that regional trends in precipitation and dry season intensity do indeed dominate over any coherent pantropical trend. The high interannual variability of precipitation makes it difficult to detect any overall trend in the American tropics, but there is a hint of a marginally significant drying trend in the Asian tropics. In African tropical rainforest regions, however, the drying trend is very strong, particularly at the northern edge of the tropical rainforest zone. This trend appears to be associated with the general drying of the Sahel region in the second half of the twentieth century. This overall drying in the tropical forest belt contrasts with the general expectation that precipitation levels will increase with global warming. However, the net pattern of precipitation change projected for tropical rainforests (typically between −1 and +2% per decade; Giorgi 2002) is highly dependent on the spatial pattern of projected drying. The current drying trend may continue, or it may simply reflect natural oscillations, and reverse in coming decades. The reduction of rainfall in specific tropical regions seems to be decoupled from the recent pantropical warming. Regions that have not experienced drying (e.g. Amazonia) show similar rates of temperature increase to regions that have (e.g. Africa).

A frequently cited climate modelling scenario (Cox *et al.* 2000; Cowling *et al.*, Chapter 16, this volume), suggests that north-eastern Amazonia may be vulnerable to extreme drying in response to circulation shifts induced by global warming, and that the consequent die-back of tropical rainforest could substantially accelerate global warming. To date, there is little evidence of any drying trend; in fact, the region has become marginally

wetter (precipitation trend +1.9 ± 2.2% yr^{-1}, $p < 0.10$; no significant trend in dry season index). However, this region is strongly influenced by ENSO; if the ocean–atmosphere system were to shift into a sustained 'El Niño-like' state in coming decades (as opposed to the ENSO simply increasing in frequency and amplitude), the region could be vulnerable to drying.

The apparent marginality of Africa's tropical rainforest zone stands out in this analysis. This is the driest tropical rainforest region, and in recent decades it has generally become drier. The extent of African tropical rainforest seems particularly vulnerable to small shifts in ocean–atmosphere circulation. The palaeo-record seems to confirm this: large areas currently covered by African tropical rainforest appear to have been covered by savanna in the last ice age (Morley 2001), and perhaps as recently as 2500 yrs ago (Maley and Brenac 1998), contrasting with the continuity of forest cover in most of Amazonia (Mayle and Bush, Chapter 15 and Maslin, Chapter 14, this volume).

With some caveats about biogeographical and paleohistorical differences between regions, it appears that a study of how surviving African forests have responded to the drying trends of recent decades may yield useful insights into how tropical rainforests in general may respond to future drying. If eastern Amazonia were to dry over this century, perhaps we can gain insights into the future of its surviving forests by examining what happened to Africa's forests over the last decade of the last century?

Finally, the phrase 'surviving forests' is important to keep in mind. The most important process to affect the northern African tropical rainforests over recent decades was probably not the reduction in rainfall, but fragmentation and clearance. The long belt of tropical rainforest at the southern edge of west Africa now largely consists of small, logged-over fragments. Climate and deforestation are coupled, in that drier regions are more likely to be deforested, and deforestation contributes to modifying local and global climate. The fate of many of the world's tropical rainforests are not likely to be primarily determined by climate trends, but by human actions on forest use or protection.

Acknowledgements

The climate dataset used in these analyses was developed by the University of East Anglia Climate Research Unit, and in particular we thank David Lister and David Viner for assistance in obtaining the date, and Daniel Wood for assistance with the preparation of the climatic datasets. Yadvinder Malhi is supported by a Royal Society University Research Fellowship, and James Wright was supported by the DeLemos bequest of the University of Edinburgh.

CHAPTER 2

Twenty-first century atmospheric change and deforestation: potential impacts on tropical forests

Wolfgang Cramer, Alberte Bondeau, Sibyll Schaphoff, Wolfgang Lucht, Ben Smith, and Stephen Sitch

Wet tropical forests, and the carbon stocks they contain, are currently at risk due to both deforestation and climate change. In this chapter we quantify the relative roles of carbon dioxide (CO_2), temperature, rainfall, and deforestation on the future extent and condition of rainforests, and study the magnitude of their feedbacks on atmospheric CO_2 concentrations. We apply a dynamic global vegetation model, using multiple scenarios of tropical deforestation (extrapolated from two estimates of current rates) and multiple scenarios of changing climate (derived from four independent off-line general circulation model simulations). Results show that deforestation is likely to produce large losses of carbon, despite the uncertainty concerning exact deforestation rates. Some climate models suggest that the combined change in rainfall and temperature will produce additional large carbon emissions due to increased drought stress. One climate model, however, produces an additional carbon sink. Taken together, our estimates of additional carbon emissions during the twenty-first century, for all climate and deforestation scenarios, range from 101 to 367 GtC, resulting in CO_2 concentration increases above background values by between 29 and 129 ppm. Notwithstanding this range of uncertainty, continued tropical deforestation will most certainly play a very large role in the build-up of future greenhouse gas concentrations.

Introduction

There is broad agreement about the importance of tropical forests for the global carbon cycle and hence global climate. There is still debate, however, about the area affected annually by tropical deforestation, the resulting flux of carbon to the atmosphere, and the feedbacks of this flux to the climate system (cf. Table 2.1, Houghton 1999; Fearnside 2000; Malhi and Grace 2000; Achard *et al.* 2002, 2004; DeFries *et al.* 2002). Some scenarios of future climate change show that tropical forests could become an unprecedented source of carbon during the twenty-first century, even in the absence of additional anthropogenic deforestation (Cox *et al.* 2000; Cramer *et al.* 2001). A central question is whether carbon

emissions from climatic impacts on tropical forests could become as large as those from anthropogenic deforestation during the coming decades. If such an additional flux appeared, then the rate of atmospheric CO_2 increase per unit of fossil fuel emissions would be greater than at present.

Recent assessments by the Intergovernmental Panel on Climatic Change (IPCC) have suggested a global anthropogenic deforestation flux[1] on the order of 1.6 GtC yr^{-1} (Bolin *et al.* 2000; Prentice *et al.* 2001) for the last two decades of the twentieth century. However, in their analysis of the spatial extent of

[1] In this chapter, carbon fluxes from the biosphere into the atmosphere ('sources') are given as positive numbers and sinks as negative numbers, except in the case of local plant productivity estimates which are also given as positive numbers.

Table 2.1 Recent estimates of carbon loss from tropical forests to the atmosphere attributed to deforestation (Gt C yr^{-1})

Region	Houghton (1999) (1980–90)	Fearnside (2000) (1981–90)	Malhi and Grace (2000) (1980–95)	DeFries et al. (2002) (1980s)	Achard et al. (2004) (1990–9)
America	0.55 ± 0.3	0.94	0.94	0.37	
Africa	0.29 ± 0.2	0.42	0.36	0.10	
Asia	1.08 ± 0.5	0.66	1.08	0.18	
Total	1.90 ± 0.6	2.00	2.40	0.65	1.07 ± 0.31

tropical forest cover from long-term satellite time series, DeFries et al. (2002) challenge these estimates as unrealistically high, suggesting only 0.6 Gt C yr^{-1} for the 1980s and 0.9 Gt C yr^{-1} for the 1990s. Using different sensors and methods, Achard et al. (2002) estimated a similar emission rate for humid tropical forests (0.64 ± 0.21 Gt C yr^{-1} for 1990–7) and, in a recent update (Achard et al. 2004), for all tropical forests (1.07 ± 0.31 Gt C yr^{-1} for 1990–9). If confirmed, they imply a modification of the global carbon budget that has implications for other compartments that are estimated with uncertainty, particularly the postulated Northern Hemisphere extratropical sink.

The problem of estimating deforestation-related fluxes consists of several subproblems each are associated with its own problems and uncertainties. First, estimating forest *area* is affected by the definitions of 'forest' versus 'non-forest' area, which vary widely in terms of tree size and density (e.g. Noble et al. 2000). Some discrepancies between flux estimates can also be explained by the technology that is used for estimation of the area covered by 'nearly intact forest', such as ground-based surveys, air photos, or satellite images (e.g. DeFries et al. 2002). Further, all area estimates must deal with the question whether the deforested area is left bare, or is converted to a vegetation type with different carbon density, or undergoes regrowth (Houghton 1999).

Once the area is estimated, then the density of carbon in the removed vegetation needs to be assessed. But total biomass (above- and below-ground) is known with sufficient precision for only a few locations. Natural forests are spatially heterogeneous, due to environmental conditions, management, or natural disturbance history, even in the absence of deforestation, and this creates further uncertainty in the density estimate.

Finally, the resulting carbon flux from removal of a part of this biomass, including its temporal evolution must be estimated as a comparison between the balance of the ecosystem without human interference and the deforested landscape. Direct measurements of such fluxes only exist for a few, and always undisturbed sites, from campaigns lasting up to a few years only. Even the undisturbed sites are affected by year-to-year climate variation, by trends in climate such as the anthropogenic enhancement of the greenhouse effect, and possibly by direct CO_2 'fertilization'. Inventory based approaches, in contrast, reconstruct the net balance over the time between inventories (usually many years), but they allow only very indirect estimations of the causes behind temporal and spatial variability. Reconstructing the overall carbon balance by inversion of atmospheric CO_2 concentrations is only a limited possibility for the tropics, due to sparse station network and uncertainties in estimating net uptake in undisturbed forests.

Improvement of carbon budget estimates, both for the reference value, for recent tropical deforestation and for the climate-change driven flux, requires that the quantities mentioned above are better constrained than previously. To integrate sparse information from surveys, statistics, and imagery, and to allow extrapolation of current trends into the future, a combination of process-based numerical models and parameters delivered from experiments and observations can be used.

The future role of tropical forests in the global carbon cycle and the climate system is a function of future deforestation rates and the degree to which remaining forests will be sustainable or can even increase their carbon stock. Deforestation is strongly influenced by social and economic development including international agreements about the protection of forest resources. Future productivity in pristine or sustainably managed forests is influenced by climate change. For the assessment of adaptation and mitigation options in tropical forest regions, it is of interest to examine the different uncertainties of these factors to each other and to rate their relative importance.

Methods and data

Design

Our quantitative assessment of the tropical forest carbon balance uses a scenario approach, based on a dynamic (global) vegetation model, driven by data on climate (monthly temperature and precipitation), atmospheric CO_2, soil texture, and deforestation. The model is based on quantitative understanding of relevant processes and is therefore expected to generate results that broadly capture environmental gradients and temporal trends, as well as the non-linear evolution of carbon stocks in response to perturbations. Biomass stock and carbon flux estimates from this model have been tested successfully against a wide array of observations in several biomes. For tropical forests, we present here some additional tests using different methods. As a first step of the application, we compare carbon fluxes attributable to historic deforestation with those attributed to other causes. Then, in a factorial experiment, we compare different plausible rates of future deforestation and different scenarios of regional climate change against each other, using the total carbon balance between atmosphere and biosphere as the diagnostic variable.

Previous assessments of carbon fluxes due to tropical deforestation have used inventory-based statistical approaches, sometimes supplied with additional data from remote sensing or from eddy covariance measurements at selected sites. The application of a process-based numerical model is expected to improve on such work when it can be assumed that major processes exhibit important, and known, non-linear features (such as the different rates of decline in above and below-ground

carbon pools after destruction of the canopy)—even when the parameters of these functions can be estimated only from a limited amount of empirical evidence. Such models typically allow for the study of feedbacks between changing atmospheric conditions and the fluxes of energy and matter through the biosphere (Brovkin *et al.* 2004).

The dynamic global vegetation model LPJ

Dynamic global vegetation models (DGVMs) combine representations of biogeochemical processes with representations of processes contributing to the dynamics of vegetation structure and composition. Existing DGVMs (reviewed by Cramer *et al.* 2001) differ in their degree of complexity and suitability for different tasks. One common feature is their 'generic' formulation which makes them suitable for broad-scale assessments in any biome. Typically, they use forcing data for climate and CO_2 concentration, thereby allowing the assessment of both direct and indirect (climatic) effects of CO_2 enrichment. The Lund-Potsdam-Jena Model (LPJ) is a DGVM of intermediate complexity that has been shown to produce credible results for a broad range of applications (Sitch *et al.* 2003).

The LPJ model includes, on the basis of generalized plant functional types (PFTs), the major processes of vegetation dynamics, such as growth, competition, and plant demography. For tropical ecosystems, LPJ uses three different PFTs, named 'Tropical broad-leaved evergreen trees', 'Tropical broad-leaved raingreen trees', and 'Tropical herbaceous plants'. All three PFTs share the bioclimatic requirement that the mean temperature of the coldest month, calculated for 20-yr running means, does not fall below 15.5°C (Prentice *et al.* 1992). The main parameters for the three PFTs are listed in Table 2.2.

The environment inputs to LPJ are monthly means of temperature, precipitation, and cloudiness, as well as soil texture and annually averaged atmospheric CO_2 concentration. From these, and based on available carbon pools and structural tissue from previous years, LPJ calculates daily gross primary production (GPP) for each PFT on the basis of a widely used coupled photosynthesis

Table 2.2 Parameter values for tropical PFTs in the LPJ model

PFT	W/H	z_1 (−)	z_2 (−)	g_{min} (mms^{-1})	r_{fine} (−)	a_{leaf} (yr)	f_{leaf} (yr^{-1})	$f_{sapwood}$ (yr^{-1})	f_{root} (yr^{-1})	S_{GDD} (°C)
Tropical broad-leaved evergreen	W	0.85	0.15	0.5	0.12	2.0	0.5	0.05	0.5	—
Tropical broad-leaved raingreen	W	0.70	0.30	0.5	0.50	0.5	1.0	0.05	1.0	—
Tropical herbaceous	H	0.90	0.10	0.5	1.00	1.0	1.0	—	0.5	100

Note: W/H indicates woody or herbaceous stature; z_1 and z_2 are the fraction of fine roots in the upper and lower soil layers, respectively; g_{min} is the minimum canopy conductance; r_{fine} is the fire resistance; a_{leaf} is the leaf longevity; f_{leaf}, $f_{sapwood}$, f_{root} are the leaf, sapwood, and fine root turnover times, respectively and S_{GDD} is the growing degree day requirement to grow full leaf coverage.

Source: Sitch et al. (2003).

and water balance scheme (Farquhar et al. 1980; Haxeltine and Prentice 1996 a,b; Sitch et al. 2003). From GPP, autotrophic (maintenance and growth) respiration R_a is subtracted, as well as a cost for reproduction. Allometric relationships are used to determine whether the annually remaining photosynthate is turned into long-lived (e.g. wood) or short-lived (e.g. leaves, fine roots) plant material. Litter and soil organic matter decomposition are driven by seasonal temperatures and soil moisture status (calculated internally in the model). From this partitioning of GPP, LPJ derives total fluxes such as net primary productivity (NPP = GPP − R_a), heterotrophic respiration (R_h), emissions from natural fire disturbances (excluded for the present application), and as a result net ecosystem exchange (NEE = GPP − R_a − R_h − disturbance). Since the water balance is tightly coupled to carbon fluxes, the model also yields actual evapotranspiration and local runoff. Comparison of total carbon stocks at different times gives net ecosystem productivity (NEP).

The LPJ responds realistically to changing water availability in a context where atmospheric CO_2 and temperature are also changing. The model generally optimizes water use as real plants do, thereby adjusting leaf area and stomatal conductance, such as to maintain the highest possible NPP. Despite these regulatory mechanisms, growth under water-stressed conditions includes an increased respiration cost, and NPP is reduced. Increased ambient CO_2 concentrations allow for reduced canopy conductance and therefore conserve water with the result of higher NPP. The magnitude of this effect,

at the ecosystem level, over the longer term, and in significantly higher-than-present ranges of CO_2 concentrations is still poorly understood because of the constrains on empirical and experimental observations. Earlier studies seem to indicate, however, that the CO_2 response in LPJ is more likely to be overestimated. For the climate change response studied here, it follows that reduced water availability possibly should lead to even higher carbon releases than those presented below.

The LPJ model has been tested successfully against a wide range of observations, including short-term flux measurements of carbon and water, satellite-based observations of leaf phenology and photosynthetic activity, inversions of atmospheric CO_2 measurements, and others (e.g. Dargaville et al. 2002; Lucht et al. 2002; Sitch et al. 2003; Gerten et al. 2004). The version of LPJ presented here incorporates an improved representation of the water cycle compared to the model described by Sitch et al. (2003). The most important changes encompass the inclusion of additional processes, such as interception and soil evaporation, and the daily distribution of precipitation (Gerten et al. 2004). For tropical forests, results of this LPJ version also compare well with estimates of carbon pools and fluxes made with a variety of other approaches (Cramer et al. 2004).

Similar to other DGVMs, LPJ in its basic mode produces potential natural vegetation, that is, the structure and carbon content of the biosphere in the absence of human impact. To account for deforestation in this study, we remove parts or all of the biomass in the affected grid cells, adding the loss to a total of all losses. The deforestation loss is estimated as follows: from a time series of historical deforestation, we determine the amount of (additional) forest destruction in any given year and each grid cell. Based on the input data, the deforested area is turned either into 'intensive agriculture' or 'extensive agriculture'. The resulting carbon pools under extensive agriculture are estimated to correspond to a remaining 20% tree cover plus the carbon that would be stored in a natural grassland (the LPJ vegetation type that most closely resembles an agricultural cropping system). For intensive agriculture, all trees are removed and replaced by grassland. This approach to the accounting of vegetation processes after deforestation, represents the best estimation currently achievable, since insufficient parameter sets exist to account for process-based estimates of different land use types.

Baseline data

Climate and CO_2

For this study, we used monthly fields of mean temperature, precipitation, and cloud cover taken from the CRU05 (1901–98) monthly climate data on a $0.5° \times 0.5°$ global grid, provided by the Climate Research Unit (CRU), University of East Anglia (New et al. 1999, 2000). Monthly cloud cover data for 1997 and 1998, not so far compiled by CRU, were taken as the average over the previous 30 yrs. Monthly data were interpolated linearly to provide 'quasi-daily' time series of climate. Atmospheric CO_2 concentrations from 1901 to 1998 from various sources (ice-core measurements and atmospheric observations), were used following Sitch et al. (2003). Soil texture data were as in Haxeltine and Prentice (1996a), based on FAO data as in Zobler (1986) and FAO (1991).

Areas with potential and actual tropical forest

Estimating historical losses of carbon due to tropical deforestation requires assessment of the area that once may have been occupied by tropical trees prior to large-scale land clearing, combined with a map of the present distribution. At the global level, the area of potential distribution is

well determined by climate (Prentice et al. 1992), but the actual area is more difficult to assess. Data sources for the present distribution are regional and national forest inventories, aggregated by FAO to a global dataset (FAO 2003), and remote sensing data from several platforms. A common feature of both data sources is that they depend on a consistent definition of 'forest' versus 'non-forest', usually requiring a more or less arbitrary cut-off value to be applied at the continuum between dense forest and open land. Remote sensing offers the advantages of greater consistency, shorter time intervals between observations, although cloud cover and sub-pixel forest fragmentation present problems that are resolved only by approximate methods.

In order to achieve both a realistic estimate of current tropical forest cover and a framework which allows the application of deforestation scenarios for the future, we have combined a long-term reconstruction of deforestation rates (HYDE, Klein Goldewijk et al. 2001) with the potential forest area determined by LPJ. For the global forest area, and for the three tropical regions Africa, Asia, and America, the total areas resulting from our approach are similar to those found by the most recent FAO assessment.

Deforestation rates

As an estimate of region-specific long-term historical deforestation rates, we used the HYDE dataset (Klein Goldewijk et al. 2001), which provides land cover classifications on a $0.5° \times 0.5°$ longitude/latitude grid for the years 1700, 1750, 1800, 1850, 1900, 1950, and 1970.[2] From HYDE, we assumed that deforestation had occurred on all grid cells classified as 'intensive agriculture' or 'extensive agriculture'. Between the time-slices, we interpolated deforestation linearly for each grid cell to provide a quasi-continuous yearly historical dataset. The resulting present-day tropical forest distribution is shown in Fig. 2.1.

---

[2] The HYDE dataset provides numbers for 1990 as well, but for consistency with the scenario calculations, we used the values from Achard et al. (2002) and Malhi and Grace (2000) for this purpose.

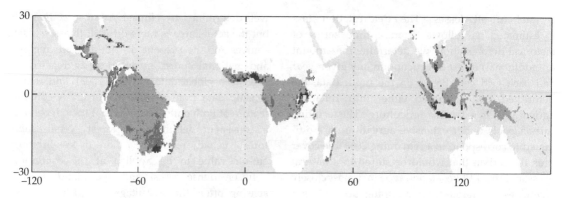

Figure 2.1 Global distribution of potential tropical forest area, simulated by LPJ on the basis of 1980–90 climate conditions (all colours) and of deforested areas according to the HYDE database (light grey: forest; intermediate grey: marginal croplands; dark grey: intensive croplands).

Scenarios

Carbon dioxide

For the period 1991–2100, all climate and DGVM simulations were based on the IPCC IS92a emission scenario (Houghton *et al.* 1992), which includes intermediate assumptions about human demography and economic activity and causes an increase of CO_2 emissions into the atmosphere of about $1\%\ yr^{-1}$ CO_2 equivalent throughout the twenty-first century. For a long time, this scenario was considered a 'business as usual' scenario of carbon emissions and associated atmospheric CO_2 concentrations. Here its use is mainly justified by the fact that outputs are available for this scenario from several climate modelling centres, thereby allowing comparison between different climate models.

Climate

The climate change scenarios are derived from output from four different climate models, provided by the IPCC Data Distribution Center, covering the period from 1901 to 2100. To generate climate information for future conditions, we first derived anomalies of monthly means (temperature, precipitation, cloudiness) from four different climate

model outputs (CGCM1, CSIRO, ECHAM4, HadCM3[3]) by calculating the differences between each modelled monthly value and a present-day baseline average for that month. These anomalies were then applied to a gridded baseline climate dataset (average 1969–98 from the CRU-dataset). This procedure retains the broad spatial features of climate change, as well as the monthly variability as simulated by the climate model, while ensuring that the spatial pattern, and the general nature of present-day climate remain as in the observations. Because the interannual variability of the resulting time series is driven by the climate model only, no specific year in the calculations can be compared to a given true calendar year.

Indicative averaged anomalies of temperature and precipitation for these climate scenarios, averaged for tropical land areas are given in Fig. 2.2. During the twenty-first century, all scenarios show warming on all continents, ranging from 1.9°C (CSIRO, America, 2041–60) to 7.8°C (CGCM, Africa, 2081–2100). Scenarios differ considerably with regard to precipitation. While most models indicate increased precipitation in tropical areas, magnitudes of the signal vary, and two models show notable exceptions: HadCM3 in America with a reduction of 17% by the end of the century, and CGCM with a similar reduction in Africa.

Remarkably, climate scenarios not only differ in the broad spatial trends of precipitation, but also with respect to the temporal variability in the tropical regions. Figure 2.3 illustrates this for the

[3] CGCM1: First version of the Canadian Global Coupled Model; CSIRO: Commonwealth Scientific and Industrial Research Organisation global coupled ocean–atmosphere–sea-ice model; ECHAM4: ECMWF/HAMBURG model v. 4; HadCM3: Coupled atmosphere–ocean GCM developed at the Hadley Centre, Bracknell, UK; further details to be found at http://ipcc-ddc.cru.uea.ac.uk/dkrz/dkrz_index.html

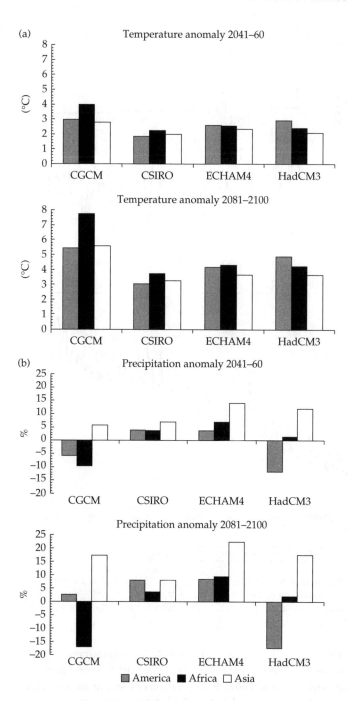

Figure 2.2 (a) Temperature anomalies for tropical regions (30°S–30°N), for all four climate models, expressed as difference *relative* to the period 1969–98. (b) Precipitation anomalies for tropical regions (30°S–30°N), for all four climate models, expressed as difference *relative* to the period 1969–98.

four climate models and the zone between 30°S and 30°N in America. The HadCM3 model clearly demonstrates much larger interannual variability than the other three models.

Deforestation

Scenarios of future deforestation were derived from two recent estimates of current trends. Based on the assessment of a variety of sources, Malhi

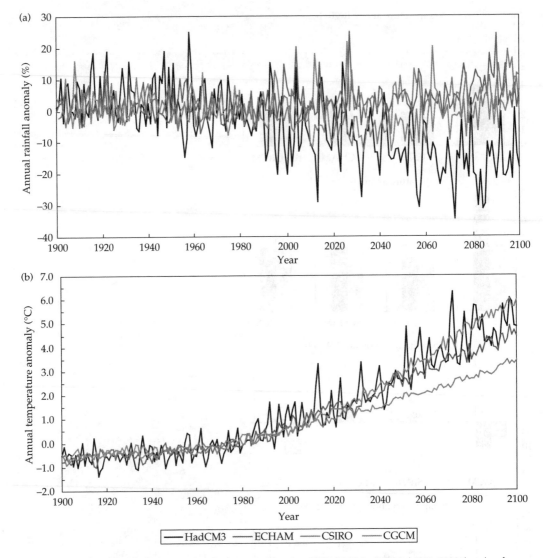

Figure 2.3 Climate reconstruction and scenario data for the tropical Americas (30°S–30°N) in the period 1901–2100, based on four different general circulation models. Upper panel (a) change in annual total rainfall (per cent) relative to the mean over 1969–98. Lower panel (b) change in annual mean temperature relative to the mean over 1969–98. Model simulations were made with monthly data. (See also Plate 2.)

and Grace (2000) suggest that, between 1990 and 1995, $129 \times 10^3 \, km^2$ of tropical forests have been deforested each year. As one scenario, we extrapolate this value (differentiated for the three main areas America, Africa, and Asia) linearly into the future. Directly based on a remote-sensing analysis covering a large sample of tropical forests, Achard et al. (2002) found much lower deforestation rates

for the period 1990–7, for total tropical forests $49.0 \times 10^3 \, km^2 \, yr^{-1}$ (or $58.0 \times 10^3 \, km^2 \, yr^{-1}$ in Achard et al. 2004). Extrapolation of these numbers yielded what is labelled 'low deforestation' (i.e. based on Achard et al. 2002) and 'high deforestation' (i.e. based on Malhi and Grace 2000), subsequently. The temporal evolution of the remaining forest decline is shown in Table 2.3.

Table 2.3 Forest areas, deforestation areas, and remaining forest areas as used in this study. The trend in future deforestation is used with a low and a high scenario, by direct extrapolation of trends reconstructed by Achard *et al.* (2002) ('low deforestation') and Malhi and Grace (2000) ('high deforestation')

| | Forest area in 1900 (Klein Goldewijk *et al.* 2001) | | Deforestation rates | | Remaining forest area in 2100 | | | |
| | 10^7 km^2 | % of pot.[a] | 1990–7 (Achard *et al.* 2002) 10^3 km^2 yr^{-1} | 1990–5 (Malhi and Grace 2000) 10^3 km^2 yr^{-1} | Low deforestation | | High deforestation | |
					10^7 km^2	% of pot.	10^7 km^2	% of pot.
America	9.66	94	22	56.9	4.92	48	2.07	20
Africa	4.39	89	7	37.0	2.15	44	0.65	13
Asia	3.24	93	20	35.1	0.43	12	0.02	<1
Total	17.28	92	49	129.0	7.49	40	2.74	15

[a] Potential area as defined by the climatic limits in LPJ.

Deforestation scenarios were applied year by year, with uniform linear rates in each of the three regions, across all grid cells as long as these are covered by forests. For comparison, we also calculated scenarios with zero further deforestation after 1995. This way of allocating deforestation estimates is deliberately simplistic, because alternative methods would require many assumptions that could not be supported by data.

Results

The response of tropical forests to changing CO$_2$, climate, and deforestation

At any location, ecosystems are simultaneously affected by radiation, temperature, moisture, and nutrient availability, and direct interference such as anthropogenic deforestation. To illustrate the capacity of the model to adequately reflect these multiple forcings, we compare calculations for the site Manaus (central Brazilian Amazonia) in two contrasting climate simulations (Fig. 2.4).

The left column of the diagram shows that, in the ECHAM4 simulation, LPJ exhibits moderate interannual variability in productivity as well as soil respiration, reflecting slight changes in water availability due to precipitation changes. In the twenty-first century, temperature increase leads to slightly accentuated drought stress, but interannual variability hardly changes. The productivity and

respiration effects of increased drought are moderated by the direct effect of CO$_2$ increase (not shown, cf. Cramer *et al.* 2001, Sitch *et al.* 2003). The remaining drought stress reduces NPP slightly after about 2030 and yields increased mortality rates. Vegetation carbon pools are slightly reduced, but the litter and soil carbon pools do not increase, since the additional litter is rapidly consumed by soil respiration. Around the model year 2042, the slight continuous change in temperature and precipitation appear to pass some threshold, giving a more marked increase in drought stress, this is followed by a reduction in vegetation carbon which is not recovered during the remainder of the century.

Similar simulations with the HadCM3 model indicate differences already for the present day. Small differences in GPP and NPP, probably mostly due to lower moisture stress, produce a forest with significantly more live biomass. This climatic difference is the result of the technique we used for climate data production. For consistency, monthly anomalies from the climate model were applied throughout the time series from 1900 to 2100. This results in identical long-term means for the period 1969–98, but the inter-annual variability is determined by the climate model alone, and turns out to be quite different between the two climate models. The comparison between ECHAM4 and HadCM3 for the model years 1969–98 therefore illustrates that tropical forests in LPJ are significantly affected by climate variability.

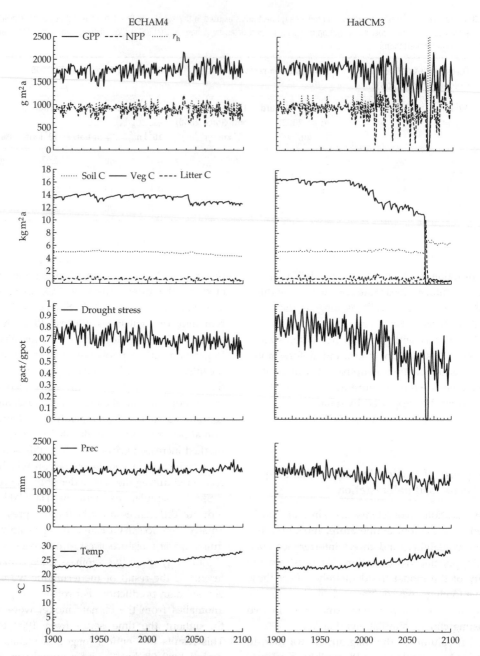

Figure 2.4 Carbon fluxes, pools, and environmental factors in LPJ simulations for the Manaus site, driven by output from two climate models ECHAM4 (left) and HadCM3 (right), without deforestation. Top row: gross primary productivity (GPP), net primary productivity (NPP), and heterotrophic respiration (r_h) [g C m^{-2} a^{-1}]; second row: soil, vegetation, and litter carbon pools [kg C m^{-2}] (vegetation carbon contains all structural biomass, above- and below-ground); third row: ratio between potential (unstressed) and actual canopy conductance, indicating the drought stress simulated by LPJ (high values indicate low drought stress); fourth row: monthly mean temperature (deg C), fifth row: monthly total rainfall (mm).

Climate simulated by HadCM3 in the twenty-first century undergoes dramatic changes in the American tropics. While temperature increases at a rate similar to that of the ECHAM model, precipitation is reduced significantly, causing a dramatic increase in drought stress. In addition, interannual variability increases strongly. Productivity and respiration respond to this variability in a synchronized manner, that is, the reduced litter fall in low productivity years generates reduced soil respiration, despite the high temperatures. It therefore appears as if it is mostly the reduction in GPP which causes the large losses in vegetation carbon, which occur in the peak years of drought stress—and these losses hardly ever recover in more favourable years. The series of model years 2070–2, with significantly higher temperatures and reduced rainfall than before, then lead to a total collapse of vegetation at the Manaus site, with some increase in soil carbon and a very weak recovery towards the end of the century. In the model simulations, the loss of biomass appears to be a function of both dramatically increased climate variability and the increase in drought stress brought about by temperature increase and precipitation decline. Some shifts in plant functional type composition (favouring the more drought-adapted seasonal tropical trees) occur in response to climatic change and are already embedded in the quantities shown in the diagram.

Neither timing nor magnitude of climate fluctuations, as simulated by the climate models in this particular grid cell, may be taken as reliable forecasts of future environmental conditions at Manaus. However, they may serve as illustrations for the changing nature of the overall climatic regime that can be anticipated if greenhouse gas forcings and the model process formulations are accurate. The response of the ecosystem model to these fluctuations represents an extrapolation beyond experimentally proven model behaviour, but it is founded on the formulation of physiological processes that are checked against experimental evidence. In conclusion, the probability of a vegetation collapse in the central Amazon cannot be determined with high accuracy—rather, these results must be seen as an illustration of the general rate and magnitude of response that tropical ecosystems could show if the physical environment would change as prescribed by the scenarios. Present day analogues to such dry climates exist on the fringes of the Central American tropical rainforest, and LPJ simulations are broadly in agreement with ecosystem conditions at these fringes.

The carbon balance of global tropical forests

Carbon fluxes due to past tropical deforestation

Table 2.4 summarizes our estimates of recent annual tropical carbon losses, as well as the total carbon loss to the atmosphere, caused by the combination of deforestation and climate change as it occurred during the twentieth century, and compares them with the extrapolations made by

Table 2.4 Model-based estimates of carbon loss from tropical forests to the atmosphere attributed to deforestation

	Annual flux 1980–1995 [Gt C yr^{-1}]			Total flux 1901–1998 [Gt C]		
	Houghton (1999)[a]	This study		Houghton (1999)[b]	This study	
		Low def.[c]	High def.[d]		Low def.	High def.
America	0.55	0.40	0.70	30.5	18.61	24.38
Africa	0.29	0.18	0.32	9.5	8.99	11.21
Asia	1.08	0.30	0.49	38.6	11.06	13.83
Total	1.90	0.89	1.51	79.0	38.65	49.36

[a] Flux estimate averaged for 1980–90.
[b] Total for 1850–1990.
[c] Deforestation rate from Achard *et al.* (2002).
[d] Deforestation rate from Malhi and Grace (2000).

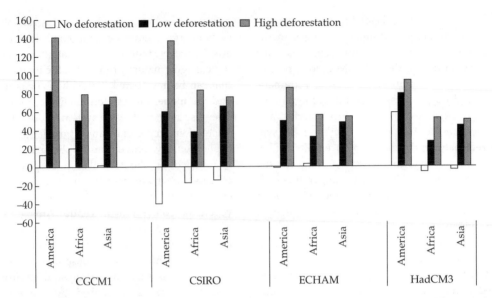

Figure 2.5 Total carbon balances [Gt C] for the twenty-first century, four different climate scenarios and three tropical forest regions, with no deforestation, low deforestation (Achard *et al.* 2002), and high deforestation rates (Malhi and Grace 2000), estimated by LPJ, including effects of increased atmospheric CO_2 and climate on ecosystem processes.

Houghton (1999). LPJ estimates are separated according to the 'low' (Achard *et al.* 2002) and 'high' (Malhi and Grace 2000) deforestation estimates for the most recent period, but are identical for the period until 1990 where the HYDE data were used. The main difference concerns Asian forests, which are estimated to produce much larger fluxes by Houghton. For Africa and America, the bookkeeping model and LPJ with high deforestation rates yield broadly comparable numbers. The low deforestation case consequently gives lower flux estimates.

The table also shows that the uncertainty concerning the actually deforested area (i.e. the difference between 'low' and 'high' deforestation) is rather large for the recent period. The comparison between recent fluxes and the century total, demonstrates that this uncertainty also carries over to the total amounts.

Carbon fluxes due to tropical deforestation in the twenty-first century
Extrapolation of high and low deforestation trends into the twenty-first century, assuming changing

atmospheric CO_2 according to IPCC IS92a, and changing climate as simulated by the four different climate models, yields widely differing carbon balances for each of the eight resulting scenarios, as well as for the case of 'no deforestation' (Fig. 2.5). Reflecting the reconstructions, the difference between deforestation scenarios is comparatively low in Asia (17–21%) and much higher in the Americas (26–105%) and Africa (50–117%).

In the 'no deforestation' scenario calculated with the CSIRO model, CO_2 increase and climate change generate a much more productive tropical forest zone, sequestering about 50 Gt of additional carbon. This is due to the much lower temperature increase simulated for the Americas, combined with the persistence of wet conditions, both of which distinguish the CSIRO model from the other climate simulations (Fig. 2.3). In contrast, HadCM3 and CGCM1 show carbon sources caused by climate change in some tropical forest regions, while ECHAM4 produces no significant sinks or sources. The scenarios from HadCM3 and CGCM1 differ greatly: while HadCM3 locates the climate-driven source exclusively in the Americas (and generates sinks in both Africa and Asia), the source in CGCM1 is mostly in Africa but is also significant in the Americas. ECHAM4 and CSIRO exhibit much less difference between continents.

The climate-driven sink in the CSIRO simulation is greatly affected by the remaining forest area at each point in time, and the simulation therefore shows the largest difference between no deforestation, low deforestation, and high deforestation. For all models, the largest 'high deforestation' flux comes from America.

Discussion

Overall, our comparison shows that anthropogenic deforestation is the most important factor in determining the role of tropical forests in the global carbon cycle. The sources or sinks produced by climate change are also significant components, however, and they strongly affect the spatial pattern of associated ecosystem changes.

Present rates of tropical forest destruction, even if they are confirmed to be at the low end of recent reconstructions, remain an important reason for concern, not only due to the negative impact the deforestation has on the environment of tropical regions, but also due to the amount of additional climate change that may be caused by the continuing release of carbon. In the most optimistic case, this carbon flux (if calculated as the difference between 'low deforestation' and 'no deforestation') still amounts to about 100 GtC in the course of the twenty-first century (HadCM3)—the most pessimistic case ('high' minus 'no' deforestation) gives more than 360 GtC (CSIRO) (Table 2.5). Following the accounting scheme proposed by House

Table 2.5 Total twenty-first century carbon loss from tropical forests, for different climate and deforestation scenarios, calculated as the difference between the net balance of forests in the absence of deforestation and the scenario value [GtC]

	Low def.	High def.
CGCM1	−166.6	−261.0
CSIRO	−235.5	−367.1
ECHAM4	−128.5	−194.3
HadCM3	−101.3	−148.1

et al. (2002), this range could lead to additional increase in atmospheric CO_2 concentrations between 29 and 129 ppm.

Furthermore, the comparisons illustrate that the current uncertainty concerning global precipitation patterns as projected by climate models has large implications on the tropical forest zone and hence on feedbacks between climate change and the global carbon balance. More detailed analysis of the regional behaviour of climate models should reveal whether the large Amazon dieback as shown by some of the models is indeed likely to occur in the coming decades.

Acknowledgements

The CRU05 climate data were kindly supplied by the Climate Impacts LINK Project (UK Department of the Environment Project EPG 1/1/16) on behalf of the Climatic Research Unit, University of East Anglia, UK. We gratefully acknowledge discussions with Frédéric Achard, Franz-W. Badeck, Martin Heimann, Jon Lloyd, Yadvinder Malhi, Carlos Nobre, Oliver Phillips, I. Colin Prentice, and Kirsten Thonicke.

CHAPTER 3

Forest–climate interactions in fragmented tropical landscapes

William F. Laurance

In the tropics, habitat fragmentation alters forest–climate interactions in diverse ways. On a local scale (<1 km), elevated desiccation and wind disturbance near fragment margins lead to sharply increased tree mortality, altering canopy-gap dynamics, plant-community composition, biomass dynamics, and carbon storage. Fragmented forests are also highly vulnerable to edge-related fires, especially in regions with periodic droughts or strong dry seasons. At landscape to regional scales (10–1000 km), habitat fragmentation may have complex effects on forest–climate interactions, with important consequences for atmospheric circulation, water cycling, and precipitation. Positive feedbacks among deforestation, regional climate change, and fire could pose a serious threat for some tropical forests, but the details of such interactions are poorly understood.

Introduction

The fragmented landscape is becoming one of the most ubiquitous features of the modern world. Nowhere is habitat fragmentation occurring more rapidly than in the tropics, where several hundred million hectares of forest have been destroyed during the last few decades (Lanly 1982; Achard *et al.* 2002). The correlated processes of habitat loss and fragmentation are probably the greatest single threat to tropical biodiversity (Laurance and Bierregaard 1997) and alter many ecosystem functions such as carbon storage, biogeochemical cycling, and regional hydrology (Lean and Warrilow 1989; Kauffman *et al.* 1995; Fearnside 2000).

Here I synthesize available information on the impacts of habitat fragmentation on forest–climate interactions in the tropics. Although much is uncertain, it is apparent that fragmentation alters such interactions in diverse ways and at varying spatial scales. Understanding these interactions and their effects on forest functioning is essential both for interpreting the effects of global climate change on tropical ecosystems, and for assessing the impacts of rapid forest conversion on the physical and biological environment.

Size and shape of fragments

The processes of deforestation and forest fragmentation are inextricably linked. As land conversion proceeds, remnant forest patches almost always persist (by happenstance or design) within a matrix of drastically modified land, such as cattle pastures, slash-and-burn farming plots, or scrubby regrowth. Most human-dominated landscapes are numerically dominated by small (<100 ha) forest fragments (Fig. 3.1), although a sizeable fraction of the remaining forest may persist in a few large (>1000 ha) fragments (Ranta *et al.* 1998; Gascon *et al.* 2000; Cochrane and Laurance 2002).

One of the most critical consequences of habitat fragmentation is a drastic increase in the amount of abrupt, artificial forest edge. Prevailing land uses, such as slash-and-burn farming and cattle ranching, typically create irregularly shaped fragments with large amounts of edge (Fig. 3.2). In the Brazilian Amazon, for example, the area of forest in 1988 that was fragmented (<100 km² in area) or vulnerable to edge effects (<1 km from forest edge) was over 150% larger than that which had actually been deforested (Skole and Tucker 1993). Remote-sensing analyses suggest that because of rapid deforestation, almost 20,000 km of new forest edge is being

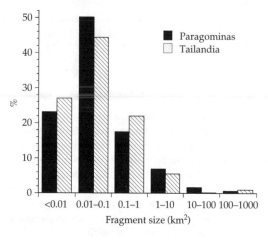

Figure 3.1 In human-dominated landscapes, most forest fragments are small (<1 km²). Data shown are for two fragmented areas in eastern Amazonia (after Cochrane and Laurance 2002).

created each year in Brazilian Amazonia alone (M. A. Cochrane, personal communication).

Microclimate and wind

On a local scale (<1 km), deforestation reduces plant evapotranspiration, humidity, effective soil depth, water-table height, and surface roughness; and increases soil erosion, soil temperatures, and surface albedo (Wright *et al.* 1996; Gash and Nobre 1997). Thus, the cleared lands that surround forest fragments differ greatly from forest in their physical and hydrological characteristics.

The forest edge is the interface between fragments and their adjoining clearings, and the proliferation of edge has major impacts on many ecological processes. Undisturbed rainforests are dark and humid, with stable temperatures, little wind, and nearly continuous canopy cover (Lovejoy *et al.* 1986; Laurance *et al.* 2002a), but when adjoined by clearings these conditions are sharply altered. On newly created edges, elevated temperatures, reduced humidity, and increased sunlight and vapour pressure deficits can penetrate

Tailândia   Paragominas

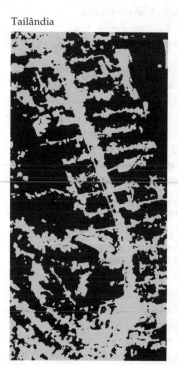
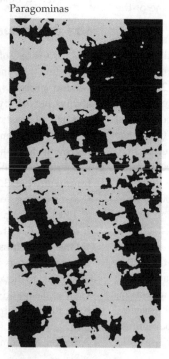

Figure 3.2 Habitat fragmentation leads to a proliferation of forest edge, as shown in two landscapes in eastern Amazonia (each 600 km² in area; dark areas are forest and light areas are mostly pastures). Tailândia, a government-sponsored colonization project, shows the characteristic 'fish-bone' pattern of deforestation, whereas Paragominas is a cattle-ranching and logging frontier. For each square kilometre of cleared land in these two areas, an average of 1.5 km of forest edge was created (after Cochrane and Laurance 2002).

at least 40–60 m into fragment interiors (Kapos 1989; Didham and Lawton 1999; Sizer and Tanner 1999). Such changes increase evapotranspiration in understorey vegetation, leading to depleted soil moisture and creating stresses for drought-sensitive plants (Kapos 1989; Malcolm 1998).

In addition, the edges of habitat remnants are exposed to increased wind speed, turbulence, and vorticity (Bergen 1985; Miller *et al.* 1991). Wind disturbance is an important ecological force in the tropics, especially in the cyclonic and hurricane zones from about 7° to 20° latitude (Webb 1958; Lugo *et al.* 1983), and also in equatorial forests affected by convectional storms (Nelson *et al.* 1994) and prevailing winds (Laurance 1997). Winds striking an abrupt forest edge can exert strong lateral-shear forces on exposed trees and create considerable downwind turbulence for at least 2–10 times the height of the forest edge (Somerville 1980; Savill 1983). Greater windspeeds increase the persistence and frequency of wind eddies near edges that can heavily buffet the upper 40% of the forest (Bull and Reynolds 1968).

These physical alterations lead to sharply increased tree mortality and damage within 100–300 m of fragment margins (Fig. 3.3; Laurance *et al.* 1998a). In central Amazonia, large (>60 cm

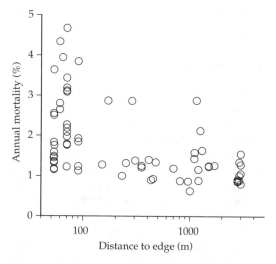

Figure 3.3 In rainforests, tree mortality rises sharply near forest edges. Data shown are from a long-term experimental study of forest fragmentation in the central Amazon (after Laurance *et al.* 1998a).

diameter) trees are especially vulnerable, dying nearly three times faster near edges than in forest interiors (Laurance *et al.* 2000). Some trees near edges simply drop their leaves and die standing (Lovejoy *et al.* 1986; Sizer and Tanner 1999), apparently because sudden changes in moisture, temperature, or light exceed their physiological tolerances. Many others are killed by winds, as evidenced by the fact that trees near edges are significantly more likely to be snapped or uprooted than those in forest interiors (D'Angelo *et al.* 2004).

Chronically elevated tree mortality has myriad effects. It fundamentally alters canopy-gap dynamics (Laurance *et al.* 1998a; Lewis 1998), which influences forest structure, composition, and diversity. Smaller fragments often become hyper-disturbed, leading to progressive changes in floristic composition (Laurance 1997). New trees regenerating near forest edges are significantly biased towards disturbance-loving pioneer and secondary species and against old-growth, forest-interior species (Viana *et al.* 1997; Laurance *et al.* 1998b). Lianas—important structural parasites that reduce tree growth, survival, and reproduction—increase in density near edges and may further elevate tree mortality (Oliveira-Filho *et al.* 1997; Viana *et al.* 1997; Laurance *et al.* 2001b). Leaf litter accumulates near edges (Sizer *et al.* 2000; Vasconcelos and Luizão 2004) as drought-stressed trees shed leaves, and may negatively affect seed germination (Bruna 1999) and seedling survival (Scariot 2001). Finally, fragmented forests exhibit a marked decline of biomass (Laurance *et al.* 1997), increased necromass, and accelerated carbon cycling (Nascimento and Laurance 2004), and are probably a non-trivial source of atmospheric carbon emissions (Laurance *et al.* 1998c).

Accelerated tree mortality directly affects forest–climate interactions by increasing the density of treefall gaps and altering canopy structure. Recurring canopy damage exacerbates edge-related changes in microclimate, increasing daytime temperature (Malcolm 1998) and vapour pressure deficits (Camargo and Kapos 1995) and altering the amount and spectral quality of light reaching the forest floor (Turton 1992). Such changes create physiological stresses for sensitive plant species

(Kapos *et al.* 1993; Bruna 2002). Gaps in the canopy are also prone to wind vortices that can kill or damage adjoining trees (Bull and Reynolds 1968) and can become foci for recurring canopy disturbances (Laurance 1997). Thus, edge-related changes in microclimate can be substantially magnified by elevated canopy damage near edges.

Edge and landscape structure

The physical structure of edges strongly influences forest–climate interactions. In the tropics, newly formed edges (<5 yr old) are structurally open and thus more permeable to lateral light and hot, dry winds than are older edges, which tend to become 'sealed' by a proliferation of vines and second growth (Kapos *et al.* 1993; Camargo and Kapos 1995; Didham and Lawton 1999). Wind damage, however, is unlikely to lessen as fragment edges become older and less permeable, as downwind turbulence increases when edge permeability is reduced (Savill 1983). Nevertheless, edge structure influences the intensity of many edge effects, and land use practices that repeatedly disturb fragment margins (such as regular burning) can have severe impacts if they prevent natural edge closure (Cochrane *et al.* 1999; Didham and Lawton 1999; Gascon *et al.* 2000).

In addition, edge orientation (aspect) can affect microclimatic parameters that influence plant germination, growth, and survival (Turton and Freiberger 1997). For example, heat and desiccation stress are highest on edges facing the afternoon sun (Malcolm 1998), whereas wind disturbance and atmospheric deposition of pollutants (Weathers *et al.* 2001) are greatest on edges exposed to prevailing winds.

Fragment size and isolation also influence forest–climate interactions. Large clearings surrounding fragments have greater 'fetch' than small clearings, resulting in higher wind velocities and increased structural damage to adjoining forest stands (Somerville 1980; Savill 1983). Desiccation and temperature extremes are also likely to increase with clearing size. Fragments that are small or have irregular boundaries are especially vulnerable to edge effects, because any point within the fragment will be influenced by multiple nearby edges, rather than a single edge (Malcolm 1994, 1998). Empirical and modelling studies in central Amazonia suggest that the impacts of edge-related tree mortality will rise sharply once fragments fall below 100–400 ha in area, depending on fragment shape (Laurance *et al.* 1998a).

Finally, the structure of modified vegetation surrounding fragments can clearly affect forest–climate interactions. Fragments surrounded by regrowth forests are partially buffered from damaging winds and harsh external microclimates, and suffer lower edge-related tree mortality than do those encircled by cattle pastures (Mesquita *et al.* 1999). The hydrology of different vegetation types also varies considerably. For example, in the eastern Amazon, degraded cattle pastures (which are dominated by shrubs and small trees) contain deep-rooted plants that absorb deep soil-water and thereby maintain moderately high rates of evapotranspiration during the dry season. This is in sharp contrast to managed pastures (grass monocultures), which contain virtually no deep roots and exhibit little evapotranspiration during dry periods (Nepstad *et al.* 1994). Hence, fragments surrounded by managed pastures may experience greater desiccation stress than those surrounded by degraded pastures or regrowth, because the former fail to recycle water vapour into the atmosphere during the critical dry-season months.

Edge-related fires

Except when subjected to strong droughts (Leighton and Wirawan 1986; Peres 1999; Gudhardja *et al.* 2000), large, unbroken tracts of humid tropical forest are usually highly resistant to fire, both because the dense canopy maintains humid, nearly windless conditions and because fine fuels such as leaf litter, which can be highly flammable, decompose rapidly (Nepstad *et al.* 1999a). When fragmented, however, tropical forests become drastically more vulnerable. Fragments tend to have dry, fire-prone edges with large amounts of litter and wood debris (Cochrane *et al.* 1999; Nascimento and Laurance 2004; Vasconcelos and Luizão 2004). They are also frequently juxtaposed with cattle pastures, which are regularly burned to control weeds and promote new grass. In addition, fragments are particularly vulnerable to periodic droughts, which increase already-high tree mortality and litter production (Laurance and Williamson 2001) and thereby augment forest fuels at a time when conditions are driest. Finally, forest fragments are frequently disturbed by logging (Laurance and Cochrane 2001), which further exacerbates forest desiccation, fuel loading, and vulnerability to fire (Uhl and Kauffman 1990; Holdsworth and Uhl 1997; Siegert *et al.* 2001).

In the eastern Amazon, surface fires that originate in adjoining pastures can penetrate large distances into fragment interiors (Fig. 3.4; Cochrane and Laurance 2002). Although confined to the forest floor, such fires are highly destructive because rainforest plants are poorly adapted for fire, having thin bark and no underground buds from which to resprout (Uhl and Kauffman 1990). Even light fires kill up to half of all trees and virtually all vines (Cochrane and Schulze 1999; Barlow *et al.* 2003). Subsequent fires are far more intense because dying plants increase fuel loads and reduce canopy cover, promoting forest desiccation (Cochrane *et al.* 1999). Forest fragments affected by such recurring fires may 'implode'

over time as their margins collapse inward (Gascon *et al.* 2000). Because surface fires can penetrate up to several kilometres into forest interiors, even very large (>100,000 ha) forest fragments may be vulnerable (Cochrane and Laurance 2002).

Spatial and temporal variability in rainfall have major effects on fire frequency and intensity. Fires are especially problematic in tropical regions affected by strong dry seasons or by periodic droughts— such as those occurring during El Niño events throughout large areas of the neotropics, Southeast Asia, and Australasia (Leighton and Wirawan 1986; Kinnaird and O'Brien 1998; Cochrane *et al.* 1999; Curran *et al.* 1999; Nepstad *et al.* 1999a; Barlow *et al.* 2003). In the Brazilian Amazon, more than one-third of the closed-canopy forest experiences soil-water deficits during strong droughts (Nepstad *et al.* 2001) and some 45 million ha (13% of the total forest area) may already be vulnerable to edge-related fires (Cochrane 2001).

Regional effects

Insights into land–atmosphere interactions at landscape to regional scales (ca. 10–1000 km) derive from simulation models that may integrate data from weather satellites, Doppler radar, radiosonde sites, micrometeorological studies, and other observations. Conclusions of larger-scale models (including global-circulation and meso-scale models) are more tenuous than those focusing on local processes, given the potentially great complexity of interacting factors at widely varying spatial scales. Rather than attempt a detailed review, I focus here on a few key findings.

A number of regional modelling studies have attempted to project the future climatic impacts of severe tropical deforestation. To simplify the models, several studies have assumed complete conversion of Amazonian (Nobre *et al.* 1991; Dickinson and Kennedy 1992; Lean and Rowntree 1993; Gash and Nobre 1997) or Southeast Asian (Henderson-Sellers *et al.* 1993) forests to pasture or savanna. Although results have varied, most studies predict that uniform deforestation will lead to markedly decreased regional rainfall (on the order of 20–30%) as well as lower evaporation,

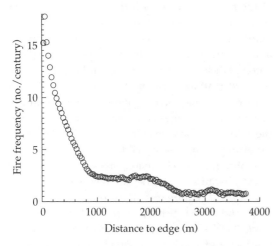

Figure 3.4 Estimated fire frequencies as a function of distance from forest edge for the Tailândia region in eastern Amazonia. Data are based on remote-sensing imagery spanning a 14-yr period (after Cochrane and Laurance 2002).

cloud cover, and soil moisture; and higher albedo and surface temperatures (Lean and Rowntree 1993; Sud *et al.* 1996; Gash and Nobre 1997).

However, regional patterns of forest loss are never uniform. For example, tropical deforestation is much greater in south-eastern Amazonia than elsewhere in the basin and more severe in western Africa than in the Congo Basin (Laurance 1999). To better approximate reality, some investigators modelled actual (ca. 1988) deforestation patterns in Brazilian Amazonia. Predicted effects have been less dramatic than for uniform clearing, with deforested regions experiencing limited (6–8%) declines in rainfall, moderate (18–33%) reductions in evapotranspiration, elevated surface temperatures, and greater wind speeds (due to reduced surface drag) that could affect moisture convergence and circulation (Walker *et al.* 1995; Sud *et al.* 1996). Although of limited magnitude, for the large expanses of Amazonia that have pronounced dry seasons and are strongly influenced by El Niño droughts (Nepstad *et al.* 1994), even modest drying and warming trends could potentially cause marked increases in forest vulnerability to fire, especially where forests are fragmented or logged.

Forest clearing and fragmentation create mosaics of land with different physical properties. One potentially important effect of increasing fragmentation is the 'vegetation breeze', whereby moist air is pulled away from forests into adjoining pastures and clearings (Silva Dias and Regnier 1996; Baidya Roy and Avissar 2000). The humid air over forests is drawn into the clearing and condenses into rain-producing clouds, then is recycled—as dry air—back over the forest. This effect has been observed in clearings as small as a few hundred hectares, but extensive clearings spanning roughly 100 km or more apparently cause much larger-scale forest desiccation (Avissar and Liu 1996). In Rondônia, Brazil, Silva Dias *et al.* (2002) describe a 20-km-wide zone of reduced rainfall surrounding large forest clearings. Thus, by drawing moisture away from adjoining forests, large clearings might increase the vulnerability of forests to fire.

The deforestation process itself can also increase forest desiccation. Smoke from tropical forest fires

has been shown to reduce rainfall and possibly cloud cover (Rosenfeld 1999). This occurs because burning hypersaturates the atmosphere with cloud condensation nuclei (microscopic particles in aerosol form) that bind with water molecules in the atmosphere, inhibiting the formation of raindrops. As a result, large-scale forest burning can create rain-shadows that have been observed to extend for hundreds of kilometres downwind (Freitas *et al.* 2000). Aerosols from forest burning may also affect the thermodynamic stability of the atmosphere, by absorbing and scattering incoming solar radiation and altering cloud formation, but the consequences of such changes for the regional hydrology and climate are poorly understood (Martins *et al.* 1998; Andreae 2001). During the dry season, large expanses of the Amazon (ca. 1.2–2.6 million km^2) exhibit significantly elevated levels of atmospheric aerosols from forest burning (Procopio *et al.* 2004).

The diverse environmental changes that affect cleared and fragmented landscapes might interact in complex ways. Of particular concern is that positive feedbacks may arise (Fig. 3.5), in which large-scale deforestation increases local or regional desiccation and thereby renders remaining forests more vulnerable to fire, promoting further deforestation (Laurance and Williamson 2001; Nepstad *et al.* 2001). Such feedbacks could be driven by many of the mechanisms described above—edge-related fires, the vegetation breeze, the moisture-trapping effects of smoke plumes, and regional drying and warming from declining evapotranspiration, among others. The net effect is that large expanses of forest that are currently too humid or intact to burn readily may become more vulnerable in the future. Such changes could profoundly influence the rate and spatial pattern of forest destruction; for example, more-seasonal forests often face considerably higher conversion pressure than wetter forests, because the former are both easier to burn initially and require less effort to maintain as pastures and farms (Schneider *et al.* 2000; Laurance *et al.* 2002b). A greater incidence of fire in the tropics could also have important global effects, by increasing greenhouse-gas emissions and thereby exacerbating global warming (Fearnside 2000; Houghton *et al.* 2000).

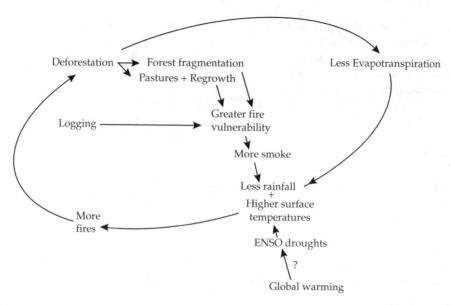

Figure 3.5 Potential positive feedbacks among forest fragmentation, logging, fires, and climate change in the Amazon (after Laurance and Williamson 2001).

Larger-scale effects

In addition to local and regional effects, tropical deforestation could have important remote impacts on other regions. In Costa Rica, extensive deforestation of the nearby Caribbean lowlands has apparently led to marked downwind reductions of cloud cover, rainfall, and mist at Monteverde Cloud Forest Reserve (Lawton *et al.* 2001). Modelling studies suggest that rapidly increasing deforestation in the Indonesian Archipelago may have a strong impact on the broader regional climate as a result of feedbacks among the biosphere, atmosphere, and ocean (Delire *et al.* 2001). Some simulations suggest that heavy Amazonian deforestation will alter precipitation in areas south of the basin (Henderson-Sellers *et al.* 1993), in Central America and the Caribbean, and even at middle and higher latitudes (Gandú and Silva Dias 1998; Gedney and Valdez 2000; Avissar and Nobre 2002; Avissar *et al.* 2002).

Finally, the regional climatic effects of deforestation could potentially interact with global warming. Costa and Foley (2000) concluded that global warming would exacerbate the effects of Amazonian deforestation, which by reducing evapotranspiration limits the capacity of the land surface to cool itself. The net effect could be markedly higher surface temperatures and a 20% reduction in regional rainfall (Costa and Foley 2000). In addition, some models suggest that El Niño droughts and tropical storms may increase in frequency or severity as a result of global warming (IPCC 1996; Timmerman *et al.* 1999). At the least, the frequency of warm weather events should rise, and the likelihood of cool weather events should decline, as a consequence of higher mean temperatures (Mahlman 1997). Thus, by increasing the incidence of droughts and hot weather, global warming could potentially promote alarming positive feedbacks among deforestation, regional desiccation, and fire (Fig. 3.5).

Conclusions

Much remains unknown about the influence of land-cover change on tropical forest–climate interactions. Local-scale processes have been best characterized, but important questions remain. For example, virtually nothing is known about the effects of altered microclimatic conditions near forest edges on plant and soil respiration rates.

Higher temperatures near edges should generally increase respiration whereas reduced humidity near edges could have an opposite effect (e.g. Chambers *et al.* 2001d). Given that tropical soils contain more carbon (in soil organic matter and root biomass) than the above-ground vegetation (Davidson and Trumbore 1995; Moraes *et al.* 1995), altered soil respiration rates could potentially have a large impact on the carbon balance of fragmented forests. On larger scales, our understanding of the effects of deforestation on regional climates is still rudimentary, despite many indications that such effects will be deleterious to forests.

For those attempting to assess the effects of global-change phenomena on intact tropical forests, it must be emphasized that edge-related alterations can penetrate large distances into forest tracts. Diverse physical and biotic changes often occur within the first 100–300 m of edges, and other phenomena, such as surface fires, may penetrate up to several kilometres inside forest margins. Where forest tracts adjoin major clearings, alterations in atmospheric circulation might infiltrate even farther into forests, perhaps 20 km or more. Finally, markedly increased concentrations of atmospheric aerosols, which have poorly understood effects on cloud cover and atmospheric stability, can occur up to several hundred kilometers downwind of major areas of biomass burning. Given the rapid pace of forest conversion in the tropics, care must be taken to distinguish the consequences of global-change phenomena from the ever-increasing effects of landscape- and region-scale alterations.

PART II

Atmospheric change and ecosystem processes

CHAPTER 4

Predicting the impacts of global environmental changes on tropical forests

Simon L. Lewis, Yadvinder Malhi, and Oliver L. Phillips

Recent observations of widespread changes in mature tropical forests such as increasing tree growth, recruitment and mortality rates, and increasing above-ground biomass suggest that 'global change' agents may be affecting tropical forests far from deforestation fronts. However, consensus has yet to emerge over the robustness of these changes and the environmental drivers that may be causing them. This chapter focuses on the second part of this debate. In previous work we have identified 10 potential widespread drivers of environmental change: temperature, precipitation, solar radiation, climatic extremes (including El Niño–Southern Oscillation (ENSO) events), atmospheric CO_2 concentrations, nutrient deposition, O_3/acid depositions, hunting, land use change, and increasing liana numbers. We note that each of these environmental changes is expected to leave a unique 'fingerprint' in tropical forests, as drivers directly force different processes, have different distributions in space and time, and may affect some forests more than others (e.g. depending on soil fertility). We present testable *a priori* predictions of forest responses to help ecologists attribute particular changes in forests to particular causes. Finally, we discuss how these drivers may change and the possible future consequences for tropical forests.

Introduction

Over the past century virtually all ecosystems on Earth have come under increasing human influence. This has been through direct contact and transformation (e.g. for farming, through hunting, or the use of fire), the effects of habitat fragmentation, the production of pollutants or the substantial alteration of major biogeochemical cycles, such as the global carbon, water, and nitrogen cycles. However, for tropical forests that are far from most direct human impacts, the question as to whether these ecosystems have been substantially altered and what may be causing these changes is actively debated (Lewis *et al.* 2004a).

Whether tropical forests are showing widespread secular changes in dynamics, and why, is of broad interest as they store ca. 40% of the carbon residing in terrestrial vegetation and annually process about six times as much carbon through photosynthesis as humans release to the atmosphere through fossil fuel combustion (Malhi and Grace 2000). In addition, tropical forests harbour more than 50% of the world's species (Groombridge and Jenkins 2003). Thus, relatively small yet consistent changes *within* remaining tropical forests as a biome could have global consequences for global climate, biodiversity, the carbon cycle, the rate of climate change, and hence human welfare.

Two widespread changes in tropical forests have received attention: increases in tree stem turnover and increases in the above-ground biomass of forest stands (Phillips and Gentry 1994; Phillips *et al.* 1998b). A case can be made that these trends have not been caused by widespread changes in environmental drivers, but are artefacts caused by the compilations of disparate datasets containing methodological errors coupled with the use of inappropriate statistical techniques (summarized in Phillips *et al.* 2002a). In response, much work has gone into expanding the datasets, particularly across South America (Malhi *et al.* 2002a), and addressing these 'artefactual' explanations of the trends (Phillips *et al.* 2002a; Chapter 10, this volume; Baker *et al.* 2004; Chapter 11, this volume; Lewis *et al.* 2004b,c, Chapter 12, this volume). These newer analyses confirm the qualitative findings of the initial two analyses, and have shown additional trends in South American tropical forests:

(1) stem turnover is rising owing to simultaneous increases in both recruitment and mortality;
(2) recruitment rates are greater than mortality rates, causing a net increase in stem density;
(3) stand-level growth and biomass mortality rates have both increased; and
(4) growth rates, on average, exceed mortality rates, leading to the documented increase in above-ground biomass (Baker *et al.*, Chapter 11, this volume; Lewis *et al.* 2004b, Chapter 12, this volume; Phillips *et al.*, Chapter 10, this volume).

Thus, one obvious question is: what may be causing such a suite of changes across large areas of tropical forest? To answer this question we first need to know which physical, chemical, and biological changes to the environment have occurred over recent decades and what their likely effects on tropical forest trees have been. In a recent paper, we presented a framework to systematically review the evidence that various widespread environmental changes have occurred across the tropics and assess the evidence of the most likely effects of each of these changes on tropical forests (Lewis *et al.* 2004a).

Our framework consists of four parts:

(1) systematic identification of mechanisms that might cause change;
(2) assessment of how much each potential driver has changed over the late twentieth century;
(3) assessment of whether the level of change in the driver is likely to be ecologically meaningful; and
(4) use of the knowledge that the different types of environmental change directly force different

ecological processes, have varying distributions in space and time, and may affect some forests more than others (e.g. depending upon soil fertility), to generate unique a priori predictions for each hypothesized change.

In the future, new hypotheses can be added to the framework, or refinements made to existing hypotheses and predictions, as necessary.

In this chapter we focus solely on part (4), generating predictions of change in growth, recruitment, mortality, and biomass in tropical forest stands from a systematic review of which environmental drivers have changed over recent decades (Lewis *et al.* 2004a).

For the purposes of identifying possible mechanisms of change, we assume that tropical tree populations are in a three-phase, time-lagged circular system that links growth, recruitment, and mortality processes (Fig. 4.1). Changes in these process rates result from forcing one of these processes, which in turn may alter the other two processes. We consider all potential drivers of change, regardless of whether they may be caused by anthropogenic changes or decadal-scale environmental oscillations. We do not

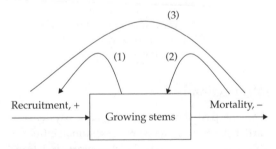

Figure 4.1 Schematic diagram of the relationships between stem mortality, recruitment, and growth of stems in a forest stand. Recruitment adds and mortality subtracts (the fluxes) stems from the standing stock of growing stems in the stand (the pool). The processes are linked since (1) growth affects recruitment, as a change in growth will change the rate of arrival of stems at the minimum threshold to join the stem population, (2) mortality affects growth, as changes in mortality change the number of remaining stems altering competition for resources, and (3) mortality affects recruitment, through the same process as (2). The system is time-lagged: a change to either growth, recruitment, or mortality will take time to percolate through the system, for example, an increase in mortality will not cause an instantaneous increase in recruitment, as juveniles will take time to respond and grow to become newly recruited stems. The same applies to biomass: a pool (standing stock) and input (growth increments) and output (losses of biomass from mortality) fluxes are also linked in a time-lagged circular process.

include an exhaustive list of possible drivers: if very implausible we ignore them (e.g. significant decreases in tropical stratospheric O_3, and hence increases in UV-B radiation).

Environmental changes over the past three decades

We have previously identified four potential physical drivers of change: air temperatures, precipitation, solar radiation, and climatic extremes (including ENSO); three potential chemical drivers: atmospheric CO_2 concentrations, nutrient depositions, and tropospheric O_3/acid deposition; and one potential biological driver: hunting pressure (Table 4.1, see also Lewis *et al.* 2004a). Two further drivers were identified: first, land use changes that can simultaneously alter the physical, chemical, and biological environment, and second, increases in lianas (woody climbing plants) of which the cause is unknown.

Altogether we identified 10 distinct drivers. Only one of these drivers, changes in hunting pressure, is expected to initially drive changes in tree recruitment (Table 4.1). Increasing liana density and changes in climatic extremes/ENSO events are expected to initially drive changes in mortality. Land use changes are expected to change both recruitment and mortality via different mechanisms. The other seven drivers are all expected to initially drive changes in tree growth.

Spatially, the drivers have distinct patterns: some have relatively uniform changes at the pantropical scale (CO_2, air temperatures), some at regional scales (rainfall, solar radiation, climatic extremes, nutrient and acid depositions) and some at only local scales (hunting, land use change; Table 4.1). In addition, the drivers can be grouped as to the type of change. Both pantropical drivers (CO_2 and air temperatures) are *point-change drivers*, as are rainfall and solar radiation (driver increases by Y at a point X where it was initially present). Land use change is an *extension-change driver* (driver increases by Y at a point X where it was initially absent). All the remaining drivers integrate both point and extension changes (Table 4.1). Combining the scale at which changes are relatively uniform and the mode of change—point or extension—gives the extent of the expected changes

in the driver. Thus, an extension-change driver that causes uniform changes at a local scale can have a near-global extent if that process, like increasing hunting pressure, is occurring over almost all areas of the biome. From the available data we can state that over the past two decades that:

(1) air temperatures and CO_2 concentrations have increased globally and fairly uniformly across all tropical forest regions;
(2) hunting and land use change have increased globally, but not uniformly at all locations;
(3) rainfall, N deposition and O_3/acid depositions have changed regionally;
(4) there was no documented global increase in climatic extremes/ENSO events; and
(5) incoming solar radiation may or may not have increased across some regions, but the diffuse: direct ratio probably has changed regionally.

This exercise shows that rainfall, N deposition, and O_3/acid depositions have only altered unidirectionally at a regional scale and are thus very unlikely to account for any widespread changes documented in tropical forests. More specifically different continents and regions are likely to be exposed to very different suites of changing drivers over the past two decades. Here, we draw attention to the main trends in the changes of drivers by continent, highlighting key differences between (but note that there are uncertainties associated with some of these trends, see Lewis *et al.* 2004a). Tropical South American forests are likely to have experienced:

(1) increasing temperatures, from a higher baseline than tropical African forests;
(2) increasing CO_2 concentrations;
(3) no consistent rainfall trends;
(4) ENSO droughts, but with a strong spatial pattern running approximately north-east (strongly affected) to south-west (weakly affected and often in antiphase);
(5) relatively small increases in N deposition and O_3/acid depositions compared with tropical Asian forests; and
(6) the lowest levels of fragmentation of any continent.

Solar radiation may also have increased across the tropical South America, although there is

Table 4.1 Summary of changes in the physical, chemical, and biological environment that may be driving temporal changes in the structure and dynamics of the tropical forest biome

Driver	Hypothesis	Description of mechanism	Level of driver	Impact of driver	Scale of change[a]	Type of change[b]	Extent of change[c]	Absolute annual change[d]	Theoretical consistency of effects?[e]	Mechanism experimentally demonstrated?	Key prediction
Air temperature	Air temperature	Long-term temperature increases affect photosynthesis, increasing/decreasing growth rates	Physical	Growth	Regional	Point	Global	$+0.024°C$	No	Yes	Growth rate changes correlate with local temperature trends
Air temperature	Respiration costs	Long-term temperature increases increase respiration rates decreasing growth rates	Physical	Growth	Regional	Point	Global	$+0.024°C$	Yes	Yes	Growth rate changes correlate with increases of minimum temperatures
Air temperature	Soil warming	Long-term temperature increases soil nutrient availability, increasing or decreasing growth rates	Physical	Growth	Regional	Point	Global	$+0.024°C$	No	Partially[f]	Growth rate changes correlate with local temperature trends with highest relative increases on nutrient-poor soils
Rainfall	Rainfall	Long-term rainfall changes change growth where water is a limiting resource	Physical	Growth	Regional	Point	Regional	-0.067 mm	Yes	Yes	Only water-limited sites where rainfall has decreased show dynamic change
Solar radiation	Global dimming	Long-term decreases in insolation affects photosynthesis increasing/decreasing growth rates	Physical	Growth	Regional	Point	Regional/or near-global	-0.30 W m^{-2}	No	Partially[f]	Growth rate changes correlate with local insolation trends
Solar radiation	Changing energy budget	Recent increases in solar radiation due to decreased cloudiness increases growth rates	Physical	Growth	Regional	Point	Regional	$+0.13$ W m^{-2}	Yes	Yes	Growth rate changes correlate with local insolation trends
Climatic extremes/ENSO	Climatic extremes/ENSO	Increased frequency and/or severity of extreme weather events increases tree mortality	Physical	Mortality	Local/regional	Point + extension	Regional/near-global	-0.020 index units	Yes	Yes	Dynamics altered where extremes for example, ENSO drought are known or suspected to have occurred

Driver	Process	Description	Type	Component	Scale			Magnitude	Detectable		Notes
CO_2	light use efficiency	Long-term atmospheric CO_2 increases photosynthesis, increasing growth rates	Chemical	Growth	Global	Point	Global	+ 1.53 ppm	Yes	Yes	Growth rate increases across most forests with greatest absolute increase in nutrient-rich aseasonal forests
CO_2	water use efficiency	Long-term atmospheric CO_2 increases water use efficiency, increases growth where water is a limiting resource	Chemical	Growth	Global	Point	Regional/or near-global	+ 1.53 ppm	Yes	Yes	Growth rate increases at nutrient-rich seasonal sites even where no change in rainfall
Nutrient depositions	N deposition	Changes in N deposition change growth rates	Chemical	Growth	Regional	Extension + some point	Regional	+ 0.013 kg N ha^{-1}	Yes[g]	Yes[g]	Changes in growth rates where deposition has increased
Nutrient depositions	P depositions	Changes in P deposition change growth rates	Chemical	Growth	Regional	Extension + some point	Regional	Unquantified	Yes[g]	Yes[g]	Changes in growth rates where deposition has increased
O_3/acid deposition	O_3/acid deposition	Increases in acidity of rainfall and tropospheric ozone decreases growth and increase mortality	Chemical	Growth	Regional	Extension + some point	Regional	+ 1.1 million ha > 60 ppb O_3	Yes	Yes	Changes where changes in driven known or suspected to occur, for example, downwind of biomass burning/industrial activity
Hunting	Competitive release	Hunting of frugivores impacts seed dispersal of large-seeded species, benefiting small-seeded animal and wind-dispersed seeds which have higher growth and stem turnover rates	Biological	Recruitment	Local	Extension + point	Near global	+ 0.13 million tonnes bushmeat	Yes	No	Only defaunated sites would show changes
Land use change	Edge effects	Tree mortality increases due to increasing encroachment of human-modified habitat	Biological	Mortality	Local	Extension	Near global	Unquantified increase	Yes	Yes	Mortality increases correlated with distance from edge

Table 4.1 (Continued)

Driver	Hypothesis	Description of mechanism	Level of driver	Impact of driver	Scale of change[a]	Type of change[b]	Extent of change[c]	Absolute annual change[d]	Theoretical consistency of effects?[e]	Mechanism experimentally demonstrated?	Key prediction
Land use change	Fragmentation-pioneer effects	Habitat fragmentation increases the number of pioneer trees in degraded areas which flood remaining forest with their seed, increasing pioneer numbers increasing recruitment	Biological	Recruitment	Local	Extension	Near global	+0.59 million ha	Yes	No	Disregarding pioneer species from data removes effects
Second-order	Lianas	Tree mortality increases due to increased liana loading	Biological	Mortality	Unknown[h]	Unknown[h]	Unknown[h]	+0.28 lianas ha^{-1}	Yes	No	Mortality increases where significant increases in liana populations detected or suspected

[a] Scale at which changes are relatively uniform (local, regional, global).

[b] Either a driver increase by Y at a point X where it was initially absent (extension-changes), or a driver increase by Y at a point X where it was initially present (point-change).

[c] Product of scale at which changes are relatively uniform and the extent of those changes (regional, near global, global).

[d] Using linear regression on available annual means for each driver from 1980 to 2000, where possible.

[e] That is, is there consensus in the published literature that if a driver changes the process driven will occur in a specified direction?

[f] Shown for a temperate forest.

[g] Additions of nutrients increases growth if that nutrient is limiting; but which nutrient is limiting is unknown in most forests.

[h] Depends on the driver of the increases in liana density and dominance.

uncertainty over this (see Lewis *et al.* 2004a for a discussion).

African tropical forests, in contrast, are likely to have experienced:

(1) relatively low initial air temperatures, which are increasing;
(2) increasing CO_2 concentrations;
(3) a strong trend of decreasing rainfall, which is rarely affected by ENSO droughts;
(4) relatively small increases in N deposition and O_3/acid depositions, compared with tropical Asian forests;
(5) very high levels of bushmeat extraction (hunting) compared with tropical South American and Asian forests; and
(6) moderate levels of fragmentation compared with tropical Asia.

Asian tropical forests are likely to have experienced:

(1) increasing temperatures, from a higher baseline than tropical African forests;
(2) increasing CO_2 concentrations;
(3) no overall trends in rainfall;
(4) high and rapidly increasing N deposition and O_3/acid depositions, compared with tropical African and South American forests; and
(5) highly fragmented and often intensively logged forests compared with the other continents.

In addition there is some evidence of a decrease in solar radiation, but an increase in the diffuse fraction owing to relatively higher air pollution compared with tropical African and South American forests.

Clearly, these broad differences between continents will contain regional variation. For example, Brazil's Atlantic forest—highly fragmented and near intensive agriculture and large industrialized areas—may experience drivers more similar to forests in Southeast Asia than much of tropical South America.

Table 4.1 and Fig. 4.2 give an indication of how much each driver has changed across the tropical forest biome. However, note that the quality of the data varies widely and is frequently very sparse both spatially and temporally. Very careful inter-

pretation is required: see Lewis *et al.* (2004a) for details.

While we identified 10 possible drivers of change that may be causing shifts in tropical forest structure and dynamics, some drivers potentially affect multiple processes, giving a total of 16 distinct hypotheses (Table 4.1). Thus, increasing temperatures may affect either photosynthetic or respiration rates, or accelerate soil nutrient cycling; solar radiation changes, whether increases or changes in diffuse/direct proportions, may increase growth, but via different mechanisms; increasing CO_2 concentrations may affect both light use efficiency and water use efficiency; nutrient deposition may increase nitrogen inputs alone and/or inputs of other nutrients; and land use change creates forest edges, and alters the population dynamics of pioneer trees, both of which may alter forest dynamics. In Lewis *et al.* (2004a) we review the relevant theoretical, laboratory and field experiments, and observations relating to each hypothesis, and below merely state the hypotheses, followed by predictions of impacts on tree growth, recruitment, and mortality. These include the unique predictions, or fingerprints, from the combination of a changing environmental driver (with its spatial and temporal signal), coupled with its forcing of a particular forest process, and the expected differential changes in different forest types (e.g. forests growing on low versus high fertility soils).

Predictions of changes in tropical forests

Different drivers of change initially force different processes, and often have different distributions in time and space. These properties allow us to state testable predictions for each hypothesis, which, when compiled as suites of predictions, demonstrate that each driver should leave a unique fingerprint or 'signature' in permanent tree sample plot data. Of course once likely drivers and mechanisms are established, experimental tests will be required to increase confidence that correlative patterns are also causal. Below we give sets of testable predictions in response to each driver (see Table 4.2 for a summary).

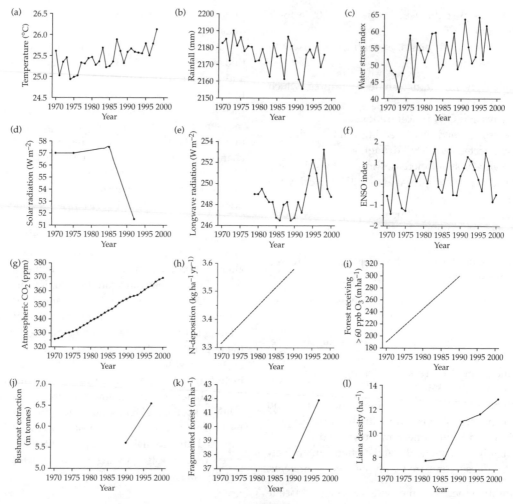

Figure 4.2 Environmental changes across the tropical forest biome, 1970–2000. Dots represent data points, solid lines are linear interpolation between data points. Dashed lines represent linear regression from numerical model output. The data are from multiple sources of variable quality. Cautious interpretation is required; see Lewis *et al.* (2004a) for further details. Data represent mean changes for the biome unless otherwise stated. (a) Mean annual air temperature, (b) mean annual rainfall (note: most regions show no trend; the decline is driven largely by one area, northern Congo), (c) the water stress index of Malhi and Wright (Chapter 1, this volume) (note: most regions show no trend; the decline is driven largely by northern Congo), (d) total solar radiation from ground measurements from across the Southern Hemisphere, (e) estimated annual 'top-of-atmosphere' longwave radiation, correlated with cloudiness and hence incoming solar radiation, from 20°N to 20°S, (f) annual multivariate ENSO index anomaly, (g) annual atmospheric CO_2 concentration measured at Mauna Loa, Hawaii, (h) estimated mean annual nitrogen deposition onto tropical forests, from model output, (i) estimated area of forest receiving high O_3 depositions, from model output, (j) estimated total wild mammal meat extraction from tropical forests, (k) total tropical forest in small fragments, estimated from satellite data, (l) large (>10 cm diameter) liana density in western Amazonian forest plots.

Temperature

Increasing air temperatures increase the rates of virtually all chemical and biological processes in plants and soils, until temperatures reach the point where enzymes are denatured. A rise in temperature may affect four major processes that may in turn alter tree growth: photosynthesis, respiration, soil nutrient availability, and ontogenetic development. Hence we can make two sets of predictions,

Table 4.2 Predictions for 16 hypotheses that may have caused changes in tropical forest dynamics during the latter part of the twentieth century

	1[a]	2	3	4	5[a]	6	7	8	9	10[a]	11	12	13	14	15	16
Multiple tests																
Growth, rect, and biomass ↑ especially where physical factor ↑	?	n	y^b	n	?	y^b	n	n	n	n	n	n	n	n	n	n
Growth, rect, and Biomass ↑ especially where physical factor ↓	?	y	n	n	?	n	y^b	n	n	n	n	n	n	n	n	n
Growth, rect, and Biomass ↑ especially where chemical factor ↑	n	n	n	n	n	n	n	y^b	y^b	?	y	n	n	n	n	n
Growth, rect, and Biomass ↑ especially where chemical factor ↓	n	n	n	n	n	n	n	n	n	?	n	y^b	n	n	n	n
Recruitment rate ↑ only where biological factor ↑	n	n	n	n	n	n	n	n	n	n	n	n	y^b	y^b	y^b	y^b
Mortality rate ↑ especially where physical factor ↑	y^b	y^b	y^b	n	n	y^b	y^b	n	n	n	n	n	n	n	n	n
Mortality rate ↑ especially where chemical factor ↑	n	n	n	n	n	n	n	y^b	y^b	?	n	y^b	n	n	n	n
Mortality rate ↑ especially where biological factor ↑	n	n	n	n	n	n	n	n	n	n	n	n	y^b	y^b	y^b	y^b
Single tests																
Stand biomass ↑	?	n	y?	n	?	y	n	y?	y?	?	y	n	n	n	n	n
Stand biomass ↓	?	y	y	y	?	n	y	n	n	?	n	y	n	y	y	y
Stand stem density ↑, as recruitment lags mortality	?	n?	y	n	?	y	n	y	y	?	y	n	n	n	n	y?
Stand stem density ↓, as recruitment lags mortality	?	y?	n	y	?	n	y	n	n	?	n	y	n	y	n	y
Mortality change and growth change positively correlated	?	y	y?	y?	?	y?	n	y?	y?	?	n	y	y	n	y	n
Recruitment change and growth change positively correlated	?	y	y	y	y	y?	n	y	y	?	y	n	y	n	y	n
Structurally similar sites respond similarly	y	y	y	y?	n	y?	n	y	y	n	n	n	n	n	n	?
Recruits larger and faster-growing in later censuses	y	n	y	y	n	y	n	y	y	n	n	n	y	y	y	n
Changes correlate with distance from the nearest edge	n	n	n	n	n	n	n	n	n	n?	n?	n?	n?	y	y	?

Table 4.2 (Continued)

	1[a]	2	3	4	5[a]	6	7	8	9	10[a]	11	12	13	14	15	16
Removal of pioneers removes effects	n	n	n	n	n	n	n	n	n	n	n	n	n	n	y	n
Changes increasingly dominated by large trees	n?	n?	n?	n?	y?	n?	?	n?	n?	n?	n	?	n	y[b]	n	y?
Increase in mortality dominated by deaths of earlier recruits	n	n	n	n	n	n	n	n	n	n	n	n	y	n	y	n
Large trees increasingly prone to death	n?	n?	n?	n?	?	n?	y?	n?	n?	y?	n	y?	n	y	y	y
Small trees increasingly prone to death	n	n?	n	n	n	n	?	n	n	n	n	n?	y	n	y	n
Mortality ↑ restricted to a few species	n	n	n	n	n	n	n?	n	n	y?	n	n?	y	n	y	n
Magnitude of change correlates with soil fertility	y[b]	n?	y[b]	y[b]	y[b]	y[b]	n	y[b]	y[b]?	y[b]	y[b]	n?	y?	n	?	y?
Magnitude of change correlates with soil water	y[b]	n?	y[b]	y[b]	y[b]	y[b]	n?	y[b]	y[b]	n	n	n?	n	n	n	n
Recruitment ↑ correlates with level of defaunation	n	n	n	n	n	n	n	n	n	n	n	n	y	n?[a]	n?[a]	n

[a] There is uncertainty whether changes in the driver cause increases or decreases in growth.

[b] These hypotheses can be further distinguished, as different mechanisms predict different responses, for example, maximum growth rate increases under hypothesis 1 will be at nutrient-rich aseasonal sites that warmed, while under hypothesis 3 maximum growth rate increases will be at nutrient-poor aseasonal sites.

Notes: Abbreviations: y, yes; n, no; ?, response unclear. Multiple tests are applicable to more than one external forcing factor, for example, chemical factor CO_2, or N deposition. Each hypothesis gives a unique 'fingerprint' of expected responses to the tests.

one for if rising temperatures increase growth and one for if rising temperatures decrease growth (Table 4.2; Lewis *et al.* 2004a).

The unique prediction is that forest growth and biomass increases/decreases would correlate with long-term trends in temperature. If growth rates changes were best correlated with mean daily minimum temperatures, this would suggest that changes in respiration were dominating the pattern. If correlations were best between growth rates and mean daily maximum temperatures, then the impacts of temperature on photosynthetic rates would be the most promising candidate. If growth rates were correlated with temperature changes, but effects are disproportionately large and positive in forests growing on nutrient-poor soils, then soil warming is the most likely mechanism. Differing regional-scale responses may provide further information. Central African forests, in particular, are consistently cooler than many other tropical forest regions (Malhi and Wright, Chapter 1, this volume), so if changes in growth are significantly different in African forests compared with Southeast Asian or South American forests, this may indicate that temperature changes are important. Also, some regions (notably western Amazonia) experienced cooling in the 1950s and 1960s owing to local climatic oscillations. Few plot data are available from this period, but if these cooling trends recur, then patterns of temperature response may be reversed. Spatially and temporally consistent results like these scenarios could provide good evidence of a 'fingerprint' of increasing temperatures (cf. Parmesan and Yohe 2003).

Precipitation

Annual rainfall and the length and the intensity of the dry season have not altered consistently across the tropical forest biome, but rainfall appears to have decreased markedly across the northern Congo Basin over at least the past two decades (Malhi and Wright, Chapter 1, this volume). Thus, changing precipitation cannot drive biome-wide shifts in dynamics, but may dominate patterns in some areas. The unique prediction from this hypothesis is that changes in growth and possibly mortality will correlate with long-term changes in rainfall patterns, but only in seasonal forests and only where changes in rainfall reduce or extend the intensity of water shortage in the dry season.

Solar radiation

The amount of incoming solar radiation may have increased over much of the tropics over the past two decades, owing to a decrease in tropical cloudiness as measured by satellites (Wielicki *et al.* 2002). By contrast, ground-based measurements suggest that significant reductions in solar radiation reaching the Earth's surface have occurred during the past 50 yr ('global dimming'), apparently caused by increased anthropogenic aerosol production (Stanhill and Cohen 2001). If satellite data showing an increase in incoming solar radiation are correct, then increases in forest growth are predicted, and will follow the spatial pattern of the satellite data, if trees are light-limited (e.g. see Graham *et al.* 2003). However, theory predicts two opposite outcomes for tropical forests in response to global dimming, if it is occurring. First, a long-term reduction in solar radiation would decrease photosynthesis, and hence growth and stand biomass, if trees are light-limited. Alternatively, the reduction in total solar radiation is likely to be accompanied by an *increase* in diffuse solar radiation. This is expected to increase total canopy photosynthesis, as diffuse light can penetrate deeper into a canopy by reducing the self-shading of leaves in the canopy, as has been shown experimentally (Gu *et al.* 2003). It is currently unclear exactly what changes in solar radiation have occurred, and whether changes in direct/diffuse components have altered over recent decades. Any changes in solar radiation patterns can be expected to have a spatial and temporal pattern—we can predict that changes in forest growth will track these changes.

Climatic extremes/ENSO

Climatic extremes across much of the tropics are strongly correlated with the ENSO events, a quasi-regular internal oscillation of the equatorial

Pacific ocean and the atmosphere (Fedorov and Philander 2000). The El Niño (positive) phase of ENSO can cause intense drought across many, but not all, tropical areas, whereas the opposite La Niña (negative) phase can cause very heavy rainfall and flooding in these areas. The regions where precipitation is most affected by ENSO extremes are north-eastern Amazonia, Central America, and insular Southeast Asia, whereas precipitation in African tropical forests shows little consistent relationship with ENSO (Malhi and Wright, Chapter 1, this volume). Over the past 30 yr there has been little discernible trend in the number, frequency, or intensity of ENSO events across the tropics (Fig. 4.2). However, two of the strongest events in the twentieth century occurred in the past two decades, in 1982–3 and 1997–8.

The unique prediction for the climatic extremes/ El Niño drought hypothesis is that mortality rate increases and stand biomass decreases will correlate with the spatial distribution of ENSO effects and/or other known extreme climatic events. For example, areas such as south-western Amazonia, which are largely unaffected by ENSO events, would be expected to show little or no changes in forest dynamics and biomass. By contrast, in north-eastern Amazonia, which is more strongly affected by ENSO events, larger increases in mortality rates and reductions in biomass would be predicted. In addition, by comparing African forests with Amazonian and Southeast Asian forests, the effects of temperature and drought may be teased apart as ENSO events cause warmer, but not drier, conditions in African forests.

Carbon dioxide

Carbon dioxide is a key substrate of photosynthesis. Atmospheric CO_2 concentrations are increasing by 1.5–2.5 ppm yr^{-1}. The general response of plants to this increase may be an increase in resource use efficiency, especially higher rates of photosynthesis, which improves light use efficiency, and decreasing stomatal conductance and transpiration, which improves water use efficiency. In addition, any acclimation of photosynthesis to higher CO_2 concentrations would reduce key enzyme usage,

notably Rubisco, increasing whole-plant nutrient use efficiency (e.g. Saxe *et al.* 1998). Hence if trees increase their light use efficiency in response to rising atmospheric CO_2 concentrations, increases in net growth from forest stands are predicted, with the greatest increases in forests where other factors are least limiting (aseasonal forests growing on nutrient-rich soils). Growth increases in forests growing on nutrient-poor soils are also expected, although these forests would be expected to show smaller absolute increases than forests on higher fertility soils. Within a given forest stand, those individuals least limited by other resources may respond most; thus, the largest relative increases may be from juvenile trees, especially in disturbed areas (where light and nutrients may be more abundant) as they may be least limited by light, water, and nutrients. This may increase stem and biomass turnover, as gap durations decrease and average tree lifespans shorten, creating a positive feedback that promotes further forest disturbance and hence increased growth. Increasing light use efficiency will decrease the light compensation point of trees, which we predict will lead to an apparent increase in stem recruitment in forest plot data. Furthermore, as more trees are at the minimum threshold for measurement (usually 10 cm diameter at breast height) in a given light environment, this would also increase stem density above the same threshold. If the above-mentioned changes occur, more rapid increases in growth, recruitment, and mortality rates may be expected than the changes in stand-level biomass and stem density, which may change more modestly.

If plants increase their water use efficiency the unique prediction is for increases in growth in forests only where water is limiting for part of the year (i.e. no change in the growth or biomass of aseasonal forests is predicted), and increased growth would occur even when no changes in rainfall had occurred. If trees increase their light use and water use efficiency simultaneously in response to rising CO_2 concentrations, the greatest absolute increases in growth would be expected from forests growing on nutrient-rich soils where water is limiting for part of the year. Also, increasing water use efficiency is predicted to cause changes in mature-tree canopy height wherever hydraulic stress is a limiting factor on tree height.

Nutrient deposition

Globally, human activities have more than doubled inputs of nitrogen, as both oxidized nitrogen compounds, NO_X, and the reduced N compound, NH_3, to terrestrial ecosystems (Matson *et al.* 2002). While most of these increases have been in temperate regions (owing to most sources being at mid-latitudes and the short lifespan of NO_X and NH_3 in the atmosphere), N deposition rates have increased substantially across the tropics over recent decades, because of increases in fertilizer use, fossil fuel combustion, and biomass burning (Matson *et al.* 2002). It is likely that deposition of other nutrients, notably phosphorus, may have also increased downwind of biomass-burning regions, but, to our knowledge, global-scale estimates of increases in P and other nutrient depositions are unavailable.

If increases in nutrient deposition occur, we can predict that forest growth and biomass will alter in forests only where the deposition has occurred *and* where that particular nutrient is limiting. In general, elevated nutrient deposition is expected downwind of biomass burning and near areas of industrialization and intensive agriculture. Rates of nutrient deposition are likely to have a strong spatial pattern: N deposition rates are likely to be large in Southeast Asia compared with South America and Africa owing to the much higher levels of intensive agriculture and industrialization; thus, if changes in growth show large continental-scale differences this may be evidence that nutrient deposition is causing changes in some areas. If, in general, forests are not N-limited, increased soil acidification may occur, so decreasing growth rates and biomass. Overall, until it is known which nutrient or nutrients are limiting a given forests growth, robust predictions of changes in forest growth with additional nutrient inputs are very difficult.

Ozone/acidic depositions

Forest near areas of industrialization and biomass burning may experience high tropospheric O_3, SO_2, and other acidic depositions. Plants grown under high O_3 levels exhibit decreased growth and visible damage. If tropospheric O_3 and acidic depositions are altering forest dynamics then we can uniquely predict that declines in growth rates and increases in mortality will correlate with increases in tropospheric O_3 and acid deposition levels. These effects will be greatest immediately downwind of biomass burning regions and industrial activity. Like nutrient deposition, while locally and regionally important, changes in O_3 and acid depositions are unlikely to explain biome-wide shifts in dynamics over recent decades.

Hunting

Hunting of wild mammal meat (bushmeat) is the single most geographically widespread form of resource extraction in tropical forests, affecting even remote areas (Robinson and Bennett 2000). The preferred prey of hunters are overwhelmingly herbivores, which disperse seeds and eat seeds and seedlings (Robinson and Bennett 2000). Their removal may therefore affect the regeneration of some species and affect plant regeneration more generally by favouring the competitive release of some groups of species. Specifically, in terms of altering tropical forest dynamics, the hunting of frugivores may reduce the dispersal of large-seeded animal-dispersed species, benefiting smaller-seeded animal- and wind-dispersed trees, which on average have higher growth and turnover rates (Phillips 1997). We predict an increase in recruitment and growth rates only at sites where hunting pressure has increased and the competitive release of smaller-seeded species has occurred.

Land use change

Across the tropics between 6 and 14 million ha of tropical forest are destroyed annually, the differences being due to different definitions of tropical forest and assessment methodologies (FAO 2000; Achard *et al.* 2002; DeFries *et al.* 2002). The resulting forest fragmentation and isolation lead to a variety of impacts on tropical tree populations that affect dynamics, by (1) changing the microclimate that trees grow in, notably decreasing soil moisture (water availability) and increasing air temperatures and evaporative demand because of the penetration of warmer air from non-forested areas; and (2) increasing the wind speeds that trees encounter, increasing tree fall rates.

Habitat fragmentation also affects tree recruitment patterns both directly (as a result of increasing tree fall disturbances, which increase light levels in the understorey) and indirectly (as a result of altered faunal communities, which can affect herbivore and seed predation). These effects may occur over large spatial scales, on the order of at least several kilometres from forest boundaries (Laurance, Chapter 3, this volume). In addition, habitat fragmentation may alter forest dynamics indirectly, by increasing the numbers of pioneer and gap-demanding non-pioneer trees, which, being good dispersers, may flood the remaining forest with their seeds, increasing the proportion of pioneer recruits in the remaining forest. As pioneers have faster growth and stem turnover than non-pioneers, these changes would increase stem turnover and stand-level growth rates.

If habitat fragmentation, and hence edge creation, is the cause of changes in forest dynamics we predict that changes in recruitment, mortality, and biomass would correlate with distance from forest edges. However, if habitat fragmentation leads to pioneer trees flooding the remaining forest, and thus increasing dynamic rates, then these changes should be directly attributable to an increase in the density of individuals of pioneer and gap-demanding non-pioneer species within long-term forest plots. Removing these 'functional groups' from plot datasets would remove the trends.

Lianas

Increases in the number and biomass of large lianas over the late twentieth century in tropical forests have been documented in undisturbed forest plots across western Amazonia, but the cause of this increase is unclear (Phillips *et al.* 2002b). Lianas are structural parasites that decrease tree growth and increase mortality, and are mostly disturbance-adapted. If lianas are

driving changes in forest dynamics or are themselves responding to changes in a driver, then uniquely, changes in dynamics should be correlated with changes in the density of large lianas and their net biomass.

Discussion

We focus on two questions. First, can the suites of predictions made for each of the 10 potential environmental drivers and 16 potential hypotheses help narrow the range of likely causes of recent changes in South American tropical forest structure and dynamics? Second, how might the environmental drivers, and hence tropical forests, alter in the future?

Potential drivers and forest plot data

Data from a network of long-term plots across the tropical forests of South America (Malhi *et al.* 2002a) have shown increases in stem recruitment and mortality, and hence stem turnover (Phillips *et al.*, Chapter 10, this volume), with recruitment exceeding mortality leading to an increase in stem density (Lewis *et al.* 2004b). In addition, stand-level forest growth has increased, as has mortality on a stand-level biomass basis, and hence biomass turnover has increased (Lewis *et al.* 2004b, Chapter 12, this volume), with growth exceeding losses causing an increase in stand-level above-ground biomass (Baker *et al.*, Chapter 11, this volume). Overall, these forests appear to have become more productive, more dynamic, and to hold more biomass. What driver or drivers may be causing this suite of simultaneous changes within these forest plots?

An environmental driver that directly increased mortality could not produce a net increase in above-ground biomass, so could not be driving changes in the dynamics documented. Similarly, a driver increasing recruitment could not cause the observed increases in biomass losses from mortality, nor the increases in growth or net above-ground biomass by the amounts documented, as recruited stems are a very small proportion of total above-ground biomass and stem numbers. However, a driver that increases the levels of resource availability could cause an increase in net primary productivity (NPP), and hence stem growth and the other observed trends (Lewis *et al.* 2004b).

Three environmental drivers may have increased resource availability across South America over the past 20 yr and hence increased forest growth are air temperatures, solar radiation, and atmospheric CO_2 concentration. Air temperatures and CO_2 have certainly increased, whereas more data on changes in solar radiation and cloudiness are needed to clarify possible increases in incoming solar radiation and possible shifts in direct and diffuse components. In addition, there is no consensus over whether CO_2 increases, air temperature increases, or changes in solar radiation actually increase forest growth. Further analyses of forest plot data, data on solar radiation, and experiments on the effects of temperature and CO_2 increases on tropical forest stand growth are required to move beyond our current working hypothesis that an increase in resource availability across South America is driving increases in NPP and hence increasing growth and accelerating forest dynamics.

Further work may help to distinguish the three possible drivers and seven hypotheses for increasing resource availability and hence NPP (Table 4.2). First, determining whether these trends are pantropical or are just occurring across one continent should be a research priority for ecologists. This requires expanding the network of plots, particularly across the Congo Basin and Borneo. Second, there are several potential tests to separate the different mechanisms. For example, the increase in incoming solar radiation across the tropics is likely to have a spatial pattern. Southern Borneo, the wettest parts of west Africa and northern Queensland are all tropical forest areas where NPP has been predicted to decline in a recent study by Nemani *et al.* (2003). By contrast, northern Borneo, Central America, the central Congo Basin, and most of Amazonia are predicted to have increased NPP over the 1980s and 1990s (Nemani *et al.* 2003). This spatial signature can be tested using long-term forest plot data. Likewise, examining spatial differences in background temperature, notably the generally lower temperatures

in African forests, and any correlations between changes in stem growth and changes in temperature may be illuminating.

The future

Across South America, at least, tropical forests appear on average to be increasing their year-on-year growth rates, and likewise their mortality and overall dynamism. In addition, growth is exceeding mortality, so these forest stands are carbon sinks for the time being, buffering the increase in atmospheric CO_2 concentrations, and hence the rate of climate change.

Under the simplest scenario of a steady rise in forest productivity over time, it is predicted that relatively slow-growing and undynamic forests would remain a carbon sink for a century or more (Chambers *et al.* 2001c), while others have suggested that the tropical forest carbon sink may reverse and become a substantial source (Cox *et al.* 2000; Körner 2004). The drivers documented in this chapter will all change in the future, most probably causing further alterations in forest structure and dynamics. In general, the drivers expected to cause elevated *mortality* are very likely to increase, as

(1) predicted increases in climatic extremes are likely to increase tree mortality;
(2) increasing N deposition, O_3/acid deposition, and other pollutants may increase tree mortality as industrialization continues to spread across the tropics; and
(3) habitat fragmentation and fires increases mortality rates in the remaining forest.

In addition, the current increases in the rates of *growth* (apparently caused by year-to-year better conditions for growth) cannot continue indefinitely, as

(1) if CO_2 is a cause, the biomass of old-growth forests may reach an upper limit because of physiological CO_2 saturation or architectural constraints;
(2) if the possibly large increases in incoming solar radiation have occurred and are the cause, this may simply reflect a decadal-scale oscillation, and

thus may not continue for long (Wielicki *et al.* 2002); and

(3) if temperature increases are the cause, points of inflection will more often be reached, decreasing net photosynthetic rates and carbon gains, and/or respiration costs will rise, both substantially reducing growth rates.

In addition, we predict increases in either stem recruitment and/or mortality rates under all 10 drivers (either directly or via effects on growth), so stem turnover rates are also very likely to continue to rise in the future. Overall, future changes are likely to include further increases in stem mortality and recruitment rates, with mortality overtaking recruitment, leading to declines in stem density at an increasing number of sites. Simultaneously, stand-level growth is expected to asymptote or decrease and biomass losses from mortality continue to rise. As a result stand biomass may decrease in surviving old-growth forests well within the current century. While there is considerable uncertainty on the future trajectory of the drivers and the responses of forests, the expected changes in the drivers plausibly predict that the current carbon sink contribution of mature tropical forests to slowing the rate of climate change is almost certain to diminish and quite possibly reverse in the future (e.g. Cox *et al.* 2000). This accelerated climate change could stimulate carbon loss from tropical forests, generating a positive feedback loop.

The drivers of change we document can be grouped into four categories on the basis of their underlying causes

(1) decadal-scale natural climatic oscillations;
(2) global greenhouse-gas emissions and resulting climate change;
(3) increasing industrialization in the tropics; and
(4) the further integration of forest products and land into expanding market economies.

To slow or halt widespread changes in remaining tropical forests will require large reductions in global fossil fuel burning, and a different form of development in tropical countries than has occurred in temperate nation-states, critically including the management of current and future logging/deforestation/settlement frontiers. Under 'business as usual' conditions, rapid global changes will more severely impact the world's remaining tropical forests with global consequences for biodiversity, climate, and human welfare.

Acknowledgements

The authors thank T. Baker, W. Laurance, J. Lloyd, and M. Williams for valuable comments on the manuscript. The work was funded by NERC (NER/A/S/2000/01002). S. Lewis and Y. Malhi are supported by Royal Society University Research Fellowships.

CHAPTER 5

Ecophysiological and biogeochemical responses to atmospheric change

Jeffrey Q. Chambers and Whendee L. Silver

Atmospheric changes that may affect physiological and biogeochemical processes in old-growth tropical forests include (1) rising atmospheric CO_2 concentration, (2) an increase in surface temperature, (3) changes in precipitation and ecosystem moisture status, and (4) altered disturbance regimes. Elevated CO_2 is likely to directly influence numerous leaf-level physiological processes, but whether these changes ultimately reflect in altered ecosystem carbon storage is unclear. The net primary productivity (NPP) response of old-growth tropical forests to elevated CO_2 is unknown, but unlikely to exceed the maximum experimentally measured 25% increase in NPP with a doubling of atmospheric CO_2 from pre-industrial levels. In addition, evolutionary constraints exhibited by tropical plants adapted to low CO_2 levels during most of the late Pleistocene, may result in little response to increased carbon availability. To set a maximum potential response for a central Amazon forest, an individual-tree-based carbon cycling model was used to carry out a model experiment constituting a 25% increase in tree growth rate that was linked to the known and expected increase in atmospheric CO_2. Results demonstrated a maximum carbon sequestration rate of only $0.05\,\mathrm{Mg\,C\,ha^{-1}\,yr^{-1}}$ for an interval centred on calendar years 1980–2020. This low sequestration rate results from slow-growing trees and the long residence time of carbon in woody tissues. In contrast, changes in disturbance frequency, precipitation patterns, and other environmental factors can cause marked and relatively rapid losses and gains in ecosystem carbon storage. It is our view that observed changes in tropical forest inventory plots over the past few decades are more likely being driven by changes in disturbance regimes, and other environmental factors, than by a response to elevated CO_2. Whether observed changes in tropical forests are the beginning of long-term permanent shifts or a transient response is uncertain, and remains an important research priority.

Introduction

Atmospheric greenhouse gases are rapidly increasing as a result of human activities and CO_2 alone represents about 60% of the total global warming potential of well-mixed greenhouse gases (Hansen *et al.* 1998). A vast amount of research indicates that climate change is an inevitable consequence of increasing greenhouse gases, although the magnitude, timing, precise location, and direction of these changes remain uncertain (IPCC 2001). Considerable effort has been directed towards determining the role of tropical land use

change in the production of CO_2 (Houghton 1991), but less research has focused on the response of old-growth tropical forests to changes in atmospheric chemistry and climate. Recent work using a network of tropical forest inventory plots demonstrated increases in (1) tree recruitment and mortality (Phillips *et al.*, Chapter 10, this volume), (2) total tree biomass (Baker *et al.*, Chapter 11, this volume), and (3) climbing woody plant (liana) abundance (Phillips *et al.* 2002b). Whether or not these observations are evidence of response to atmospheric change, or a transient phenomenon, is an interesting and important question.

Forest response to atmospheric change can occur at a number of different phenomenological scales. Some of the physiological responses of leaves to changes in atmospheric CO_2 concentration ($[CO_2]_{atm}$) or light intensity, for example, can be almost instantaneous, whereas changes in tree community structure brought about by a gradually shifting climate may take decades to centuries. In addition, some responses may be transient, and not reflect long-term ecosystem acclimation, and not all changes have impacts on forest carbon balance and net ecosystem productivity (NEP). For example, photosynthetic carbon assimilation often responds strongly to elevated CO_2, whereas the corresponding growth response may be considerably less marked or non-existent. Conversely, changes in carbon allocation to long-lived woody tissues, or slow-cycling soil organic matter (SOM), result in relatively long-term shifts in forest carbon balance (Chambers et al. 2001c; Telles et al. 2003).

There have been a large number of studies on the effects of elevated CO_2 on physiological, ecological, and biogeochemical processes. However, the relevance of most of these studies to old-growth forests remains unclear in many cases, and tropical studies are quite limited (Chambers and Silver 2004 and references therein). Elevated CO_2 experiments of plants undergoing stress or changes in resource availability may offer some insight, and studies in natural (unmanaged) ecosystems may be most relevant for understanding tropical forest response. Recently, free air CO_2 enrichment (FACE) technology has allowed manipulation of ambient CO_2 concentration while minimizing experimental artefacts (Allen et al. 1992), although FACE experiments of forested ecosystems are few (e.g. DeLucia et al. 1999; Norby et al. 2001). Natural CO_2 springs provide additional opportunities to study tree response to lifetime elevated CO_2 exposure (Hättenschwiler et al. 1997; Tognetti et al. 2000).

This review summarizes ecophysiological and biogeochemical responses of tropical forests to atmospheric change, and addresses the following questions: Will tropical forest ecosystems accumulate additional carbon under elevated CO_2, and if so, at what rate? How will tropical forests respond to altered disturbance regimes expected under a changing climate? Is there evidence that tropical forests are already responding to elevated CO_2 or other atmospheric changes? An important consideration is not to predict what the precise changes will be, but to highlight the important processes that may lead to alterations in forest structure and function.

Leaf gas exchange

Predicting the response of tropical forests to elevated atmospheric CO_2 is impeded by the absence of experimental studies for intact tropical ecosystems. Existing elevated CO_2 studies for tropical forest plants are limited to seedlings and saplings (Würth et al. 1998a), and model communities (Körner and Arnone 1992; Winter et al. 2001a,b). Generalizing results from these studies is further confounded by differences in container (pot) size, soil nutrient status, and competitive interactions among species. Nevertheless, results from these and innumerable extra-tropical studies have shed considerable light on leaf-level responses to elevated CO_2, higher temperatures, and changes in moisture status.

Photosynthesis responds differently to CO_2 concentration at different temperatures (Farquhar et al. 1980). This is primarily due to competitive carboxylation and oxygenation reactions catalyzed by Rubisco. The rate of the oxygenation reaction (photorespiration) increases more rapidly than the carboxylation reaction as temperature increases, and CO_2 is also less soluble than O_2 at higher temperatures (Farquhar and von Caemmerer 1982; Lambers et al. 1998). Thus, under elevated CO_2, higher temperatures are less inhibitory of carbon assimilation because the relative substrate concentration for the carboxylation reaction is increased. Kirschbaum et al. (1994) modelled these direct photosynthetic responses and predicted that, compared with rates at pre-industrial atmospheric CO_2 concentration in the year 2000 photosynthesis would be about 20% higher at 35°C but only 5% higher at 5°C.

One of the most consistent responses of plants to elevated CO_2 is an increase in water use efficiency (WUE). Generally, the amount of carbon assimilated per unit water loss (i.e. WUE) increases because stomatal conductance and transpiration are reduced under elevated CO_2, whereas internal CO_2 concentration remains relatively constant, so that more carbon is assimilated at a given transpiration rate (Hsiao and Jackson 1999). Thus, an increase in WUE is similar in some respects to a decrease in water stress (Amthor 1999). The magnitude of this effect, however, is dependent on the experimental conditions. Field *et al.* (1995) found that stomatal conductance was reduced by 31% for plants in growth chambers, 17% for plants in open-top chambers, and 4% for plants grown in the ground. An increase in WUE is probably the most common leaf-level response to elevated CO_2, although changes in WUE are not necessarily linked with proportional changes in plant growth.

Tropical trees often experience midday and post-midday declines in photosynthetic carbon assimilation rates that may be driven by moisture stress (Williams *et al.* 1998). If moisture stress is the causal factor, midday depression may be partially ameliorated under elevated CO_2, although experimental studies are lacking. However, other factors may also cause reductions in post-midday carbon assimilation rates. For example, photosynthesis is one of the most heat-sensitive of plant physiological processes, and temperatures in the range of 35–45°C tend to inhibit photosynthetic rates (Berry and Björkman 1980). Because direct solar radiation in the tropics often increases midday leaf temperatures beyond 40°C (Koch *et al.* 1994), protective mechanisms may be frequently invoked, resulting in relatively lowered post-midday photosynthetic rates under environmental conditions similar to those that occur during morning hours. Daily variation in photosynthetic assimilation may also be driven in part by intrinsic circadian rhythms that are probably insensitive to changes in $[CO_2]_{atm}$ (Goulden *et al.* 2004).

Plant respiration

Two different effects of elevated CO_2 on plant respiration have been suggested and are often referred to as direct (short-term) and indirect (long-term) effects. Although these studies were not carried out on tropical plants, many results from extra-tropical studies are probably relevant. First, initial studies indicated that a direct reduction in respiration was caused by inhibition of the activity of a key enzyme (cytochrome *c* oxidase) by CO_2 in the mitochondrial electron transport chain (Gonzalez-Meler *et al.*

1996). However, recent studies suggest that flux control cannot be attributed to enzymatic constraints (Gonzalez-Meler and Siedow 1999). It appears that elevated CO_2 has little if any direct effect on plant respiration, and that CO_2 effects found in earlier studies were primarily experimental artefacts caused by chamber leaks and other factors (Chambers and Silver 2004 and references therein). Although small direct inhibition of plant respiration is still plausible (Hamilton *et al.* 2001), this effect is certainly much less important than suggested in earlier studies.

The indirect (long-term) effects of elevated CO_2 on plant respiration are also contradictory. A recent meta-analysis found that growth in elevated CO_2 reduces leaf (dark) respiration by about 18% (per unit mass) (Curtis and Wang 1998). However, other studies have found increases or no long-term response of leaf respiration to elevated CO_2, and the response of woody tissue respiration also demonstrates conflicting results (Chambers and Silver 2004).

It appears that the long-term response of plant respiration to elevated CO_2 is complex, and further studies are needed to clarify contradictory results. In ecosystems where wastage (e.g. non-functional, futile cycles, alternative oxidase) respiration is an important process, the response of respiration to elevated CO_2 may be even more complex. For example, in addition to the cytochrome pathway which produces relatively large amounts of ATP, plants also have an alternative respiratory pathway of questionable utility (Lambers 1997). No studies have looked at how carbohydrate partitioning between the cytochrome and alternate oxidase responds to elevated CO_2. Because total respiratory efflux is comprised of construction (R_c), maintenance (R_m), and wastage (R_w) components, studies that consider only R_c and R_m may be obscuring relative changes in these key processes. Chambers *et al.* (2004a) found that about 70% of assimilated carbon was lost as autotrophic respiration for central Amazon trees growing on nutrient deficient soils. If much of this respiration is non-functional (i.e. R_w), central Amazon trees may already be capturing carbon in excess of immediate metabolic needs, and may not be particularly responsive to increased carbon availability.

Adaptive and evolutionary constraints

Recent plant exposure to elevated $[CO_2]_{atm}$ is a relatively short-term phenomenon compared to the past ~500,000 yr when atmospheric CO_2 levels ranged between 180 and 280 ppmv (Petit *et al.* 1999, Chapters 15 and 16, this volume). Thus, much of the world's flora may be adapted to the relatively low CO_2 levels of the late Pleistocene, and adaptive traits conferring survival at low CO_2 may constrain plants from exploiting increased carbon availability (Sage and Coleman 2001). This evolutionary legacy may have resulted in plants that are overly conservative in terms of stress response, growth potential, allocation patterns, and storage investment. Thus despite the current and expected increase in $[CO_2]_{atm}$, many plants may respond as if carbon is still a limiting resource. There may be species or guilds (e.g. lianas) that respond favourably to elevated CO_2, but changes in species composition driven by the small annual increase in CO_2 are likely to be commensurately slow and not observable in a few decades of forest inventory plot data (e.g. Laurance *et al.* 2004).

Due to substrate limitation and a relative increase in photorespiration under low CO_2, photosynthetic rates for C3 species are predicted to have been 20–30% less in the late nineteenth century and 40–60% less during the last glaciation (Sage 1995). Vegetation from tropical regions would have been more negatively impacted because the effect of low $[CO_2]_{atm}$ on both photorespiration and substrate limitation is much greater at higher rather than lower temperatures (Sage and Coleman 2001). Also, in the same way that CO_2 enrichment can attenuate the inhibitory effects of other stresses, CO_2 depletion can aggravate stresses (Sage 1995), and these stresses may have been heightened in many tropical trees. Conditions that are not harmful today may have been unfavourable under lower CO_2, and if these conditions increased the probability of mortality, they may have acted as strong selective pressures (Sage and Cowling 1999). Thus, although temperature and moisture stresses are expected to be relieved under elevated CO_2, due to evolutionary constraints, many plants may be conservative in response to this relief. Tropical tree species may exhibit particularly conservative adaptations conferring elevated photosynthetic capacity, and more carbohydrate storing mechanisms, than needed under elevated CO_2.

Tropical trees often experience a number of environmental and ecological stresses including chronically low nutrient availability, strong seasonal variability in precipitation, large diurnal changes in leaf temperature and plant water status, and intense competition for resources with other plants. Some of these stresses may have been aggravated during recent geological time. In addition to lower $[CO_2]_{atm}$, for example, precipitation may have been reduced in many tropical regions during Pleistocene glaciations (Mora and Pratt 2001). Given these conditions of the evolutionary past, how would tropical trees have adapted to effectively balance the acquisition of numerous resources with differing availabilities? In general, plants living under stressful conditions often acquire limiting resources in excess of metabolic needs, and then exploit these stored resources during stressful periods (Chapin 1980). Plant response to stress involves complex interactions among resource availabilities and acquisition strategies under the prevailing conditions (Grime 1988), and these responses may be maladaptive under changing conditions (Sage and Cowling 1999). To avoid stresses associated with CO_2 depletion during the Pleistocene, genetically fixed limits on growth rate, restricting competition between new tissue production and storage, and ensuring adequate carbohydrate reserves during stress conditions, may have been advantageous (Sage and Cowling 1999). Thus, developmental controls over maximum meristematic activity may restrict the growth potential for many tropical tree species, limiting growth even when carbon availability increases.

Net primary productivity

Although many studies show an increase in productivity under elevated CO_2, the magnitude of this response is often dependent on the experimental conditions. First, changes in productivity found under elevated CO_2 are usually smaller and more variable than stimulation of leaf photosynthesis (Koch and Mooney 1996). Second, enhanced productivity is generally lower for natural as compared with managed ecosystems (Field 1999). Third, productivity stimulation is often reduced under conditions of low soil nutrient supply (Kirschbaum *et al.* 1994; Curtis and Wang 1998), although stimulation may be enhanced under low water supply (Poorter and Perez-Soba 2001). Winter *et al.* (2000) found that biomass accumulation for a model community of two species grown in open-top chambers in a forest clearing in Panama was not significantly affected by elevated CO_2. In contrast, with the addition of large amounts of fertilizer (Winter *et al.* 2001b), biomass accumulation was enhanced by more than 50%.

Free air CO_2 enrichment technology (Allen *et al.* 1992) alleviates some of the problems associated with other methods for experimentally increasing ambient CO_2 concentration. In a loblolly pine plantation, DeLucia *et al.* (1999) found that NPP was stimulated by about 25%, and that this increased productivity was maintained for at least 4 yr (Hamilton *et al.* 2002). In contrast, in an adjacent longer-running non-replicated FACE experiment, Oren *et al.* (2001) found that productivity during years four through seven returned to rates found in the control plots although growth enhancement was similar to that found in the replicated experiment for the first 3 yr (Hamilton *et al.* 2002). Short-term and long-term responses to elevated CO_2 are, in many cases, likely to differ substantially, underscoring the need for a cautious approach to broad extrapolations from short-term studies.

The residence time of carbon in different tissues varies considerably in tropical forests. Leaves and fine roots, for example, are short-lived and decompose relatively quickly, whereas slow-growing trees of the central Amazon can live for more than 1000 yr (Chambers *et al.* 1998), and the trunks and branches of large dead trees can take many decades to decompose (Chambers *et al.* 2000). Thus productivity gains that primarily increase fast-cycling tissues will have little effect on ecosystem carbon storage, but the partitioning of additional photosynthate to woody tissue

production results in relatively large, long-term increases in carbon storage.

In summary, experimental and theoretical evidence suggests that NPP in forested ecosystems may increase by anywhere from 0% to 25% under doubled pre-industrial $[CO_2]_{atm}$. Further constraining this range for tropical forests is hindered by a lack of appropriate large-scale manipulative studies. However, to at least quantify the maximum long-term response, we carried out a modelling experiment. Briefly, forest carbon cycling was modelled by simulating recruitment, growth, death, and decomposition of individual trees. The model was parameterized with, and tested against, extensive field data from the central Amazon, and model details are provided in Chambers *et al.* (2004b). To test carbon accumulation response to increasing $[CO_2]_{atm}$, an increase in NPP was linked to the known and expected rise $[CO_2]_{atm}$, with the NPP response following a so-called beta (β) function (Amthor and Koch 1996). Results indicated a remarkably long lag-time before total tree biomass reached a new dynamic equilibrium, with a maximum sink capacity for calendar years 1980–2020 of only 0.05 $MgC ha^{-1} yr^{-1}$ (Fig. 5.1), or about an order of magnitude less than observed by Baker *et al.* (Chapter 11, this volume) in the RAINFOR forest inventory plots. Thus, at least for this central Amazon forest, we would argue that it appears unlikely that recent increases in biomass (Baker *et al.*, Chapter 11, this volume) are being driven by elevated CO_2.

Tree mortality and succession

Changes in disturbance frequency and tree mortality in tropical forests can have large impacts on forests structure and functioning, and there is evidence that changes are occurring in tropical forests. Phillips *et al.* (Chapter 10, this volume) found that rates of both tree mortality and recruitment have increased since the 1950s, with most of the increase since the 1980s. Although the cause of this observed increase in forest turnover is unclear, strong El Niño and La Niña (ENSO) events have been pronounced during the latter part of the twentieth century (Fedorov and Philander 2000), and the frequency and intensity

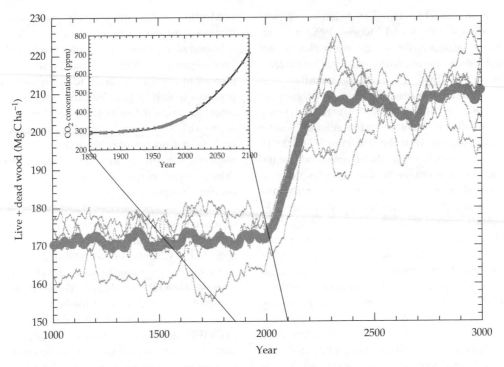

Figure 5.1 Modelled response of total wood carbon (trees and woody surface litter) in a central Amazon forest to a 25% increase in NPP with five-model runs at 10 ha each (thick line mean response). The increase in NPP was tied to the known and expected increase in atmospheric CO_2 from 1850 to 2100 (insert graph) using a beta function (see text). Near the year 2200, total wood carbon storage reached a new dynamic equilibrium with a 25% increase. Response was remarkably slow with an accumulation for the period of 1980–2020 of only 0.05 $MgC ha^{-1} yr^{-1}$.

of ENSO events may be linked to climate change (Timmermann *et al.* 1999).

Phillips *et al.* (2002b) documented an increase in the dominance of woody climbing plants (lianas) in Amazonian forests using a series of permanent inventory plots. Because lianas thrive in disturbed environments, this change may be partially due to increased forest turnover rates (Phillips *et al.*, Chapter 10, this volume) which would increase the forest fraction in the gap regeneration phase. Although lianas may also respond more favourably to elevated $[CO_2]_{atm}$, the increase in $[CO_2]_{atm}$ over the few decades that the inventory plots have been monitored is probably not yet sufficient to drive noticeable changes in species composition. Conversely, increased disturbance can result in rapid shifts in ecosystem processes and tree species composition (Laurance *et al.* 1997, 1998a). CO_2-driven changes in species composition may become

increasingly important as $[CO_2]_{atm}$ continues to rise throughout the twenty-first century, but it is quite unlikely that current subtle shifts in species composition (Laurance *et al.* 2004a, Chapter 9, this volume) are being driven by increasing $[CO_2]_{atm}$.

In many tropical forests the vast majority of gaps are relatively small and the abundance of true pioneer species (e.g. *Cecropia* spp., *Vismia* spp.) is low. For example, gaps smaller than 200 m^2 were 76% of all canopy openings at La Selva Biological Station (Sanford *et al.* 1986), and daily integrated light levels in these small gaps are less than 20% full sunlight (Denslow *et al.* 1990). A number of ecological processes are thought to vary as a function of gap size. The species composition of trees that fill gaps, for example, is often quite similar to that of the surrounding forest, and only once a gap passes a critical large size do pioneer species become abundant (Hubbell *et al.* 1999).

Thus tree mortality and gap formation typically result in relatively small changes in biogeochemical cycling, although once a critical threshold is passed, forest structure and functioning can change dramatically.

Changing atmospheric conditions may result in synergistic effects that ultimately result in marked changes in the structure and functioning of tropical forests. First, changes in disturbance frequency such as ENSO events and the intensity of convective storms may alter the size and abundance of canopy gaps. Increased tree mortality will not only initiate regeneration in gaps, but will also result in increased growth for mature canopy trees due to the release of local competition for resources (Chambers et al. in press). Elevated mortality will be particularly important if the largest trees are impacted more heavily. In many tropical forests a large fraction of the above-ground biomass is stored in a small number of the largest trees (Brown et al. 1995), and Laurance et al. (2000) found that mortality rates for the largest trees increased more than smaller trees in response to forests fragmentation.

Nutrient uptake

Nutrient availability is well known to influence plant response to elevated CO_2 (Arnone and Körner 1995; Winter et al. 2001b). In temperate ecosystems, low nitrogen(N) availability often limits NPP and has the potential to inhibit growth responses to CO_2 fertilization (Finzi et al. 2002). In highly weathered soils typical of tropical forests, N is relatively abundant and P is the most commonly limiting element to NPP. Phosphorus can also become limiting to vegetation due to P occlusion with Fe and Al oxides and hydroxides, and the formation of resistant secondary minerals. Because highly weathered tropical soils typically have abundant reactive Fe and Al minerals, the removal of P from the rapidly cycling labile pool through these mechanisms is thought to be a dominant fate of P in these ecosystems (Uehara and Gilman 1981).

The question of whether P limitation will inhibit the growth response to elevated CO_2 in tropical forests is complex. Plants have evolved several mechanisms to alleviate low P stress including associations with mycorrhizal fungi (Went and Stark 1968): the ability to produce enzymes that may help release P adsorbed on exchanged sites or held in organic molecules, the ability to resorb P from foliar tissue prior to leaf senescence (Vitousek 1984; Silver 1994), and a strategy of allocating significant C to root biomass on or near the soil surface to efficiently cycle P from decaying organic matter (Stark and Jordan 1978). The response of these processes to elevated CO_2 is more likely to control nutrient interactions with elevated CO_2 than direct changes in nutrient uptake in tropical forests.

In a recent review, Lloyd et al. (2001) suggest that tropical forests are not strongly limited by P, and that P limitation will not inhibit a potential growth response under elevated CO_2. Their argument stems from two primary assumptions: (1) that the labile P pool is in equilibrium with the non-labile pool over time periods relevant to current atmospheric changes, and (2) that P mineralization and uptake in soils will keep pace with P demand due to the production of organic acids associated with root and mycorrhizal activity. The Lloyd et al. (2001) proposal that tropical forests are not truly limited by P contradicts widely held beliefs that low P pools strongly influence rates of tropical forest growth.

Factors controlling rates of P mineralization are not well understood in tropical forests. For rates of P mineralization to increase commensurately with an increase in growth rates under elevated CO_2, stimulation of microbial activity and/or the production of phosphatase enzymes is required. A study using a cultivated wheat species found that root phosphatase production increased under elevated CO_2 when soil P levels were low (Barrett et al. 1998). We know of no similar studies on tropical forest tree species, and too little is known about the behaviour of phosphatase production in tropical forests to predict the response to elevated CO_2. There is good evidence that mycorrhizal associations can be maintained under elevated CO_2, and as Lloyd et al. (2001) point out, this is likely to result from more root tissue being available for infection than from direct CO_2 enhancement.

In summary, there is not yet sufficient evidence to discount that nutrients and specifically P will limit the ability of tropical forests to enhance growth under elevated CO_2. While we agree with Lloyd et al. (2001) that tropical forests have evolved numerous strategies to help offset the strong P sorption capacity of highly weathered tropical forest soils, mechanisms to increase P acquisition sufficiently to lead to a measurable growth response lack experimental evidence.

Fine and coarse litter dynamics and SOM storage

Because NEP is the difference between NPP and heterotrophic respiration, factors that change either process result in changes in ecosystem carbon storage. However, most research focuses on the input side of this equation, and much less work has been carried out on respiration of SOM and dead plant material. Both production and decomposition are likely to be directly or indirectly sensitive to changes in $[CO_2]_{atm}$, temperature, and precipitation. These processes and their degree of coupling are also likely to be affected by the frequency and severity of disturbance events.

Tropical forests growing on highly weathered soils often allocate considerable photosynthate to root biomass (Jackson et al. 1996). Roots are more buffered against climate changes than aboveground tissues by their location in the soil. In mesocosm studies, tropical plants increased their allocation to roots when exposed to elevated CO_2 and high soil nutrient conditions (Körner and Arnone 1992), but under low nutrient availability there was no statistically significant increase in root biomass with CO_2 fertilization (Arnone and Körner 1995).

The largest changes in litter dynamics under elevated CO_2 conditions will probably be linked to alterations in litter production, either from increased productivity or changes in disturbance frequency. For example, at the Duke FACE site, Hamilton et al. (2002) found a 165% increase in total heterotrophic respiration in response to elevated CO_2 (+200 ppmv), largely brought about by an increase in fine litter production. However, if most

additional productivity under elevated CO_2 is allocated to fast-cycling tissues or exudates, forest carbon balance may be changed by very little. Conversely, in response to higher-than-average tree mortality rates, changes in coarse litter (trunks and branches >10 cm in diameter) stocks and subsequent decomposition (Chambers et al. 2000) and CO_2 release (Chambers et al. 2001d) can offset gains in above-ground biomass for a number of years (Saleska et al. 2003).

Another potentially important C sink in tropical forests is SOM. The deep profiles and fine texture of many tropical forest soils, together with the high proportion of amorphous organic and mineral material result in very large soil C storage capacity. The rate of deposition may or may not be balanced by heterotrophic respiration. The mean residence time for surface soil C in tropical forests is generally relatively short (6–30 yr) (Trumbore et al. 1995) but can be very long when SOM reacts with certain amorphous minerals such as allophane common in young volcanic soils (Torn et al. 1997). Deep, highly weathered tropical soils may be especially important repositories for C. Carbon inputs from deep roots found in seasonally dry tropical forests can dramatically increase the soil C pools at depth, which tend to have a very slow turnover time (<1000 yr) when compared to surface soil environments (Trumbore et al. 1995).

Detecting changes in soil C pools with elevated CO_2, climate change, and associated environmental changes can be very difficult in tropical forests due to the large soil C pool size, although stable and radioisotope techniques are dramatically increasing our ability to identify changes in soil C pools and fluxes. The factors that are most likely to lead to greater soil C storage in mature tropical forests are increased hydrophobicity of litter through changes in plant species composition or litter quality, increased rooting depth associated with decreased soil moisture or nutrient availability, increased C allocation to root biomass and root exudates both of which can affect the amount of C inputs and mean residence times, and greater root density and fungal associations that contribute to the formation of stable aggregates. Any one or combination of these factors could contribute to greater soil C storage over relatively short time periods. However, although many tropical soils clearly have a large long-term carbon storage capacity, Telles *et al.* (2003) demonstrate that the annual sink capacity is relatively low, and not likely to be globally significant over annual to decadal timescales.

Summary

Elevated CO_2 (doubled pre-industrial) studies in natural ecosystems have demonstrated NPP increases of 0–25%, with plants growing under nutrient stress showing the lowest response. Because many tropical forests occur on nutrient deficient soils, evidence to support a maximum 25% increase in NPP for many tropical forests under elevated CO_2 is non-existent. In addition, evolutionary constraints of plants adapted to low CO_2 levels during much of the late Pleistocene, may limit tree response to additional carbon. However, to at least constrain the maximum expected carbon sequestration response for a central Amazon forest, we carried out a model experiment with a 25% increase in NPP tied to the known and expected rise in CO_2 concentration from years 1850–2100 (Fig. 5.1). During a 40-yr interval centred on the year 2000, the model predicted a sequestration rate of only $0.05 \, \mathrm{Mg\,C\,ha^{-1}\,yr^{-1}}$, which is about an order of magnitude less than biomass accumulation measured on a global network of forest inventory plots (Baker *et al.*, Chapter 11, this volume). Thus, we suspect that observed changes in tropical forests (Phillips *et al.* 2002b; Baker *et al.*, Chapter 11, this volume; Laurance *et al.* 2004a; Phillips *et al.*, Chapter 10, this volume) are more plausibly being driven by changes in other environmental factors, such as disturbance cycles (Chambers *et al.* in press), solar radiation (Gu *et al.* 2003), or precipitation, which have the capacity to cause much more rapid shifts in ecosystem carbon storage than increasing $[CO_2]_{atm}$. An important area to focus research effort is distinguishing among processes that result in short-term transient shifts in ecosystem carbon balance, from those that have the potential to cause relatively long-term permanent changes.

Acknowledgements

This work was supported by NASAs LBA-ECO and INPAs Piculus project (G7 Nations Pilot Programs, FINEP 6.4.00.0041.00).

CHAPTER 6

Tropical forests dynamics in response to a CO$_2$-rich atmosphere

Christian Körner

The fixation and storage of carbon by tropical forests, which contain close to half of the global total of biomass carbon, may be affected by elevated atmospheric CO$_2$ concentration. Classical theoretical approaches assume a uniform stimulation of photosynthesis and growth across taxa. Direct assessments of the carbon balance either by flux studies or by repeated forest inventories also suggest a current net uptake, although magnitudes sometimes exceed those missing in the global C-balance. Reasons for such discrepancies may lie in the nature of forest dynamics and in differential responses of taxa or plant functional types. In this contribution I argue that CO$_2$ enrichment may cause forests to become more dynamic and that faster tree turnover may in fact convert a stimulatory effect of elevated CO$_2$ on photosynthesis and growth into a long-term net biomass carbon loss by favouring shorter lived trees of lower wood density. At the least, this is a scenario which deserves inclusion into long-term projections of the carbon relations of tropical forests. Species and plant functional type specific responses ('biodiversity effects') and forest dynamics need to be accounted for in projections of the future carbon-storage in tropical forests.

Introduction

Among the various facets of global change, the enrichment of the atmosphere with CO$_2$, now at concentrations close to 380 ppm, is unquestioned. There are two main mechanisms through which this rapidly progressing change can affect biota. The first effect, which is indirect, operates via the climate system. This is covered by other contributions in this volume. To the extent that climatic changes would include more disturbances (extreme events, storms) these would tend to enhance biomass turnover and reduce standing C-stocks. Gradual influences, such as atmospheric warming seem to be of less significance in the tropics, but changes in rainfall patterns would have major influences.

The second effect of elevated CO$_2$ concentrations is direct and influences carbon fixation by plants. Since forest forming tropical trees use the C3 pathway of photosynthesis, their carboxylation machinery is not CO$_2$-saturated at current concentrations. Hence, rising concentrations of CO$_2$ can stimulate photosynthesis per unit of leaf area. Whether this leads to more growth, and if it does, to more net carbon storage, is an open question and will depend on a suite of factors other than CO$_2$ concentration, and in particular on differential responses of plant species, which may lead to new community composition and ecosystem functioning. The consequences of enhanced growth rates on carbon stocks are unclear. On the one hand, if there are recruitment waves with concomitant tree mortality lagging behind, then a transitory build up of C-stocks is possible (see Lewis *et al.*, Chapter 12, this volume). However, because fast growing, shorter lived species tend to produce less dense wood (e.g. in *Cecropia*, *Ochroma*, many Bombacaceae) than slow growing, longer lived species, at steady state, faster turnover may reduce the standing crop reservoir (Phillips and Gentry 1994).

Direct influences of elevated atmospheric CO_2 concentrations on tropical plants had been reviewed by Arnone 1996 and Körner 1998. Here I was given the task to discuss effects on species diversity and forest dynamics. To start with I will recall two key issues, which concern the most widespread misconceptions in the interpretation of the responses of plants to CO_2 enrichment in the published literature: the link between CO_2 assimilation and growth, and the link between growth and carbon sequestration. For the discussion of overall forest responses, these are of particular importance.

A stimulation of assimilation does not necessarily mean that there is more growth

Since it was discovered that plants take up carbon from the atmosphere rather than from soil 200 yrs ago (De Saussure 1804, transl. 1890), there has been a widespread misconception of a direct proportionality between the rate of leaf photosynthesis and growth. This is not the place to review the relevant literature, but it should be sufficient to recall that if there were such straightforward links, agricultural research programmes would not have given up selecting cultivars with high photosynthetic capacity about 30 yrs ago. In the world's most important crop, for instance, wheat, there is no such link (e.g. Evans and Dunstone 1970; Gifford and Evans 1981; Saugier 1983; Watanabe *et al.* 1994). Rather, whether plants produce a lot or little biomass over a certain period of time depends on how, for a given rate of assimilation, plants handle assimilates (investment), how dense their tissues are (costs), and how long these function, and how growth is timed (developmental controls). Moreover, it is almost trivial to recall that all organisms need to keep a certain balance of elements in their tissues, hence, the investment of one chemical element (e.g. carbon) always requires the investment of some others, which must be available. If such other elements fall short, plants may still assimilate more carbon in response to elevated CO_2, but this carbon will at

large first dilute such other elements and then channel carbon away from structural growth (Fig. 6.1 illustrates some options). It is obvious that actual plant responses to elevated CO_2 will always reflect the availability of resources other than CO_2. For experimentalists this means that their experimental design to a large extent predefines the resulting responses to CO_2 enrichment. In order to draw meaningful conclusions on plant responses to elevated CO_2, these must be studied under the relevant nutrient conditions. These may be a full nutrient solution in horticulture, or a poor depleted soil in some natural systems. Obviously, such results are not interchangeable (Körner 1995) and must strictly be statistically separated when examining data in the literature. Most available meta-analyses do not stratify data in this way (Körner 2003a) and results therefore reflect the proportions of growth conditions under which the data were collected.

In essence, this means that one cannot expect proportionality among the increase of CO_2 concentration of the atmosphere, the degree of photosynthetic stimulation, and the rate of growth of plants (Körner 2003c). In the ideal case, one would observe the growth response *in situ* over the 60–80 yrs it may take for the current, already 30% increased CO_2 concentration to arrive at a 100% increase compared to pre-industrial conditions—obviously not an option if predictions are to be made right now. Researchers have tried to circumvent this problem by adopting experimental conditions which approximate life in a tropical forest, and exposing plants to a step increase in CO_2 in the hope of accelerating responses, to obtain at least a picture of the likely direction and nature of such responses.

A stimulation of growth does not necessarily mean enhanced carbon sequestration

An equally difficult issue to be communicated to a broad concerned audience is the difference between rates and pools. Analogous to economies, rates of carbon incorporation (growth rate) do not in a simple way scale to the stock of carbon (the capital). The total carbon capital per unit land area is the net result of input (growth) and output (decay) or, more

The fate of carbon in plants

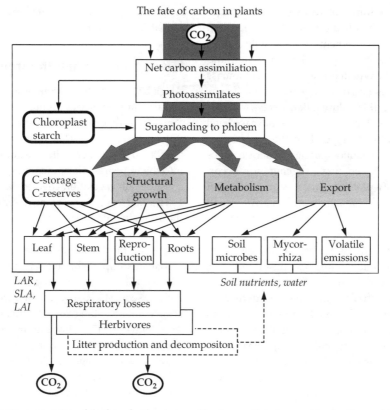

Figure 6.1 A schematic representation of the fate of carbon in trees. The diagram illustrates a variety of pathways along which photosynthates may be distributed. The resultant carbon allocation ultimately determines the rate of biomass accumulation (Körner 2003c) (reproduced from *J. Ecol.*, with permission by Blackwell Science).

precisely (as long as no carbon is lost through run-off), the difference between net CO_2 fluxes into and out of the system, termed NEE, the net ecosystem carbon exchange, as is measured in carbon flux studies from towers. The key point here is time. Over certain periods of time the accumulation of biomass by a tree (growth) enhances its carbon capital, the reason why we have foresters. On a landscape-wide scale the capital would only change if all trees were growing continuously. In reality, all trees eventually die or get harvested. If birth and death rates are balanced and the demography of trees is such that a steady forest cover is ensured, the overall biomass carbon pool would be constant, despite the fact that most trees are growing. Natural regeneration waves or forest plantation management can cause growth to be synchronized over large areas so that at landscape scales carbon pools

may cycle in regular patterns. In this case, whether a net carbon uptake or a release is recorded depends on when one makes the measurements. As the accumulation of biomass is slow, but collapse and breakdown are often very fast, there is a far greater likelihood to find forests with net carbon uptake than forests with a net carbon release although, at the landscape level, the balance may be zero (Körner 2003b). Given average tree life spans of 200 yrs in a natural forest and 30 yrs in a vigorous plantation, the ratio between areas with a net C-fixation to areas with a net C-release (death of a tree or groups of trees) could be 200:1 in a uniform natural forest, 30:1 in a plantation forest. A key aspect is the size of the regenerating patches. The larger they are, the more severe the problem becomes. Very small patches (gaps) may be nested at random positions in a hectare size test plot. However, their dynamics must also be random, not to distort the picture. Such conditions are difficult to ascertain. A random positioning of measurement stations will always reflect the spatial and temporal regeneration and mortality pattern. Whether findings of net biomass increase in forests based on inventory data (Lewis *et al.*, Chapter 12, this volume, Baker *et al.*, Chapter 11, this volume) will scale to realistic landscape level responses will always depend on the representativeness of patch dynamics and age structure. Once breakdown waves and wave regeneration occur, the probability of a plot being installed in an area that is accumulating carbon is many times greater than the probability of a plot being installed in an area in which there is net carbon release. If carbon release is not by microbial activity but by fire, there may be only two days in a millennium to witness the release, and during most of the rest of the time one would detect accumulation. For gap-mediated dynamics, the larger the plot and the smaller and more random in time the gaps, the more representative the data obtained will be. Given that carbon release will always remain a very stochastic phenomenon, the integration over time remains a challenge and the 'space for time model' (e.g. using larger measurement towers) still has its limits.

These constraints are critical for ascertaining CO_2 enrichment effects on tropical forests. Answers will remain scale-dependent, as trivial as this may sound. The question is, what sort of answers can we expect at which types of scales. I see three domains:

(1) The plant-based study of growth and development (scale $<10 \text{ m}^2$);
(2) The plot-based study of community dynamics (scale $<1 \text{ km}^2$);
(3) The landscape-based study of carbon balance (scale $>100 \text{ km}^2$).

For scaling reasons, (1) and (2) will never produce direct answers concerning the landscape carbon balance, and (3) will not provide answers on biodiversity effects. After a short account on the global dimension of missing carbon, I will focus on experimental data obtained for (1), and I will try to draw conclusions for (2), which may have implications for (3). Needless to say, no experi-

mental CO_2 enrichment data (beyond the ongoing global 'experiment') exist for (2) and (3).

Missing carbon and the tropical forest

At global scales, the carbon cycle cannot be closed, because human society annually emits more CO_2 than can be detected in either the atmospheric or the upper oceanic pools. This 'missing carbon' of 1–2 Gt (gigatons) C is generally assumed to disappear into the terrestrial biota, including the tropics. If we consider all vegetation-covered land on earth (ca. 100 million km^2, disregarding deserts and cropland), the annually missing C is 10–20 g m^{-2}. This is a very small signal to detect, in view of the 10–20 kg C m^{-2} stored in most soils and the similar amount of carbon tied up in biomass in the case of forests (Körner 2000). If the 'missing carbon' would exclusively be sequestered to forests and woodlands, which is very unlikely, the relevant land area would be ca. 48 million km^2 (17 million km^2 in the humid tropics) and the required net flux per unit area would be roughly twice as high (i.e. 20–40 g m^{-2}).

Because of their longer growing season, tropical forests (which store ca. 42% of the globe's biomass carbon; Brown and Lugo 1982; Soepadmo 1993) may suck up a relatively greater fraction of this missing carbon, but their productivity per unit of time in the growing period (per week, per month) is similar to any other humid forest in the world (Körner 1998). Recent estimates of carbon release from tropical forest destruction suggest lower rates than previously thought (Achard *et al.* 2002), reducing the globally missing carbon fraction to 1 Gt C or less, of which more than half can perhaps be explained by forest regrowth in parts of the temperate zone (Canadell and Pataki 2002). This would reduce the remaining mean carbon sequestration per unit of tropical forest area to potentially immeasurable quantities. In the following, I will try to explain why it may be more promising and significant for the understanding of the consequences of CO_2 enrichment for tropical forests to search for plant and community responses to CO_2, than to search for missing atmospheric carbon. I will illustrate that such biotic responses are likely to outweigh classical concepts of ecosystem carbon binding which are based on photosynthesis responses. My prediction is that 'Rio' will determine what 'Kyoto' had hoped to solve: it will in large part be biodiversity, that is, species-specific or functional type responses to atmospheric change, which will determine the ecosystem carbon balance. The humid tropical forest is likely to be among the first systems where this will be seen, simply because the 12 month growing season and sufficient moisture should permit significant effects to materialize faster than elsewhere.

Differential responses by species or plant functional types to elevated CO$_2$ co-determine carbon stocking

Experimental CO$_2$ enrichment work conducted in the tropics or with tropical plants until recently has been summarized by Arnone (1996) and Körner (1998). In essence, CO$_2$ has always been found to stimulate photosynthesis, leading to greater carbohydrate accumulation in plant tissues (e.g. Würth et al. 1998b) and often also to the depletion of nitrogen concentrations. Growth of tropical plants was found to be stimulated only when they were grown with ample nutrients or in isolation and/or deep shade. Under a fertile but highly competitive regrowth situation, a complex model community of tropical plants showed enhanced carbon turnover (greater litter production and carbon release from soil) but little growth stimulation (Körner and Arnone 1992). Microcosms composed of a suite of tropical tree seedlings also revealed no overall growth response, but species-specific responses in height growth, which ultimately would alter community structure (Reekie and Bazzaz 1989). When seedlings of two commonly dominant tropical tree species were grown to a height of almost 2 m in a high light environment in open top chambers in the tropics, they either showed no growth response to CO$_2$ when grown in unfertilized substrate (Winter et al. 2000), or showed a massive stimulation, when fertilized (Winter and Lovelock 1999; see also Winter et al. 2001b). A broad screening for CO$_2$-responsiveness across early and late successional tropical tree species revealed, that under favourable growth conditions, it is the early successional

species with high relative growth rate which take advantage, and not the late successional species (Winter and Lovelock 1999).

Deep shade has been found to be the situation under which the CO$_2$ responses of plants are most dramatic. The explanation is very simple. Plants can use low light intensities far better when exposed to a CO$_2$-enriched atmosphere, and the light compensation point of photosynthesis shifts to lower light intensities. *In situ* tests have revealed relative stimulations to growth of seedlings in Panama (Barro Colorado Island) by up to 50% in only 15 months of growth under ca. 0.5% of maximum midday photon flux density (a mean of 11 μmol m^{-2} s^{-1}; Würth et al. 1998a).

Responses among the tested species differed, but did not reveal any relationship to plant functional types such as shrub or tree, or nitrogen-fixing or not. Remarkably, there was no indication of a greater stimulation in the leguminous tree *Tachigali versicolor* (which was visibly nodulated) compared with the lauraceous tree *Beilschmiedia pendula*. It is still very doubtful whether enhanced nutrient availability (assuming this was the case for the legume) can stimulate tree recruitment under such deep shade conditions. A cross-continental test with a common treatment design revealed very little or no effect of nutrient addition to the understorey in a Swiss temperate and montane forest, a laurophyll forest in the Canary Islands, and a humid tropical forest in Panama (S. Hättenschwiler, M. S. Jiménez, A. Gonzalez, Ch. Körner, unpublished data). The one exception to this in deep shade was one tropical liana species which took advantage from nutrient addition (Hättenschwiler 2002). Responses may also depend on the degree and type of mycorrhizal infection, as was shown *in situ* for the above-mentioned *Beilschmiedia* by Lovelock et al. (1996).

CO$_2$-stimulated growth should improve seedling survival in the understorey, but could also cause young lianas to more likely escape the forest shade. We now have evidence that this is a plausible scenario. Three species of tropical lianas grown competitively in their natural substrate in a dynamic climate simulator, which mimics the tropical understorey temperature and humidity, yielded greater CO$_2$ responses (in relative terms)

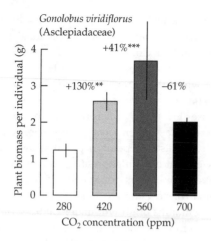

** $P < 0.01$ *** $P < 0.001$

Figure 6.2 The biomass response of a tropical liana species at seven months growth under three different levels of atmospheric CO_2 concentrations, starting from pre-industrial CO_2 concentration. Note the non-linear response and the largest relative stimulation in the range of current CO_2 concentrations (from data by Granados and Körner 2002).

when exposed to very low light (one example is shown in Fig. 6.2).

The growth of these lianas was strongly stimulated (>100%), particularly when responses at elevated CO_2 concentrations were compared with growth rates at 280 ppm, the pre-industrial CO_2 concentration. The most dramatic effect occurred between 280 and 420 ppm, that is, within the range of concentrations plants are experiencing now and during the last 100 yrs. At higher CO_2 concentrations, the responses were smaller and even reversed, above 560 ppm. These data suggest that lianas are likely to take a significant advantage from currently ongoing atmospheric CO_2 enrichment. Have lianas become more successful in recent decades? According to new analyses by Phillips *et al.* (2002b) and Wright *et al.* (2004) the answer seems yes. These demographic data suggest enhanced growth and biomass of tropical lianas, and the greater vigour of lianas is a strong candidate for explaining faster tree turnover (Phillips and Gentry 1994; Laurance *et al.* 2004a; Phillips *et al.* 2004). It is well known that lianas play a key role in tropical forest dynamics (Laurance *et al.* 2001b; Schnitzer and Bongers 2002).

If lianas as a single functional type would take greater advantage from atmospheric CO_2 enrichment, this biodiversity effect could potentially shift tropical forests from later to earlier successional stages, irrespectively of whether trees also become stimulated by elevated CO_2 or not. Early successional forests store much less carbon than old growth, late successional forests.

Paradoxically then, growth stimulation can potentially lead to a reduction in forest carbon storage. This example underlines the significance of differential effects on species or functional plant types under CO_2-enriched atmospheric conditions. The next experimental step needed would be *in situ* tropical forest CO_2 enrichment, using the technical tools developed and successfully tested for mature temperate forests (Pepin and Körner 2002). As far as the first two seasons' results of these tests in a highly diverse, 35 m tall forest near Basel show, species do indeed respond differently to elevated CO_2. Hence, the overall ecosystem response will depend on the species present.

CO2 and forest water relations

In this last section I am entering almost unexplored terrain. Beyond its potential influence on the climate and its established direct effects on plant photosynthesis, the concentration of CO_2 in the atmosphere has a third effect: elevated CO_2 reduces the width of stomatal pores in many plants. This is commonly explained by an optimization strategy of plants, keeping CO_2 uptake and vapour loss in balance, so that carbon gain per unit of water lost is maximized (e.g. Farquhar and Caemmerer 1982). A reduction of leaf transpiration in response to CO_2 enrichment would have far ranging consequences at landscape level, with more moisture retained in the ground and less used to humidify the atmosphere. Both these effects create positive feedback. High soil moisture and lower vapour pressure deficit could enhance transpiration and offset the consequences of the leaf level response (Field 1995; Körner 2000). This is a theoretical prediction. The reality for temperate forests so far tells us that there is no uniform behaviour among different tree species in this respect. Some species exhibit the expected response

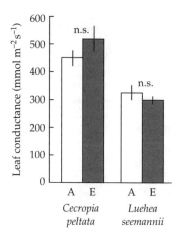

Figure 6.3 Stomatal responses to *in situ* CO$_2$ enrichment in the canopy of a tropical forest in Panama. Leaves or branches were continuously exposed to CO$_2$-enriched air for several days to weeks, and stomatal conductance was measured with a porometer (data from Körner and Würth 1996 and Lovelock *et al.* 1999).
A ambient, E elevated CO$_2$; n.s. refers to statistically non-significant differences between A and E.

(many, but not all broad-leaved deciduous species), while others (in particular evergreen conifers) do not (Ellsworth *et al.* 1995; Medlyn *et al.* 2001).

In situ data for mature tropical forest stomatal responses are available for CO$_2$-enriched canopy leaves, studied with a canopy crane in Panama (Fig. 6.3). Surprisingly, no CO$_2$ effect on stomata was observed, even though the CO$_2$-enrichment signal left a strong fingerprint of stable carbon isotopes in the CO$_2$-enriched leaves, proving that the gas did affect the leaves. *Fagus sylvatica* (beech), the most important European deciduous forest species, shows the same absence of a stomatal response, but other temperate tree species do save moisture under elevated CO$_2$ (for instance *Carpinus betulus* (hornbeam), Keel, unpublished data). It is almost certain that such inter-specific differences will occur in tropical forests as well. The consequence is a species-dependent patchiness of soil moisture, which will lead to losers and winners during periods of reduced moisture supply. It may well be that such CO$_2$-induced species-specific changes in water relations will outweigh a potential direct CO$_2$ fertilization effect on trees, as seems to be the case for grassland (Volk *et al.* 2000; Morgan *et al.* 2004). We urgently

need to address these CO$_2$-linked water relations questions, using the new technique of CO$_2$-enrichment for tall forests mentioned above. The relative abundance of savers and spenders of water in a CO$_2$-enriched world will feed back onto the climate system, causing the atmosphere either to become drier or remain unaffected. Needless to say, this has consequences for the large-scale export of moisture from humid to drier regions via water vapour. Given that species differ in their moisture response, biodiversity may also play a key role in these macroclimatic processes.

Conclusions

The consequences of atmospheric CO$_2$ enrichment for forests are commonly modeled from first principles such as, for example, photosynthetic responses. This short chapter has attempted to show the importance of accounting for the differential responses of species or plant functional types to elevated CO$_2$ when attempting to predict tropical forest responses to atmospheric CO$_2$ enrichment. I have tried to explain why I think that direct carbon accounting at the plot level can be biased towards higher than true landscape-wide C-fixation because it assumes relatively uniform uptake in space and time, rather than stochastic carbon release patterns. Once differential CO$_2$ responses of key forest taxa come into play, a greater stimulation of one type of taxa (e.g. lianas) can reduce life span in others (e.g. trees). Although there are few data to support such conclusions, currently available experimental findings make them appear plausible. Moreover, a more dynamic tropical forest seems a realistic scenario and would be in line with some long-term inventory data (Phillips *et al.*, Chapter 10, this volume, Laurance *et al.*, Chapter 9, this volume). Faster tree turnover, as documented in these studies for Amazonia, is likely to reduce the carbon stock in the medium-term even under conditions of enhanced productivity. Interactions between biodiversity and forest dynamics need to be considered when making projections about the role of tropical forests in the global carbon cycle in a CO$_2$-rich future. A more dynamic forest is not likely to sequester more carbon.

CHAPTER 7

The effects of drought on tropical forest ecosystems

Patrick Meir and John Grace

Drought stress in tropical forests can have a significant impact on global carbon, water, and energy cycles. This chapter examines drought-induced responses in the processing of carbon and water by intact tropical forest ecosystems over short (physiological) and longer (ecological) timescales. Both levels of understanding need to be represented in analyses of climate–forest ecosystem feedbacks. Although limited spatial information on the diversity of soil physical properties constrains estimates of drought vulnerability, tree functional convergence across species based on simple measures such as wood density promises to simplify how drought responses can be represented and linked to changes in forest composition through mortality indices. While insufficient on their own, satellite-derived measurements of ecosystem properties (e.g. leaf area index) and processes (e.g. mortality and photosynthesis) are expected to provide increasingly detailed information that can be used to test our understanding of short- and longer-term responses to drought.

Introduction

It has long been recognized that many tropical rainforests, paradoxically, experience periodic moisture stress, even in regions with high mean annual precipitation (Richards 1996). With a leaf area index (the mean leaf area per unit ground area) of between 4 and $6\,\mathrm{m^2\,m^{-2}}$ and a plentiful water supply, transpiration rates of about 4 mm per day are common in tropical vegetation. Clearly then, monthly rainfall of less than $\sim$100 mm may lead to moisture stress, with the effects modulated by other prevailing weather variables together with the specific physiological and hydrological properties of vegetation and soil. This value of 100 mm per month has been used widely to identify the occurrence of possible seasonal moisture deficits and to gauge the severity of climatic perturbations in rainfall at interannual or longer timescales. Even in terms of rainfall alone, however, any monthly or annual total must remain a generalization: the extent of water stress varies strongly in relation to the number of sequential months during which

rainfall is low (dry season length), and the extent to which rainfall is less than the approximated critical value of 100 mm (intensity).

Malhi and Wright (Chapter 1, this volume) discuss late twentieth century trends in tropical climate and evaluate soil moisture stress across all three main tropical forest regions by attempting to account for both dry season length and intensity. Overall, they identify a slight global reduction in rainfall over the period 1960–98 ($1 \pm 0.8\%$ per decade), but emphasize wide regional variation including both increases and decreases in rainfall, and a strong trend towards increased water stress in Africa. Long-term (multidecadal scale) fluctuations not fully captured in the 1960–98 dataset can also underlie and influence these trends (Botta *et al.* 2002), making it difficult to make unequivocal statements about changes in rainfall from 39 yrs of data. In contrast, the accompanying temperature measurements show warming in most regions, and for recent years (1976–98) strong and global increases of 0.15–0.40 °C per decade. The combined effect of increased temperature with no change (or a reduction) in rainfall leads to increased moisture stress, and this is underlined in climate and vegetation process simulations for the twentieth and twenty-first centuries (Cramer et al., Chapter 2, this volume). The latter study reports at least two leading climate models as giving very strong drying scenarios for the end of the twenty-first century, but differing in the location of the affected areas (South America, Africa; a 17% reduction in precipitation by 2100 for both), while two further models do not show patterns that are as distinctive. Such large potential effects and the uncertainty in their magnitude and location imply a strong need to understand vegetation–climate interactions under a drying climate.

So what is the significance of large reductions in rainfall to tropical rainforest vegetation, and to its interaction with the atmosphere in terms of the exchange of H_2O and CO_2? Over the short term, physiological responses constraining water loss from plants must reduce the rates of carbon assimilation and growth by trees, and the rate of CO_2 efflux from soil. Higher temperatures associated with drier weather will necessarily increase respiration rates in plant and soil, although below a particular moisture threshold soil respiration must also decrease. The immediate consequences of drought may be impacts on production and increased tree mortality, together with reductions in evapotranspiration and a change in the net gain of carbon from the atmosphere. In the longer term, differential species responses to drought may result in complete or partial changes in vegetation type.

Models of the responses by tropical rainforest to drought have previously suggested that short-term differences in rainfall associated with El Niño-Southern Oscillation events are expected to result in large switches in carbon exchange from source to sink of the order of up to ~1 $Pg\,C\,yr^{-1}$ at the regional (Amazonian) scale, with net emission rates similar to estimates from deforestation for the same region (Foley et al. 2002). Indeed, there may be a causal link between the impact of drought on rainforests and global CO_2 concentrations because of the large contribution that rainforests make to global photosynthesis (Clark et al. 2003). Coarser-scale and longer-term studies have also previously simulated strong positive feedbacks between

increased CO_2 concentration, climatic drought, and ecosystem-wide processing of carbon and water for the twenty-first century, with marked effects in South America (Cox et al. 2000). However, uncertainty in the modelled meteorology (Huntingford et al. 2004) and the modelled responses in vegetation and soil over different timescales have limited our ability to thoroughly evaluate these results.

Simulations of ecosystem–atmosphere interactions during this century generally incorporate the physiological effects of high CO_2 concentration on plant physiology (e.g. Cramer et al., Chapter 2, this volume). Increases in atmospheric CO_2 concentration above current levels (~370 ppm) up to 500–600 ppm have the potential to reduce water use by vegetation through increased carbon assimilation capacity (Jarvis 1998). The negative effects of drought on plant growth can therefore be offset through the physiological advantage accrued from higher CO_2 concentrations. However, increases in temperature associated with a higher atmospheric CO_2 concentration tend to have the opposite effect, as do increases in leaf area index that may potentially result from increased CO_2 concentration. Local and regional variations in these responses will influence the net effect in terms of water use and carbon gain, but at the forest margins both are likely to interact strongly with human activities.

Forest use by humans tends to cause fragmentation, yielding large amounts of edge and canopy gaps relative to undisturbed tracts of forest. These two features lead to greater exposure to ignition sources as well as a high probability of further drying within the litter layer and a consequent increased vulnerability to fire (Cochrane et al. 1999). The feedbacks among drought, fire, ecosystem functioning, and land use in tropical forest areas are potentially significant (Nepstad et al. 2004). However, they are not examined further here (but see Laurance et al., Chapter 9, this volume), other than to note that while the potential for forest loss and its consequent climatic impact is large in regions such as Amazonia, year-to-year variation in drought response by intact rainforest may be of a similar magnitude as deforestation in terms of CO_2 exchange (Tian et al. 1998; DeFries et al. 2002).

This review examines the physiological and ecological responses of tropical rainforest ecosystems to drought. Physiological changes in the processing of carbon and water by plants and soils are considered with the goal of understanding how it may be possible to represent these processes at short- and long timescales, in the context of climatic drying.

Hydraulic constraints in plants and soil

Plants

Carbon dioxide and water vapour pass through the same conduits—stomata—for diffusion into and out of the leaf. As a consequence, transpiration and photosynthesis are necessarily linked. Estimates of photosynthesis and production must therefore be predicated on a good understanding of water use by plants, including how water use is controlled under conditions of drought.

The response of transpiration (E) to changes in soil moisture content is determined by atmospheric demand—primarily the water vapour pressure deficit (VPD)—and hydraulic constraints to water transport in the soil–plant–air continuum. VPD at the top of rainforest canopies can be remarkably high because of the high leaf surface temperatures. Stomata tend to close in response to this high VPD (Grace *et al.* 1995). Dynamic changes in stomatal conductance (g_s) result in the diurnal control of plant water potential above a critical value below which cell damage occurs, hydraulic conductivity is lost, and transpiration ceases (Oren *et al.* 1999; Sperry *et al.* 2002). However, the rate of water transport in this continuum is partly controlled by conductivity in the rhizosphere between the bulk soil and the root surface (Passioura 1988; Kyllo *et al.* 2003) as well as by conductivity in the xylem and related tissues from roots to leaves. Changes in soil moisture and atmospheric demand therefore influence the potential range and rate of E and the total amount of photosynthesis. Responses by plants to soil moisture stress may include changes in g_s, changes in non-leaf hydraulic resistance, and changes in total leaf area or exposed leaf area. Ultimately, below a

critical minimum soil water potential, control of stomatal apertures or other loci of hydraulic resistance cannot maintain the conductivity from soil to stomata because of cavitation events ('breaks') in the water columns in the plant or the soil–plant continuum (Fig. 7.1). This results in a reduction in,

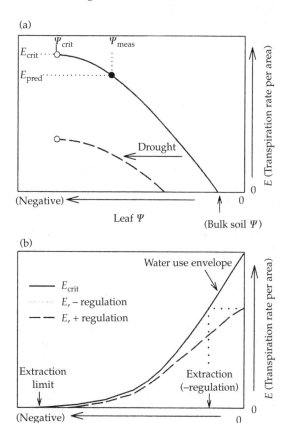

Figure 7.1 Hydraulic limitations in the soil–leaf continuum; hypothetical relationships between transpiration rate and water potential of the leaves and soil (from Sperry *et al.* 2002). (a) Steady state increases in transpiration (E) result in concomitant reductions in leaf water potential (leaf ψ). Moving from right to left, leaf water potential becomes more negative ('less moisture'); the dashed line represents the situation when the soil has been droughted, and the solid line, non-drought conditions. Beyond a maximum value of E (E_{crit}), hydraulic continuity is lost (because of cavitation of water columns), and transpiration cannot continue. (b) Stomatal regulation maintains hydraulic continuity by controlling E below a critical maximum value influenced by soil water potential (soil ψ). Moving from right to left, soil ψ becomes more negative (less moisture available). In the absence of stomatal regulation (dotted line), once E_{crit} has been passed, no further water uptake is possible.

or cessation of, growth and leads to death of the leaf or branch. *In extremis,* death of the whole plant occurs.

Measurements of E made at the leaf level provide detailed information on the response to leaf water status and atmospheric demand (e.g. McWilliam *et al.* 1996), but suffer from large spatial variability. Sap flux measurements, by contrast, provide an integrated measure of E for a single tree, and with appropriate tree size–class distribution data and information on sapwood area they have been used to successfully estimate water use at the stand level in monospecific stands (e.g. Vertessey *et al.* 1995). The high species diversity in tropical forests presents a possible challenge for this approach, but fortunately there are indications of functional convergence across tropical species in the processing of carbon and water (Meir and Grace 2002; Meinzer 2003). Using detailed measurements on 23 species in a tropical moist forest in Panama, Meinzer *et al.* (2001) showed that tree stem diameter was strongly correlated with sapwood area, maximum sap flux intensity, and the daily total flux of sap. Furthermore, apparently diverse species-based responses in E to VPD converged when the data were normalized with a measure of the branch-level ratio of leaf area to sapwood area (Meinzer *et al.* 1997), suggesting the existence of general underlying constraints to water use. Wood density has also been shown to be a good predictor of key water use parameters, including minimum daily leaf water potential, resistance to cavitation, capacitance, and total daily transpiration (Bucci *et al.* 2004). It appears that plants investing, for example, in the high mechanical strength and resistance to xylem cavitation that is associated with denser wood may face a trade off with lower water transport efficiency, lower sapwood capacitance (and hence a reduced ability to buffer changes in leaf water potential), and lower growth rates.

This apparent functional convergence among diverse tropical species based on relatively simple physical tissue properties is also consistent with a general allometric scaling scheme for water use by plants which proposes that total daily transpiration scales with the $\frac{3}{4}$ power of above-ground dry mass (Meinzer 2003). Therefore, while a limited number of studies have been made of transpiration by tropical trees and forests, it appears feasible to represent their response to drought at stand and larger scales using relatively few measurements. This has been achieved with some success (e.g. Williams *et al.* 1998; Foley *et al.* 2002), although such process-based modelling studies tend either not to address, or not to make use of, recent analyses of functional groupings.

Soils

The changes in below-ground interactions between carbon processing and water availability in drying tropical soils are less well understood. Soil physical properties dictate both water holding capacity and the availability of water to micro-organisms and plants, with consequences for the carbon cycle. For example, Tomasella and Hodnett (1997) showed that a bimodal pore size distribution in a clay-rich soil in central Amazonia strongly controlled hydraulic conductivity and soil water potential, leading to low plant available moisture. These physical properties constrain access to water for vegetation, also constraining GPP (Malhi *et al.* 1998). However, although soil water potential is relatively widely measured, soil hydraulic conductivity data are sparse for tropical rainforests and the relationship with texture is uncertain (Fisher *et al.* 2005). This lack of information on regional-scale spatial variation in soil hydraulic conductivity constrains our ability to accurately predict plant water use in drying tropical soils. As soil physical properties cannot be retrieved using remote sensing, collection of further appropriate field data should be considered a high priority.

Differences in soil porosity constrain soil respiration rates as well as plant productivity. A peaked response in soil respiration to volumetric soil moisture content is generally expected whereby respiration rates decline in very wet or very dry soils (Howard and Howard 1979), although unexpected exceptions have been reported (Davidson *et al.* 2004). The response to soil moisture becomes linear when soil moisture is expressed as soil water potential (Davidson *et al.*

1999), and is a reflection of the underlying physical constraints, such as porosity, on water availability to respiring autotrophs and heterotrophs in drying soils. A strong reduction in soil CO_2 efflux was observed by Sotta *et al.* (unpublished data) when soil moisture was experimentally reduced in an attempt to mimic soil drying during persistent El Niño conditions. Consistent with this, Saleska *et al.* (2003) also reported dry season reductions in soil respiration using eddy covariance and chamber methods. We note that if such a drought related effect was strong and widely occurring for a sufficiently long period, then a temporary storage of soil carbon could be expected through reduced losses from soil respiration, an outcome that would contradict the model results discussed above of regional-scale interannual terrestrial carbon exchange.

Over the longer term such a drought effect on respiration could also influence the total stock of moisture-labile carbon in the soil, although soil respiration rates and carbon storage will clearly depend on interacting responses to long-term changes in temperature as well. While a uniform (and global) response of increasing respiration with increasing temperature has been used previously in some models (e.g. Cox *et al.* 2000), the combined interactions among temperature, moisture, and nutrient availability are likely to moderate this response. In particular, the number of soil carbon pools and the differential changes in the size of pools that are either labile or resistant to oxidation may lower the long-term temperature response in respiration, at least at the scale of an individual ecosystem (Melillo *et al.* 2002).

Responses in soil to a drying climate also depend on the concomitant productivity of the vegetation, via the supply of assimilate to the soil. Högberg *et al.* (2001) showed that slightly more than 50% of daily GPP in a temperate forest was returned to the atmosphere via the fine roots and soil. In tropical forests there may be similar tight and dynamic links between carbon assimilation and its allocation to compartments that are above-ground ('shoot') or below-ground ('root'), and which may have long (e.g. wood) or short (e.g. fine roots, root exudate) carbon residence times.

The ratio of root to shoot (R : S) often increases with drought resistance—large R : S values are found in savanna ecosystems (Shenk and Jackson 2003), and also in some rainforest regions where extensive deep root systems provide a mechanism for seasonal drought tolerance (Nepstad *et al.* 1994). But less is known of the allocation processes and their consequent effect on soil CO_2 effluxes that accompany the response by vegetation to drought, or during transitions in vegetation, such as from rainforest to savanna. Studies from tropical plantations of *Eucalyptus* species show that R : S varies between 0.15 and 0.35 along an increasing annual precipitation gradient from 800 mm to 1600 mm (Stape *et al.* 2004).

In addition to a moisture response, soil CO_2 effluxes are affected by plant growth patterns through nutrient supply. Forests with a low nutrient status tend to be drought resistant relative to those on nutrient rich soils (Chapin 1980). The larger R : S ratio implied by this is also consistent with the variation in allocation below-ground proposed to explain observations of relatively high above-ground net primary productivity in South American forests growing on rich soils in the western Amazon (Malhi *et al.* 2004). Getting allocation responses correct in addition to direct moisture and temperature responses is perhaps more important than it might at first seem. Dufresne *et al.* (2002) showed that differences between dynamic global vegetation models (DGVMs) in the allocation of assimilate to below-ground components contributed significantly to a 3-fold difference in the size of the potential positive feedback between twenty-first century atmospheric CO_2 concentration and the terrestrial carbon cycle, an increase of 75 ppm versus 250 ppm. This positive feedback, presented by Cox *et al.* (2000), is partly dependent on the proportion of GPP contributing to the soil carbon compartment, and its subsequent release as soil respiration. The difference between the amount of GPP that ends up in vegetation or soil carbon stocks is important because of the differential climate sensitivities in each. A larger amount of soil carbon can lead to a much larger release of CO_2 into the atmosphere through increased soil respiration and a strengthened positive feedback between climate and the terrestrial carbon cycle.

Ecological change: can long-term patterns be described by process understanding?

Changes in the species composition and functional properties of tropical forest occur in response to climatic change but linking the two is difficult because few process-level measurements exist over the relevant timescale(s) of ecological change. Studies in semi-arid ecotonal landscapes suggest that vegetation change can be very rapid when precipitation declines below a threshold level (e.g. Allen and Breshears 1998). Less information is available for tropical rainforests, but documentation of the response to climatic drying resulting from the impact of the strong El Niño events of 1982/3 and 1997/8 provides some multi-year information.

Tree mortality

Research at Barro Colorado Island (BCI), Panama, using a large 50-ha permanent sample plot, has generated a rich picture of species-by-species growth patterns over more than 20 yr. The seasonal forest at BCI has experienced a long-term drying trend since the 1960s, punctuated by extreme droughts, most notably the El Niño events of 1982/3 and 1997/8. An identically sized plot at Pasoh forest in Peninsular Malaysia, where annual rainfall is slightly lower (~2000 mm versus ~2500 mm but relatively aseasonal), has experienced less extreme drought stress in recent years and provides a point of comparison. During the period 1985–95, years of 'normal' rainfall, mortality in trees with a diameter greater than 10 cm was nearly two-fifths higher at BCI than at Pasoh (2.03% yr^{-1} versus 1.48% yr^{-1}), and this pattern was also found in growth and recruitment rates (Condit *et al.* 1999). The high turnover in species population sizes at BCI showed that the community is far from equilibrium, probably because of recent drought-related disturbance. The decline

in moisture-demanding species at BCI has previously led Condit (1998) to predict an extinction rate of up to 40% in very moist areas, under the scenario of a 9 week lengthening of the dry season.

The unusual drought stress associated with El Niño events tends to increase mortality in tropical rainforests from around 0.7–2.0% yr^{-1} to 2.0–7.6% yr^{-1}, and in exceptional cases to above 10% (Table 7.1). The importance of topography and soil type in influencing drought-related mortality among forests is variable. Often, it is the largest trees that suffer the biggest increases in mortality, though in a few circumstances smaller trees have been shown to suffer more (Aiba *et al.* 2002). The higher susceptibility to drought in larger trees may simply reflect the greater exposure to solar radiation and thus evaporative demand,

Table 7.1 Impact of El Niño-induced severe drought on lowland (<1000 m a.s.l.) rainforest tree mortality (*M*), in % per year

Location	Plot no. and size (ha)	*M* (Non-El Niño)	*M* (El Niño)	Source
Kutai, Borneo	3, 0.2	1–3	14–24	Leighton and Wirawan (1986)[a]
BCI, Panama	1, 50	1.98	3.20	Condit *et al.* (1995)[a]
Manaus, Brazil	23, 1	1.13–1.91	2.44–2.93	Laurance *et al.* (2000)[b,c]
Manaus, Brazil	12, 1	1.12	1.91	Williamson *et al.* (2000)[b]
Sarawak, Borneo	2, 1.3 and 6.4	0.89	5.36	Nakagawa *et al.* (2000)[b]
Mt Kinabalu, Borneo	2, 1	0.8–1.1	2.5–2.6	Aiba *et al.* (2002)[b]
Sarawak, Borneo	3, 0.7	2.4	7.6	Potts (2003)[b]

[a] From El Niño of 1982/3.
[b] From El Niño of 1997/8.
[c] From fragmented forest.
Note: Range (of *M*) quoted if more than one plot.

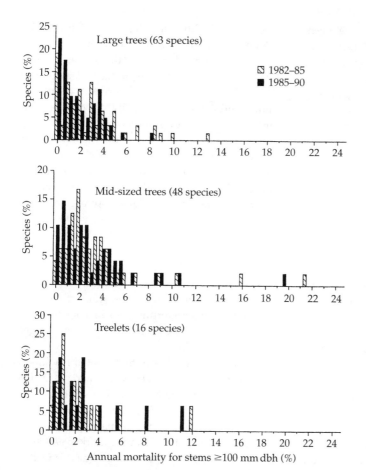

Figure 7.2 Frequency distributions for mortality rates in three different plant physiognomic forms, showing the increase in mortality under drought stress (from Condit *et al.* 1995). Data are from an interval during non-El Niño conditions (1985–90, black columns) and one that included an El Niño event (1982–5, hatched columns). Mortality is plotted in intervals of 0.5% yr^{-1}; only species with more than 20 stems of d.b.h >10 cm are included.

or may be a more complex function of differential seedling survival and species physiology. Slik (2004) reported increased drought effects in logged forest over unlogged forest in Kalimantan, predicting compositional changes as a result of high mortality in pioneer trees (e.g. *Macaranga* sp.) and increased light penetration to the forest floor. A wide range of ecological differences among species in terms of drought tolerance has been described for example, through differences in phenological timing and water use, or differences in the depth of water extraction offset against differences in the ability to tolerate low plant water potential (e.g. Stratton *et al.* 2000).

Such wide species differences are also found in studies of mortality, with a natural range from 0 to 26% yr^{-1} across species, values that correspond

to an approximate range in maximum tree age from 35 to more than 1000 yrs (Condit *et al.* 1995; Nakagawa *et al.* 2000). Representations of drought-related transitions in vegetation need to incorporate this information (Fig. 7.2), as well as the expected lower rates of mortality for more strongly seasonal rainforests (e.g. Condit *et al.* 2004). One species-level calculation of vegetation transition at BCI predicts invasion rates in response to climatic change over hundreds of years (Condit 1998). It is probably not realistic to obtain or model species-level mortality data for whole regions, but mortality functions based on tree size or growth can be used to represent compositional change. Linking changes in composition (through mortality) with changes in the physiology of species or functional groups remains a challenge. However, the identification of strong correlations between widely used measures such as wood density and drought tolerance physiology (Meinzer et al. 2003) may provide one way of making this link. We speculate that an analysis of mortality parameters, drought tolerance and wood density could facilitate the estimation of transitions in tropical forest function in response to climatic drying.

Fruiting

Fruiting responses to drought are important because of their potential impact on long-term forest composition. In Southeast Asian rainforests mast-fruiting in the dominant Dipterocarp family is sufficiently strongly correlated with El Niño events to suggest that drought, along with low temperature, may act as an environmental cue for large-scale fruiting (Williamson and Ickes 2002). Wright et al. (1999) also hypothesized that El Niño events enhance fruiting at BCI; they further show that wet and cloudy La Niña conditions following an El Niño year can lead to strongly reduced fruit production because of reduced assimilation and stored resources. It therefore seems probable that, whether or not drought triggers a fruiting response in a particular species, the capacity to perform this response is strongly controlled by the carbon resources of individual trees. If this is the case, stronger or repeated El Niño-type events could lead to failure rather than increases in fruiting. Consistent with this prediction, a strong reduction in fruiting was observed in an experimental throughfall exclusion study in the eastern Amazon (Meir et al. unpublished data).

The potential for long-term process monitoring

Liu et al. (1994) investigated the spatial and temporal trends in the Normalized Difference Vegetation Index (NDVI) derived from satellite data from South America (1981–7). They detected large changes in this index that were associated with patterns of rainfall, and in particular with the El Niño year of 1982–3. In north-east Brazil the drought was very severe and it resulted in great socio-economic difficulties for this region. The fall of NDVI associated with dry conditions presumably reflects large changes in the leaf area index (LAI), as a result of the shedding of leaves (Fig. 7.3). LAI determines the absorption of photosynthetically active radiation by the canopy, and thus the rate of photosynthesis. It also controls the rate of transpiration, and therefore the partition of energy between evaporation and sensible heat loss. If the leaf area exposed to incoming radiation is reduced, then the hydraulic stress on the vegetation will also be reduced. Many tropical rainforests contain evergreen and facultatively deciduous species, making possible fine-scale variations in LAI in response to climatic variation (e.g. Carswell et al. 2002). Although global datasets on LAI have been collated (Asner et al. 2003), the potential to continuously monitor leaf area from space provides a growing opportunity to link spatial patterns with processes that occur on longer ecological timescales.

The recent availability of high spatial resolution data ($\sim$1 m) from the Quickbird and IKONOS satellites, or very high spectral resolution data (wavelength bands of a few nanometres) from the EO-1 Hyperion sensor, has opened up significant new opportunities. For example, Clark et al. (2004) have demonstrated the potential to use IKONOS and Quickbird data to quantify canopy tree mortality rates. Asner et al. (2004) have successfully used hyperspectral data from EO-1 to detect drought effects on net primary productivity by sensing changes in two narrow-band leaf pigment reflectance indices, the photochemical reflectance index (PRI) and the anthocyanin reflectance index (ARI). These recent successes suggest the potential to monitor the future growth and productivity of remote tropical rainforests in fine ecological and physiological detail over large spatial and temporal domains. However, as with other new satellites, the datasets are still quite short term. Longer runs of AVHRR-derived values for temperature and LAI are available and have been used to drive ecosystem productivity models. Potter et al. (2004) in a recent formulation of their NASA–CASA ecosystem model used AVHRR data from 1982–98 to show correlations between NPP and interannual

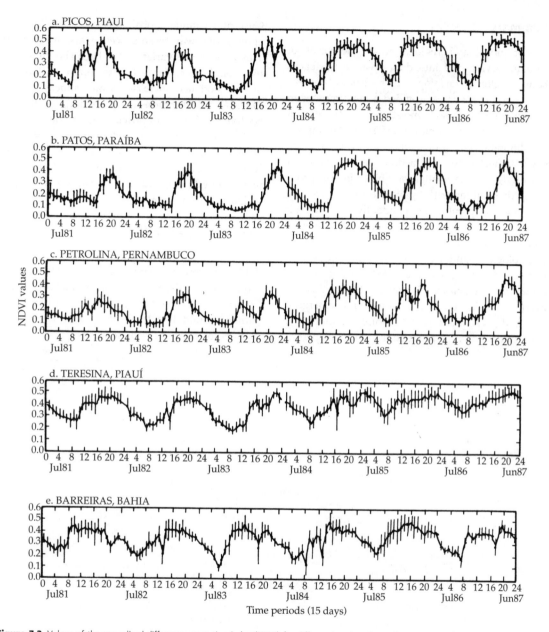

Figure 7.3 Values of the normalized difference vegetation index (NDVI) for different locations in north-eastern Brazil over a 6-yr period that included the 1982–3 El Niño event (from Liu *et al.* 1994). The plots show variation in surface greenness from the west (Picos 7°4′S, 41° 29′W) to the east (Patos 7° 1′S, 37° 16′W), and from north-west (Teresina 5° 5′S, 42° 49′W) to south (Barreiras 12° 9′ 45° 0′W), with the driest centre of the semi-arid region at Petrolina (9° 23′S, 40° 30′W).

differences in rainfall and temperature. These results are consistent with other ecosystem models that do not use satellite input data for land surface parameters, but also suggest additional climate–land surface process correlations, hence underlining the need to balance the use of remote sensing with appropriate and detailed ground observations.

Summary

Recent global and regional model scenarios of climatic drying and attendant positive feedbacks between global warming and the tropical terrestrial carbon cycle have provided a new impetus to investigations of soil moisture constraints on the functioning of tropical rainforests. An important limitation on our ability to evaluate such vegetation–climate interactions is a lack of knowledge of the spatial variability in soil properties that influence the moisture that is actually available to plants, independently of the soil moisture that is volumetrically present. However, equally important is the need to extend our physiological understanding of the short-term response to drought into the ecological processes underlying vegetation dynamics. To do this we need to make use of existing and emerging knowledge of plant growth patterns, mortality, and ultimately, reproduction and dispersal. Current dynamic global vegetation models disagree in the way these processes are represented. This inter-model variation has been shown to lead by itself to very large (up to 3-fold) differences in the modelled increase in atmospheric carbon dioxide concentration for a period as few as 50–100 yrs hence. Improving our understanding of how relatively slow, but quantitatively influential, ecological processes operate and are correlated with the capacity of vegetation to function in response to drought, will advance this element of Earth system science.

Acknowledgements

This article forms part of research funded by the EU Framework 5 Programme and the UK Natural Environment Research Council including the Centre for Terrestrial Carbon Dynamics. The authors would like to thank Y. Malhi and O.L. Phillips for helpful discussion.

PART III

Observations of contemporary change in tropical forests

CHAPTER 8

Ecological responses of Amazonian forests to El Niño-induced surface fires

Jos Barlow and Carlos A. Peres

Over the last 20 yrs the combined effects of El Niño-induced droughts and land-use change have dramatically increased the frequency of fire in humid tropical forests. Despite the potential for rapid ecosystem alteration and the current prevalence of wildfire disturbance, the consequences of such fires for tropical forest biodiversity remain poorly understood. Here we provide a pan-tropical review of the current state of knowledge of these fires, and include data from a study in a seasonally dry *terra firme* forest of central Brazilian Amazonia. Overall, this study supports predictions that rates of tree mortality and changes in forest structure are strongly linked to burn severity. Despite the paucity of data on faunal responses to tropical forest fires, some trends are becoming apparent; for example, large canopy frugivores and understorey insectivorous birds appear to be highly sensitive to changes in forest structure and composition during the first 3 yrs after fires. Finally, we evaluate the viability of techniques and legislation that can be used to reduce forest flammability and prevent anthropogenic ignition sources from coming into contact with flammable forests.

Introduction

Large uncontrolled fires linked to severe El Niño, Southern Oscillation (ENSO) events have become increasingly common across the humid tropics during the past two decades (e.g. Leighton and Wirawan 1986; Cochrane *et al.* 1999; Nepstad *et al.* 1999b; Siegert *et al.* 2001). The overwhelming majority of these fires are associated with anthropogenic influences such as logging and fragmentation (see Cochrane 2003 and Laurance 2004 for reviews of the underlying mechanisms), although sustained wildfires can also affect largely undisturbed primary forests provided that the appropriate edaphic conditions are combined with sufficiently long droughts (Peres 1999).

We examine the current state of knowledge regarding the effects of these fires on tree mortality

and faunal responses, and then relate these to our study of a large mosaic of unburned and burned forests in central Brazilian Amazonia, examining in detail the relationship among burn severity, tree mortality, and forest structure, and the effects of fires on understorey birds and large vertebrates. Finally, we review management strategies that may be adopted to prevent the spread of fire in the fire-sensitive humid tropics.

Previous Studies

Tree mortality

Several hypotheses concerning the impact of fires on tree mortality have become established as a result of studies in Amazonia and Borneo. First, tree mortality is related to burn severity (Uhl and

Kaufmann 1990; Pinard *et al.* 1999), and more severe, recurrent fires kill a much greater proportion of standing stems than do initial low-intensity fires (Cochrane and Schulze 1999). Following initially light burns, tree mortality is markedly size dependent, with smaller stems having disproportionately higher levels of mortality than larger stems (Woods 1989; Holdsworth and Uhl 1997; Cochrane and Schulze 1999; Peres 1999; Pinard *et al.* 1999; Haugaasen *et al.* 2003). However, this survival advantage of large stems does not appear to persist following more severe recurrent fires, where all stems become equally vulnerable (Cochrane and Schulze 1999). Finally, a temporal increase in tree mortality occurs over time following fire events, both within 1 yr of fires (Holdsworth and Uhl 1997; Kinnaird and O'Brien 1998; Haugaasen *et al.* 2003), and from 1 to 2 yrs after fire disturbance (Cochrane *et al.* 1999).

Faunal responses to wildfire
It is hardly surprising that terrestrial vertebrates characterized by poor climbing abilities and low mobility should be affected by surface fires, and reports of dead and injured animals in burned forests (including tortoises *Geochelone* spp. and agoutis *Dasyprocta* spp.) are not uncommon in the immediate aftermath of fires (Peres 1999; Peres *et al.* 2003). Many arboreal vertebrate species appear to succumb to smoke asphyxiation, including several primate, sloth, porcupine, arboreal echimyid rodents, and some bird species (Mayer 1989; Peres 1999). Sub-lethal injuries of more mobile animals are also common, and local hunters have reported killing several terrestrial mammals exhibiting fire-induced scars up to 4 yrs after a recurrent fire in an Amazonian forest (Barlow, unpublished data).

After the fires, surviving animals must either emigrate into nearby unburned areas (if available) or remain within the burned forest matrix. Animals moving into unburned forest can expect to face elevated interference competition through territorial aggression from conspecifics (e.g. Bierregaard and Lovejoy 1989); density-dependent reductions in fitness through exploitative competition for food, mates, or other resources; and

could suffer disadvantages from their poor familiarity with the spatio-temporal distribution of resources. All of these effects can be aggravated by overcrowding. Only where population levels in adjacent unburned forest have been artificially reduced by game hunting is it possible to envisage a relaxation of density-dependent effects, although the life-expectancy of emigrants may still be low if the local subsistence hunting pressure persists or increases (Peres *et al.* 2003).

Individuals remaining in the burned forest face a different set of problems. Initially, large-bodied animals may be hunted relentlessly by rural peoples attempting to compensate for losses of food crops. This is exacerbated by the lack of cover resulting from the scorched understorey and loss of midstorey and canopy foliage (Haugaasen *et al.* 2003), and the increasing clumping of animals around remaining fruiting trees or patches of unburned forest (Lambert and Collar 2002). Reports of large numbers or entire groups of mid-sized to large-bodied diurnal primates (including brown capuchin monkeys *Cebus apella*, howler monkeys *Alouatta belzebul*, and orangutans *Pongo pygmaeus*) being killed by hunters shortly after wildfires are not uncommon (Saleh 1997; J. Barlow personal observation).

Animals that escape or are unaffected by hunting may face severe food shortages (O'Brien *et al.* 2003), as many canopy trees abort fruit crops and shed leaves following the traumatic heat stress (Peres 1999). In some cases animals may compensate by switching to alternative dietary items. For example, primates such as howler monkeys (C. Peres, personal observation) and orangutans (Suzuki 1988) became increasingly folivorous, resorting to the post-fire regrowth of young leaves, while both pig-tailed macaques (*Macaca nemestrina*) and gibbons (*Hylobates muelleri*) took advantage of wood-boring insect outbreaks immediately after the 1982–3 fires in Indonesia (Leighton 1983). However, dietary switching may not be an option for all species: In the case of understorey birds, those with specialized dietary, foraging, or habitat requirements are also the most likely to be extirpated from burned forest 1 yr after fires (Barlow *et al.* 2002).

Information on the long-term effects of fires remain scarce, and come mainly from studies conducted following the 1982–3 fires in Southeast Asia. In general, large terrestrial browsers such as wild pigs (*Sus* spp.), Banteng (*Bos javanicus*), and deer (*Cervus unicolor* and *Muntiacus muntjak*) appeared to be able to take advantage of newly available food sources, and recovered well from the fires (Wirawan 1985; Doi 1988; Mayer 1989). However, other animals did not display the same resilience, and some nocturnal primates (Western tarsiers *Tarsius bancanus* and slow loris *Nycticebus coucang*) and the Malayan sun bear (*Helarctos malayanus*) had been either locally extirpated or drastically reduced in numbers as of 1986 (Doi 1988; Boer 1989).

Study area

Our study was located in the Arapiuns and Maró river basins, western Pará, central Brazilian Amazonia, with fieldwork extending from January 1998 to June 2002 (Barlow and Peres 2004). The results reported here are from continuous expanses of forest, with burned forest lying adjacent to unburned forest areas used as controls. The assumption that these forests were similar in structure and composition before the 1997–8 fires was supported by interviews with local people, and by the similarity in the basal areas, stem densities, and tree family abundances in unburned and once-burned plots (Barlow *et al.* 2003).

Methods

Data on tree mortality (≥10 cm d.b.h.), canopy cover, and visible char height (noted on the trunk of all measured standing stems) were obtained from a total of 44 quarter-hectare (10 × 250 m) forest plots examined between July 2000 and May 2001. Sixteen plots were placed in unburned, 22 in once-burned, and 6 in twice-burned forest. Other habitat structure variables (including understorey openness, and both woody and non-woody stem densities) were quantified as a subset of 28 plots where the understorey mist-netting of diurnal birds was conducted (Barlow and Peres in press).

Thirteen of these plots (six in unburned forest and seven in once-burned forest) had been previously sampled 1 yr after the fire (Haugaasen *et al.* 2003). The diameters of trees were converted into aboveground dry biomass values following Santos (1996).

Understorey mist-netting was conducted at 20 plots examined 1 yr after fire and in 28 of the 44 quarter-hectare plots examined 3 yrs after fire (Barlow *et al.* 2002; Barlow and Peres in press). These 28 plots were placed in unburned ($n = 10$), once-burned ($n = 12$), and twice-burned forest ($n = 6$). In total, we obtained 5543 bird captures from 34 560 net-hours of sampling effort. Information on large-vertebrate abundance is derived from semi-structured interviews with local hunters (Peres *et al.* 2003); our own observations spanning 20 field months, and line transect surveys. The surveys were conducted along four line-transects of 4 km in length, located in unburned forest ($n = 2$), once-burned forest ($n = 1$), and twice-burned forest ($n = 1$). Each transect was walked twice every month for 14 months (Feb 2001–March 2002), resulting in 448 km of outward census effort (Barlow 2003).

Results

Tree mortality and forest structure

Surface fires had decisive delayed effects on tree mortality in once-burned forest, and an additional 74 trees per hectare died between 1 and 3 yrs post-fire. While a marked temporal increase in mortality was evident among smaller stems <30 cm in diameter, the greatest increase in mortality (relative to their original abundance) was found in the largest diameter class (≥50 cm diameter; Fig. 8.1(a)). This delayed mortality reduced live tree biomass by an additional 107 Mg ha^{-1} in once-burned forest (Barlow *et al.* 2003), substantially increasing estimates of committed carbon emissions from Amazonian fires.

Burn severity (mean char height) was a strong predictor of the number of standing live stems remaining (or tree mortality) in the 44 plots sampled 3 yrs after fire (Fig. 8.2). In once-burned plots, the live-tree density (excluding regrowth of pioneers

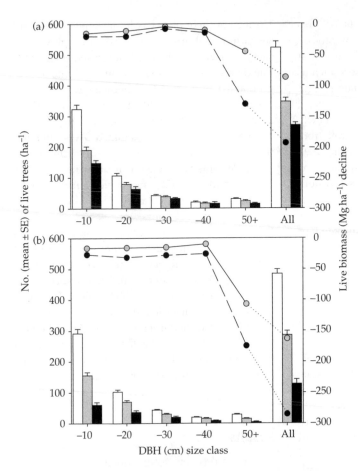

Figure 8.1 Mean number of live trees and above-ground biomass loss (relative to unburned primary forest) per hectare in different tree size classes showing (a) the temporal reponses to fires in primary forest plots (open bars; $n = 6$), and once-burned forest plots examined 1 yr (grey bars; $n = 7$) and 3 yrs (solid bars; $n = 7$) after fire, and (b) the loss of live trees and biomass 3 yrs after fire, in primary forest (open bars; $n = 16$), once-burned forest (grey bars; $n = 22$) and twice-burned forest (solid bars; $n = 6$). Circle symbols indicate the mean decline in live biomass from unburned forest levels in each disturbance treatment.

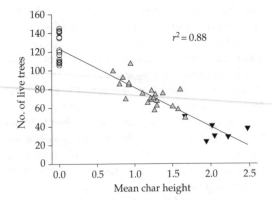

Figure 8.2 Relationship between burn severity (mean char height score) and the number of live trees in each quarter-ha plot ($F_{1,42} = 313.2$, $r^2 = 0.88$, $p < 0.001$). Unburned, once-burned, and twice-burned forests are indicated by open circles, grey triangles, and solid triangles, respectively.

after the 1997–8 fire) was, on average, only 58% of that found in the unburned forest. This was further reduced to just 26% in twice-burned plots. While trees of all size classes declined in abundance following recurrent fires (Fig. 8.1(b)), the sharpest decline relative to their initial abundance was in the largest size class (≥ 50 cm diameter; Fig. 8.1(b)).

Changes in forest structure were strongly related to the live-tree density in each plot (Fig. 8.3). The density of canopy gaps and the number of non-woody stems (such as bamboos) increased exponentially whereas understorey openness declined linearly, and the number of woody stems (< 10 cm diameter) exhibited a unimodal response (Fig. 8.3). Forest regeneration in once- and twice-burned plots was dominated by understorey and mid-storey species of shrubs and treelets typical of

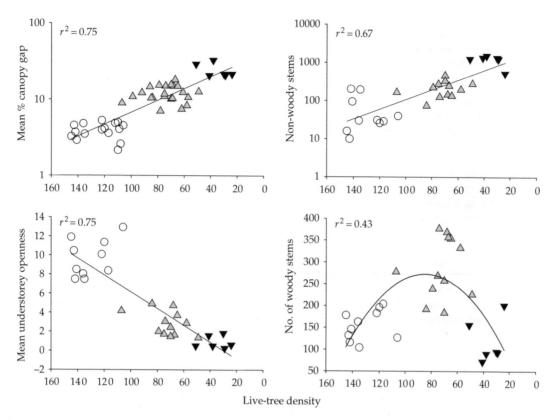

Figure 8.3 Changes in forest structure 3 yrs after fire disturbance in relation to local burn severity (here indexed as the number of live trees per plot). Unburned, once-burned, and twice-burned forests are indicated by open circles, grey triangles, and solid triangles, respectively. (Regression statisitics: canopy gaps: $F_{1,42} = 123.0$, $r^2 = 0.75$, $p < 0.001$; non-woody stems: $F_{1,26} = 52.2$, $r^2 = 67$, $p < 0.001$; understorey openness: $F_{1,26} = 79.1$, $r^2 = 0.75$, $p < 0.001$; the number of woody stems < 10 cm diameter: $F_{2,25} = 9.5$, $r^2 = 0.43$, $p < 0.001$).

disturbed forests and second growth (Table 8.1). Most saplings recorded in the understorey of unburned forest plots were regenerating canopy tree species that produce fruits dispersed by large vertebrates. Saplings dominating in once-burned forest plots were understorey and mid-storey specialists, producing fruits dispersed by small avian frugivores. Abiotic modes of seed dispersal (i.e. autochory and anemochory) became increasingly prevalent in the regeneration phase of twice-burned forest plots (Table 8.1).

Faunal responses

Primates and other mammals
Neither large frugivorous primates nor ungulates were recorded in the once-burned forest 1 yr

after the fires. However, some recovery was evident 3 yrs after the fires, and although many species were too rare to be observed during line-transect censuses, tracks around fruiting trees showed that tapir (*Tapirus terrestris*) and brocket deer (*Mazama* spp.) were at least occasional users of once-burned forest. Furthermore, almost all primate species were recorded, albeit at very low densities. Recurrent fires had far more pronounced effects, and resulted in the decline or extirpation of almost all forest species. Only small-bodied primates that are highly tolerant of second-growth, such as marmosets *Callithrix humeralifera*, titi monkeys *Callicebus hoffmannsi*, and squirrel monkeys *Saimiri ustus*, were more abundant in twice-burned forest than unburned forest.

Table 8.1 Main responses of a central Amazonian avifauna to short-term (1 yr after a single burn), medium-term (3 yrs after a single burn), and recurrent fires (examined 3 yrs after the most recent fire event)

Foraging guild	Short-term	Medium-term	Recurrent fires
Understorey species			
Dead-leaf gleaning insectivores	↓↓↓	↓↓	↓↓↓
Ant-following insectivores	↓↓	↓↓	↓↓↓
Terrestrial-gleaning insectivores	↓↓	↓	↓
Bark-searching insectivores[a]	↔	↔	↓
Arboreal-gleaning insectivores	↔	↑	↑↑
Arboreal-sallying insectivores	↓↓	↓	↓↓
Arboreal omnivore	↓↓	↔	↔
Arboreal granivore	↑↑	↑↑	↔
Arboreal nectarivore	↑	↑↑	↔
Arboreal frugivore	↔	↑↑	↔
Canopy species			
Canopy frugivores	NA	↓	↓↓

[a] Feeding superficially.

Note: ↔ = no change detected, ↓ or ↑ = slight decline or increase, ↓↓ or ↑↑ = strong decline or increase, ↓↓↓ = extremely rare or conspicuously absent from burned forest.

Table 8.2 Main responses of a central Amazonian mammal fauna (1 yr after a single burn), medium-term (3 yrs after a single burn), and recurrent fires (examined 3 yrs after the most recent fire event)

Foraging guild	Mammalian taxa included	Short-term	Medium-term	Recurrent fires
Arboreal folivores	*Bradypus, Choloepus,* and *Alouatta*	↓	↓	↓↓
Arboreal frugivores	*Cebus apella* and *Cebus albifrons*	↓	↓	↓↓↓
Small arboreal frugivores	*Callithrix, Callicebus,* and *Saimiri*	↔	↑	↑↑
Arboreal frugivores/granivores	*Pithecia* and *Chiropotes*	↓	↓↓	↓↓↓
	Sciurids	N.A.	↓	↓
Terrestrial granivores/frugivores	*Tayassu* spp., *Dasyprocta agouti*	↓	↓	↓
Terrestrial frugivores/browsers	*Mazama* spp., *Tapirus, Agouti paca*	↓	↓	↓
Carnivores	*Felis, Eira, Nasua*	↓	↔	↔

Note: ↔ = no change detected, ↓ or ↑ = slight decline or increase, ↓↓ or ↑↑ = strong decline or increase, ↓↓↓ = extremely rare or conspicuously absent from forest type.

Understorey and canopy birds

Fires had substantial impacts on the understorey forest avifauna, and once-burned forest netlines sampled 1 yr after the fires yielded fewer species and a lower abundance of birds than did unburned forest netlines (Barlow *et al.* 2002). There was no change in the species richness or capture success in six unburned plots sampled 1 yr and 3 yrs after fires (Barlow and Peres in press). In contrast, the same burned plots sampled 3 yrs after the fires ($n = 7$) displayed some signs of recovery in terms of bird species richness (mean ± SE species richness per netline: 1 yr after fire = 31.6 ± 2.7; 3 yrs after fire = 49.4 ± 2.4) and overall abundance (mean ± SE capture success per netline: 1 yr after fire = 73.1 ± 5.1; 3 yrs after fire = 168.0 ± 11.2). However, the avifaunal assemblage also became increasingly dissimilar to unburned forest over the same interval, and some species associated with second-growth habitats demonstrating 9-fold increases in abundance (Barlow and Peres in press).

Species turnover and changes in the avifaunal assemblage 3 yrs after fire were strongly associated with changes in habitat structure (such as canopy cover and understorey regeneration; Barlow and Peres in press), which were in turn determined by local burn severity (Fig. 8.3). The average (Bray–Curtis) similarity between netlines in once-burned and unburned forest was just 29%, compared to the mean background similarity of 54% between the 10 unburned plots. This decreased markedly (to only 6%) in twice-burned forest, and virtually no bird species were shared between the two extremes of the disturbance continuum.

Most insectivorous foraging guilds declined as a result of fires, although arboreal-gleaning insect-ivores increased in abundance along the gradient of burn severity. Dead-leaf gleaning and ant-following birds were extirpated from twice-burned forest, while arboreal nectarivores, granivores, and frugi-vores were most abundant in once-burned forest sampled 3 yrs after fire (Table 8.2). Large canopy birds feeding partly or entirely on fruit were affec-ted in a similar fashion as arboreal frugivorous mammals, and toucans *Ramphastos* spp., pigeons *Columba* spp., red-throated caracaras *Daptrius americanus* and oropendulas all declined appre-ciably after single and recurrent fires.

Discussion

Effects on tree mortality and forest structure

Levels of tree mortality in once-burned forest observed in this study (42%) are comparable to those in selectively logged forests in eastern Brazilian Amazonia (36–54%: Kaufmann 1991; 44%: Holdsworth and Uhl 1997; 38–66%: Cochrane *et al.* 1999; ca. 30%: Gerwing 2002), and Malaysian Borneo (53%: Woods 1989), despite substantial differences in logging intensities and natural disturbance regimes. However, these levels of mortality are all considerably higher than those reported for drier forests affected by surface fires near the south-western (23%; Pinard *et al.* 1999)

and northern (8–16%; Barbosa and Fearnside 1999) phytogeographic limits of Amazonia.

Overall, the generality of findings and predic-tions from other studies on the effects of wildfires in neotropical forests were supported here. Tree mortality was strongly related to burn severity at each forest plot (Fig. 8.2; cf. Uhl and Kaufmann 1990; Pinard and Huffman 1997), and recurrent fires had a much stronger effect than did single burns, with large-stemmed trees no longer holding a clear survival advantage (Fig. 8.1(b); Cochrane and Schulze 1999). However, by examining the temporal sequence of mortality in once-burned forest up to 3 yrs after the initial fire, we found that the short-term survival advantage initially attributed to large trees following low-intensity fires (e.g. Cochrane *et al.* 1999; Haugaasen *et al.* 2003) may be simply an artefact of research methods, since large trees damaged by blazes appear to take longer to succumb to lethal fire-stress (Barlow *et al.* 2003). Although this striking increase in large tree mortality comes from a single study, its generality is supported by the high mortality rates suffered by large trees following recurrent fires, droughts, forest fragmentation, and edge effects (see Barlow and Peres 2004).

This progressive increase in large tree mortality greatly augments previous estimates of forest biomass loss following low-intensity burns (Nepstad *et al.* 1999b). The delayed mortality of an additional 14% of trees between 1 and 3 yrs resulted in the loss of an additional 107 Mg ha^{-1} of live biomass, and increased the overall losses of live tree biomass from 23% after 1 yr to 51% after 3 yrs (Barlow *et al.* 2003). However, these estimates exclude the potential net losses from below-ground biomass and soil carbon pools (which as yet have not been quantified), or potential gains through the accumulation of biomass during post-fire regeneration, which in one study restored forest biomass to just 6% below primary forest levels 3–5 yrs post-fire (Carvalho *et al.* 2002). Moreover, high levels of uncertainty still surround estimates of carbon loss from above-ground forest biomass. Although they may increase further if tree mortality continues beyond 3 yrs after fires, or if high rates of tree damage are also considered

(Laurance *et al.* 1998a), applying an allometric model (Chambers *et al.* 2001a) that accounts for the lower living biomass of senescing larger trees may reduce the estimates.

Impact on vertebrate populations

Despite the relative paucity of studies documenting responses of tropical forest animal populations to wildfires, some patterns are beginning to emerge. First, it is evident that short-term effects of fire are frequently exacerbated by hunting, both in Amazonia (Peres *et al.* 2003) and in several parts of Southeast Asia (Saleh 1997). Wherever wildfires are caused by slash-and-burn subsistence agriculture, this situation is unlikely to change as ignition sources are often generated by the same people who subsidize their protein intake through hunting.

In the medium-term, our findings suggest that the rapid regeneration of early pioneer trees in once-burned forest lowered hunting pressure (by reducing hunter access and visibility), and providing an alternative resource pulse to terrestrial browsers and grazers. The same regeneration pulse also produced many of the small fruits, seeds, and flowers that can explain the increase in understorey birds consuming these resources (Table 8.1). However, while many terrestrial and understorey species may benefit from the vertical shift of primary productivity from the forest canopy to the understorey, canopy species (with the exception of second-growth specialists) were detrimentally affected by the sharp decline in live-tree density. This is consistent with similar responses by primate species observed in a fire-disturbed forest in southern Sumatra (O'Brien *et al.* 2003). Unsurprisingly, these declines were increasingly severe in twice-burned forest, where many primary forest specialists were extirpated, and most species that had initially benefited from, or were resilient to single burns, declined (Table 8.2). Finally, most of the burned-forest plots and transects were in forest adjacent to unburned forest, and these shifts in abundance should be seen as conservative indicators of the detrimental effects of fire.

Invertebrates

Invertebrates also exhibit substantial post-fire shifts in community structure (Fredericksen and Fredericksen 2002; Cleary and Genner 2004). Like birds in our study, the butterfly community in Borneo was strongly influenced by the recovering vegetation: all burned forests had greater numbers of herb specialists and reduced numbers of tree specialists, while liana specialists increased greatly after a single burn (Cleary and Genner 2004). Burn severity is also important, with recurrent fires resulting in much greater shifts in community composition and species loss. Crucially, the distance from unburned forest patches appears important for the recolonization of more mobile butterfly species, although less mobile endemics and habitat specialists remain restricted to these patches up to 3 yrs after the fires (Cleary and Genner 2004).

Assessing ecosystem change

The consequences of fires in tropical forest ecosystems will eventually be determined by the extent of fire coverage, fire-return intervals, and the tolerance of the biota to heat stress. Even under a relatively optimistic scenario of fire-return intervals (such as the 100–150 yrs predicted for some undisturbed forest interiors: Cochrane and Laurance 2002) the fire-sensitive biota of fire-prone forests can be expected to become increasingly impoverished and reminiscent of second-growth stands (Slik *et al.* 2002). Furthermore, in some areas a chain of interacting ecological, social, and economic factors (Cochrane *et al.* 1999; Nepstad *et al.* 1999b) threatens to decrease fire-return intervals far beyond a level that even fire-adapted or pioneer species with fast life histories can survive. Left unchecked, closed-canopy forest ecosystems can be rapidly transformed into low-biomass open forests more reminiscent of scrub-savannas (Barlow *et al.* 2003; Cochrane 2003). At present, twice- and thrice-burned forests in south-eastern and central Brazilian Amazonia are already dominated by flammable grasses and forbs (Cochrane and Schulze 1999; Barlow 2003), and fire-return intervals of less than a decade (Cochrane and Laurance 2002; J. Barlow, personal observation). This transition represents an ecosystem phase-shift from high-phytomass *terra firme* forest to fire-dominated scrublands, which is unlikely to be reversed under current climatic (Timmerman *et al.* 1999) and socio-economic conditions.

Reducing forest flammability

In both temperate and tropical forests, fire risk is closely associated with fuel humidity (Holdsworth and Uhl 1997) and fuel loads (Minnich 2001), and management options must concentrate on minimizing desiccation (Holdsworth and Uhl 1997) and increased fuel loads in the understorey (Uhl *et al.* 1991) driven by selective logging and forest fragmentation.

Where logging takes place, substantial reductions in forest flammability can be gained through reduced impact logging (RIL) practices that reduce both canopy openness and fuel loads on the forest floor. Even where the adoption of these methods does not prevent the occurrence of fires, they will at least reduce fire severity, and hence their impacts on biodiversity, forest recovery times, and the chances of a recurrent-burn regime from becoming established. Moreover, long-term forest management schemes with designated land-rights may play a major role in preventing the proliferation of illegal land clearance, poorly executed illegal logging operations (J. Barlow personal observation), and the post-logging settlements that bring ignition sources into contact with the flammable forests (Laurance 2001a). As such, the creation of 50 million ha of National Forests to be designated by 2010 in the Brazilian Amazon (Verissimo *et al.* 2002) is potentially a positive step for fire prevention.

Current patterns of frontier advance have resulted in landscapes characterized by small irregular-shaped forest patches juxtaposed with frequently burned pastures that are extremely vulnerable to fire (Cochrane and Laurance 2002). Therefore, although legislation requiring landholders to maintain 50–80% of forest within their properties has helped retain some forest cover, it

has failed to maintain forests within fire-dominated agricultural landscapes that are hydrologically viable in the long term. We recommend the encouragement of legislation changes to focus on the protection of fewer large forest areas with low edge-forest interior ratios rather than the creation of many small fragments.

Preventing combustion of flammable forests

Preventing ignition sources from reaching fire-prone forests may be achieved through small and large fire-breaks, education schemes and legislation (e.g. PROARCO), financial incentives to encourage fire-sensitive forms of agriculture (Nepstad *et al.* 2001), and the creation of substantial networks of protected areas or managed forests that can serve as fire-breaks as well as barriers to haphazard frontier advance.

Fire-breaks have been advocated to prevent flame contact with flammable forests. Proposed methods range from the removal of all fuel from 5-m strips around logged forests to the maintenance of extensive swathes of undisturbed primary forest which could act as fire-breaks between logged forests and agricultural areas (Holdsworth and Uhl 1997). Neither method, however, is without practical complications. Cleared strips would be expensive, require frequent maintenance, and may not prevent the fire continuity through sparks or tree falls. Primary forest strips would be susceptible to edge-effects that increase flammability (e.g. Laurance 2004), and without adequate protection could be affected by illegal logging and become eroded by fire over time (Cochrane and Laurance 2002). Despite these problems, fire-breaks may be successful in buffering flammable forests, and finding low-maintenance fire-resistant buffers that are financially viable should be considered a major research goal over the next decade.

Because of the huge social and economic cost of fires for rural communities (Diaz *et al.* 2002; Peres *et al.* 2003), educational schemes such as PROARCO and Programa Emergência Fogo (which inform Amazonian landholders how to minimize fire propagation into forest) are often well received in communities that have a history of fire in their region (M. Carnelutti, personal communication). Simple measures, such as dry biomass clearing along the margins of agricultural plots that will be burned and the ban on burns during the driest months of the year, can be effective in preventing fires from spreading into surrounding forests. One of the advantages of these schemes is that the causes of fires can be targeted in the most fire-prone areas. However, although financial and custodial punishments for violators infringing these rules provide an added incentive, their effectiveness is often undermined because of poor governance (Nepstad et al. 2001; J. Barlow, personal observation). Furthermore, our experience suggests that people are reluctant to invest time and energy into changing their traditional agricultural practices where fires have yet to burn large forest areas, while people investing in fire prevention often fail to accrue the benefits (Nepstad et al. 2001). Finally, such schemes are expensive to implement and enforce (the 1998–9 PROARCO campaign required a $20 million loan from the World Bank), and recent successes will not be sustained without substantial additional funding.

The simplest and most biodiversity-sensitive method for preventing fires from reaching their destructive potential is to discourage indiscriminate frontier advance (cf. Laurance and Fearnside 2002). However, with the advent of road paving, Brazilian Amazonia is likely to witness the opposite trend (Avança Brasil; Laurance et al. 2001a; Nepstad et al. 2001; Peres 2001). Although these economic developments could provide the foundation for improved governance and the institutional capacity required for the successful implementation of these measures (Carvalho et al. 2001; Nepstad et al. 2002a), this has not been the case in the past, and the successful prevention of wildfires during this latest wave of economic development will require considerable political and institutional change and unprecedented levels of regional investment (Laurance and Fearnside 2002; Nepstad et al. 2002a, b; Anon. 2004).

Conclusions

Our results concur with previous studies though the temporal increases in large tree mortality indicate that previous assessments may be conservative. Although substantial economic, social, and ecological benefits can be gained from preventing a transition from closed-canopy tropical forest to scrub savannas (Barlow et al. 2002; Diaz et al. 2002), there is still a lack of political will or consensus about how this could be achieved. Current policies protecting forest cover in private landholdings do little to prevent forest fragmentation and the creation of fire-susceptible forest edges, while the legislation and institutional capacity-building required to prevent a new wave of fires in recently developed Amazonian frontiers (Nepstad et al. 2002a) are yet to be demonstrated (Laurance and Fearnside 2002; Anon. 2004). Ultimately, success in preventing wildfire regimes from reaching their potential levels of damage in seasonally dry tropical forests will depend upon the implementation and deployment of sensible fire prevention measures and the sustained investment in institutional capability and governance required to enforce them.

Acknowledgements

Financial support for this study was provided by the Center for Applied Biodiversity Sciences of Conservation International and Jos Barlow's fieldwork was funded by a NERC PhD studentship at the University of East Anglia.

CHAPTER 9

Late twentieth-century trends in tree-community composition in an Amazonian forest

William F. Laurance, Alexandre A. Oliveira, Susan G. Laurance, Richard Condit, Henrique E. M. Nascimento, Ana Andrade, Christopher W. Dick, Ana C. Sanchez-Thorin, Thomas E. Lovejoy, and José E. L. S. Ribeiro

The rainforests of central Amazonia are some of the most species-rich tree communities on earth. Our analyses suggest that, in recent decades, forests in a central Amazonian landscape have experienced highly non-random changes in dynamics and composition. These analyses are based on a network of 18 permanent plots unaffected by any detectable disturbance. Within these plots, tree mortality, recruitment, and growth have increased over time. Of 115 relatively abundant tree genera, 27 changed significantly ($P \leq 0.01$) in density or basal area—a value nearly 14 times greater than that expected by chance. An independent, 8-yr study in nearby forests corroborates these shifts in composition. Despite increasing tree mortality, pioneer trees did not increase in abundance. However, genera of faster growing trees, including many canopy and emergent species, are increasing in dominance or density, whereas genera of slower growing trees, including many subcanopy species, are declining. Rising atmospheric CO_2 concentrations may explain these changes, although the effects of this and other large-scale environmental alterations have not been fully explored. These compositional changes could potentially have important effects on the carbon storage, dynamics, and biota of Amazonian forests.

Introduction

Are global-change phenomena altering Amazonian forests? Recent studies suggest that undisturbed Amazonian forests have become increasingly dynamic in the past few decades, with higher rates of tree mortality and turnover (Phillips and Gentry 1994; Phillips *et al.* 2004). In addition, carbon storage (Grace *et al.* 1995a; Malhi *et al.* 1998; Phillips *et al.* 1998b; Baker *et al.* 2004b) and productivity (Lewis *et al.* 2004b) in these forests appear to be increasing. Finally, lianas—climbing woody vines that often favour disturbed forest—evidently are increasing in size and abundance (Phillips *et al.*

2002). Possible evidence for such changes comes not only from plot-based studies but also from remote-sensing imagery; an indicator of primary productivity, the normalized difference vegetation index, increased markedly in South American rainforests from 1981–2000 (Paruelo *et al.* 2004).

The causes of these changes are controversial. One prominent suggestion is that the changes arise from increasing plant fertilization caused by rising atmospheric CO_2 concentrations, which is expected to increase forest dynamism and productivity (Reekie and Bazzaz 1989; Phillips and Gentry 1994; Winter and Lovelock 1999; IPCC 2001). However, other large-scale phenomena, such as alterations in regional temperature (Clark *et al.* 2003), rainfall (Condit *et al.* 1996a; Tian *et al.* 1998), available solar radiation (Wielicki *et al.* 2002), or nutrient deposition (Artaxo *et al.* 2003) might also account for some observations (Lewis *et al.* 2004a). It is also not inconceivable that local or natural phenomena, including past disturbances or sampling artefacts, could contribute to, or even generate at least some of the observed patterns (e.g. Clark 2002a, 2004; Nelson 2005).

Until recently, no studies had assessed whether Amazonian tree communities were changing in taxonomic or functional composition, in concert with observed alterations in productivity and dynamics. We recently conducted the first analysis of this nature, using a long-term (11–18 yr) dataset from permanent plots in a central Amazonian landscape (Laurance *et al.* 2004a). Here we summarize these findings and highlight their potential implications for the ecological functioning of Amazonian forests.

Methods

Study area and field methods

The study area is part of the Biological Dynamics of Forest Fragments Project (BDFFP), a long-term experimental investigation of habitat fragmentation in central Amazonia (Lovejoy *et al.* 1986). A key component of the BDFFP is a network of 66 1-ha forest-dynamics plots in fragmented and intact forest. The present study involves a subset of these plots: 18 discrete plots in lowland *terra firme* forest that span an area of about 300 km², are randomly located with respect to local topography, and are positioned at least 300 m away from any clearing to avoid edge effects (Laurance *et al.* 1997, 1998b, 2000). The plots exhibited no evidence of current or past disturbance from logging, fires, or hunting, although two plots experienced small wet-season floods that caused temporary increases in tree mortality (Laurance *et al.* 2004a).

The 18 plots were established from 1981 to 1987 and recensused at roughly 5-yr intervals for an average of 15.0 yrs (range = 11.4–18.2 yrs), with the final

census of each in 1999 or 2000. Within each plot, all trees (≥10 cm d.b.h. [diameter-at-breast-height]) were marked with permanent tags, mapped, measured for trunk diameter (above any buttresses, if present), and identified on the basis of sterile or fertile material. In total, nearly 13,700 trees were recorded (Laurance *et al.* 2004a).

Data analysis

We assessed changes in tree-community composition over time by contrasting data from the first and final censuses of each plot. We assessed changes in the abundance of tree genera, rather than species, for three reasons. First, 88% of tree species in our study area are too rare (<1 individual per hectare) to allow robust analyses of population trends. Second, within a genus of Amazonian trees, species tend to be similar ecologically (Casper *et al.* 1992; ter Steege and Hammond 2001), so analyses at the genus level capture most of the relevant information. Third, 95.3% of study trees were positively identified at the genus level, whereas a smaller percentage was identified at the level of species.

We encountered 244 tree genera in our plots, of which 115 were sufficiently abundant (initially present in at least 8 of the 18 plots) to permit rigorous analysis. For each genus, we used bootstrapping to assess changes in population density and basal area (a strong correlate of tree biomass) between the first and final censuses (see Laurance *et al.* 2004a for further explanation). This analysis makes no assumptions about the underlying statistical distribution of data. Using a conservative 1% significance level in our tests, we expected for each parameter about 1 out of 115 genera to show a significant change by chance alone. Our null hypothesis was that each tree genus exhibited no significant change in population density or basal area, which is appropriate because total density and total basal area of trees did not change significantly during our study (see below). Because this method is unreliable for genera occurring in a small number of plots, we restricted our analyses to genera present in ≥8 plots during our initial census (at this frequency, all genera exhibited reasonably stable estimates for recruitment, mortality, and growth).

We also tested whether the observed changes in density and basal area for all 115 tree genera were more similar among our 18 plots than expected by chance, using randomization tests. To do this we selected nine plots at random and determined the mean percentage change in density for each genus in the plots, and then compared these values to the mean percentage change for each genus in the other nine plots, using Pearson correlations. We repeated this 1000 times, using random combinations of plots each time. The mean and standard error for the 1000 correlations was determined, which were then used to calculate a Z statistic ($Z = \text{mean S.E.}^{-1}$). We used a one-tailed Z test to determine whether the mean value of the observed correlations was significantly greater than 0. The same procedure was used to test for changes in basal area.

Ecological traits of tree genera

For most of the 115 genera in this study, data on growth form and successional status were gleaned from published and online data sources as well as personal knowledge of the authors (see Laurance *et al.* 2004a and Supplemental Online Information). Estimates for median and maximum growth rates, mortality and recruitment rates, and mean trunk diameter were derived from demographic data from our long-term study. Distributional data on locally occurring species within each significantly changing genus, with respect to major rainfall zones in Amazonia, were mostly derived from online sources. Finally, an index of drought tolerance for 30 abundant tree species was derived from published and unpublished data from our 18 plots and from other nearby plots in the same study area.

Changes in forest dynamics and growth

Stand-level rates of annual mortality and recruitment, and the annual rate of trunk growth for individual tree genera, were generated for two largely non-overlapping intervals (ca. 1984–91 and 1992–9). For our 18 plots, the first interval averaged 7.6 ± 2.5 yrs in duration, whereas the second interval averaged 7.4 ± 0.9 yrs in duration; the first interval was more variable in length because the plots were initially established over a 6-yr period, from 1981 to 1987. Annualized mortality and recruitment data for each plot and interval were estimated using logarithmic models.

To calculate annual growth rates for each genus, the mean annual growth of each tree was determined by subtracting its initial d.b.h. from its final d.b.h., and dividing by the number of years. The median growth rate was then determined for all trees within the genus. Maximum growth rate was also calculated for each genus, and to reduce the effects of outliers the upper 10% of the rates was used as an estimate of maximum growth rate. Growth rates were calculated only for genera that had at least 10 live stems in each of the first and second intervals.

Results

Changes in tree density and basal area

A total of 27 genera exhibited significant changes during the study, with 14 genera increasing in basal area and 14 genera declining in density (Table 9.1). One genus, *Couepia*, simultaneously increased in basal area while declining in density (the result of increased tree growth but high mortality of small individuals), whereas three other genera either decreased (*Oenocarpus*) or increased (*Corythophora, Eschweilera*) in both density and basal area. Thus, excluding *Couepia*, 13 genera declined in density, and 13 genera increased in basal area, sometimes dramatically (Table 9.1). Most genera that declined in density did not also decline in basal area because of accelerated growth of the surviving trees (see below).

Mortality rates differed between the 13 increasing and 13 decreasing genera. Declining genera had much higher mortality than did increasing genera (1.57 ± 0.90 versus $0.51 \pm 0.31\%\,\text{yr}^{-1}$; $t = 4.66$, d.f. $= 24$, $P = 0.0001$), whereas recruitment rates did not differ between the two (0.50 ± 0.48 versus $0.69 \pm 0.42\%\,\text{yr}^{-1}$; $t = 1.06$, d.f. $= 24$, $P = 0.30$). Recruitment rates of increasing and decreasing genera were both on average lower than the

Table 9.1 Significantly ($P \leq 0.01$) increasing or decreasing tree genera in undisturbed Amazonian rainforests based on population density and basal area data

Genus	Family	Net change (%)
Tree density increases over time		
Corythophora	Lecythidaceae	+9.8
Eschweilera	Lecythidaceae	+4.0
Tree density decreases over time		
Aspidosperma	Apocynaceae	−13.3
Brosimum	Moraceae	−8.1
Couepia	Chrysobalanaceae	−8.9
Croton	Euphorbiaceae	−35.0
Heisteria	Olacaceae	−25.0
Hirtella	Chrysobalanaceae	−13.0
Iryanthera	Myristicaceae	−16.3
Licania	Chrysobalanaceae	−11.0
Naucleopsis	Moraceae	−17.8
Oenocarpus	Arecaceae	−32.3
Quiina	Quiinaceae	−29.0
Tetragastris	Burseraceae	−15.0
Unonopsis	Annonaceae	−15.3
Virola	Myristicaceae	−14.0
Tree basal area increases over time		
Corythophora	Lecythidaceae	+12.0
Couepia	Chrysobalanaceae	+10.8
Couma	Apocynaceae	+14.4
Dipteryx	Leguminosae	+7.2
Ecclinusa	Sapotaceae	+13.8
Eschweilera	Lecythidaceae	+7.0
Licaria	Lauraceae	+17.2
Maquira	Moraceae	+9.9
Parkia	Leguminosae	+22.0
Peltogyne	Leguminosae	+15.9
Sarcaulus	Sapotaceae	+14.4
Sclerolobium (now synonomized as Tachigali)	Leguminosae	+76.6
Sterculia	Sterculiaceae	+23.4
Trattinnickia	Burseraceae	+13.6
Tree basal area decreases over time		
Oenocarpus	Arecaceae	−29.1

stand-level rate (1.06% yr^{-1}) because they included few pioneers (which have higher recruitment).

These shifts in tree communities were not driven by large overall changes in tree density or basal area. During the course of our study, average tree density declined by 1.1%, whereas average basal area rose by 1.9%. Neither change was statistically significant ($P > 0.09$; paired t-tests).

Two lines of evidence confirm that these compositional changes reflect underlying biological processes, not sampling errors. First, randomization tests revealed that the observed changes in density ($P = 0.001$) and basal area ($P = 0.002$) for all 115 genera were consistent across the 18 plots. Second, we contrasted our results with those of a separate study (Oliveira and Mori 1999), in which trees in three 1-ha plots were censused in undisturbed forest about 6 km east of our study area, using virtually identical methods. In this study, plots were censused in August 1991 and again in September 1999. A total of 2085 trees were recorded in the plots, of which 97.1% were identified level of genus. To minimize effects of small sample sizes, we included in the analysis the 42 genera with at least 10 individuals in the 3 plots (all of these genera were present in over half of our 18 plots). Changes over time in both density and basal area were significantly and positively correlated between the two studies (Fig. 9.1). Thus, parallel studies conducted by two separate teams of investigators revealed similar patterns of change.

Differences between increasing and declining genera

Do the increasing and declining tree genera differ biologically? We reviewed available literature and Internet resources and used data from our long-term study to quantify key ecological traits for most genera (see Laurance *et al.* 2004a and Supplemental Information). The 13 increasing genera and 13 decreasing genera differed in growth form: all of the former were canopy or emergent trees, whereas six (46%) of the latter were subcanopy trees (the remainder being canopy or emergent trees), a highly significant difference ($G = 10.15$, d.f. $= 1$, $P = 0.001$; G-test). Similarly, among all 115 genera, there was a clear tendency for large trees to increase in population density (Fig. 9.2) and basal area at the expense of small trees.

Surprisingly, successional status differed little between the 13 increasing and 13 declining genera;

(a)
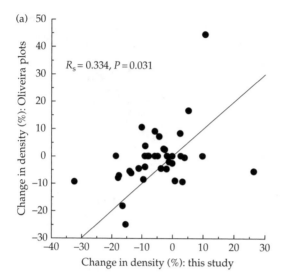

(b)
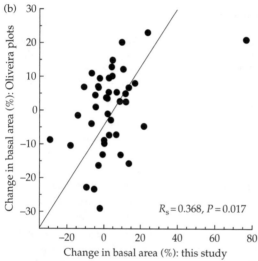

Figure 9.1 Mean percentage changes in (a) population density and (b) basal area of 42 Amazonian tree genera in two different long-term studies (correlation coefficients are for Spearman rank tests). Data are from 18 1-ha plots from the BDFFP and 3 nearby 1-ha plots studied by Oliveira and Mori (1999). The diagonal line in each figure shows $y = x$.

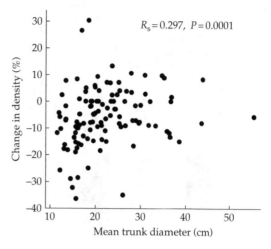

Figure 9.2 Relationship between tree size and long-term population change for Amazonian tree genera (Spearman rank correlation).

$P = 0.049$) absolute growth rates were significantly higher in the increasing than declining genera (t-tests with log-transformed data). Similar patterns were evident when all genera that increased and declined in density (not just those that changed significantly) were compared. Collectively, these trends suggest that genera with higher absolute growth rates, including many canopy and emergent trees but not pioneers, are increasing at the expense of slower growing genera, which include many smaller, old-growth subcanopy trees.

In addition, the tree community is changing in taxonomic composition. The increasing genera are dominated (57%) by three families (Leguminosae, Lecythidaceae, Sapotaceae) that are not represented among declining genera, whereas most (64%) declining genera are in families (Arecaceae, Annonaceae, Chrysobalanaceae, Moraceae, Myristicaceae) that are poorly represented among increasing genera (Table 9.1).

Changes in forest dynamics and growth

To help identify the underlying causes of these alterations, we assessed dynamical changes in the tree communities. We divided census data for each plot into two roughly equal intervals (1984–91 and 1992–9) and then contrasted overall rates of tree mortality and recruitment between the two

old-growth trees dominated (77%) both groups. In addition, none of the major pioneer genera (*Annona*, *Cecropia*, *Croton*, *Goupia*, *Jacaranda*, *Miconia*, *Pourouma*, *Vismia*) increased significantly in density or basal area, either individually or when pooled. Nevertheless, both median ($t = 2.28$, d.f. $= 24$, $P = 0.032$) and maximum ($t = 2.07$, d.f $= 24$,

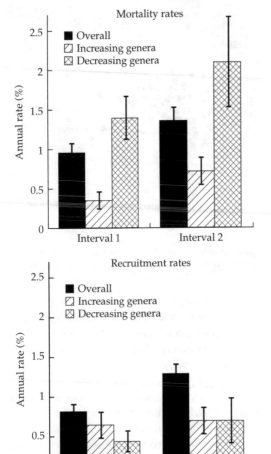

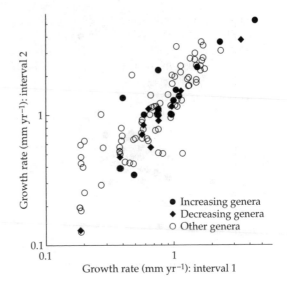

Figure 9.4 Comparison of median growth rates of Amazonian tree genera between interval 1 (ca. 1984–91) and interval 2 (ca. 1992–9). The diagonal line shows $y = x$. Growth rates accelerated markedly over time for all genera ($t = -9.74$, d.f. = 114, $P < 0.00001$; paired t-test), and accelerated significantly more for increasing than decreasing genera ($t = 2.45$, d.f. = 24, $P = 0.022$; t-test for unequal variances).

Figure 9.3 Mortality and recruitment rates (± 1 s.d.) for all trees, for 13 genera that increased in basal area, and for 13 genera that declined in density. Overall mortality ($t = -2.38$, d.f. = 17, $P = 0.03$) and recruitment ($t = -4.45$, d.f. = 17, $P = 0.0003$) accelerated from interval 1 (ca. 1984–91) to interval 2 (ca. 1992–9). However, there was no significant change over time ($P > 0.11$) in mortality or recruitment for the increasing and decreasing genera (paired t-tests).

intervals. Both rates rose markedly from interval 1 to interval 2 (Fig. 9.3); thus our forests clearly became more dynamic over time. Mortality and recruitment rates did not rise significantly for the increasing and declining genera, although the latter consistently had higher mortality than recruitment (Fig. 9.3).

Moreover, for 87% of genera, rates of trunk growth accelerated between intervals 1 and 2 (Fig. 9.4). This demonstration of enhanced growth across a wide range of tropical tree genera is consistent with stand-level increases in tree growth across South American forests (Lewis *et al.* 2004b). Notably, the average increase in absolute growth rate was higher among increasing genera than declining genera (0.55 ± 0.49 versus 0.19 ± 0.17 mm yr^{-1}); the average for genera showing no significant change was intermediate (0.41 ± 0.50 mm yr^{-1}). This difference did not occur solely because increasing genera were often large in size and decreasing genera often small: in relative terms, growth accelerated much more in increasing (57%) than decreasing (22%) genera.

Discussion

We observed three distinctive trends in this study: (1) there were positively correlated shifts in tree-community composition across geographically well separated plots, with faster growing canopy and emergent genera (but not pioneers) generally increasing at the expense of slower growing sub-canopy genera; (2) for the large majority (87%) of tree genera, incremental trunk growth accelerated from interval 1 (ca. 1984–91) to interval 2 (ca. 1992–9); and (3) tree-community dynamics (mortality and recruitment) also accelerated from interval 1 to interval 2.

There are several plausible explanations for these forest-wide changes in composition and dynamics. We discuss each of these in turn, paying particular attention to local mechanisms or sampling artefacts that could potentially influence our findings (cf. Nelson, in press).

Forest recovery from past disturbance

The forests of our study area might be in a state of disequilibrium because of ongoing recovery from past disturbance, leading to shifts over time in tree composition. The disturbance most likely to operate over such a large spatial scale as our study area (300 km^2) is a major forest fire. It is unlikely, however, that past fires could account for the suite of changes we observed. Soil charcoal is found in our study area, but the large majority of charcoal was created at least 1100–1500 yrs ago (Piperno and Becker 1996; Santos et al. 1996), and continuous forests have persisted in our study area for at least the last 4500 yrs (Piperno and Becker 1996). Detailed phytolith (plant fragment) studies suggest that the past fires were natural in origin (Piperno and Becker 1996) and, judging from the virtual absence of burnt phytoliths, that they caused relatively little forest damage (D. R. Piperno, personal communication). Moreover, the complex old-growth forest structure (Laurance 2001b), extremely high tree diversity (Oliveira and Mori 1999), and, especially, the high incidence of old (500–1000-yr-old) trees in our study plots (Laurance et al. 2004b), all suggest that fires during the past millennium had only patchy, limited effects on forest structure and composition. Finally, although recovery from past fires might plausibly promote shifts in tree-community composition (Table 9.1), it could not explain accelerating tree growth (Fig. 9.3) and forest dynamics (Fig. 9.4).

Another potentially important cause of disturbance is strong winds, especially from convectional thunderstorms (Nelson, in press). Again, these are unlikely to account for the pervasive changes we detected. First, strong winds are more likely to cause population declines of canopy and emergent trees than of subcanopy trees (Laurance et al. 2000)—the opposite of the pattern we observed. Second, pioneer trees, which increase in disturbed forest, were uncommon in our plots (<2.6% of all stems), which would seem unlikely if wind disturbance was pervasive. Third, observed changes in tree communities were not concentrated in one or a few clusters of plots; we found that, for the 115 most abundant tree genera, nearby plots did not show more similar patterns of floristic change during our study than did more-distant plots ($P = 0.92$, Mantel test), as would be expected from convectional-storm damage, which is patchy at a landscape scale (Nelson 1994; Nelson et al. 1994). Finally, wind disturbance would not cause an acceleration of tree growth, as was observed in our study.

A third possible cause of disturbance is forest flooding and soil saturation (Nelson, in press). Flooding was especially severe in 1989, which had the heaviest wet season recorded in the Manaus area since 1910 (Mori and Becker 1991). However, this is also unlikely to explain observed trends. Wet-season rains in 1989 were indeed heavy (1887 mm), but the pattern is less striking than it might initially seem. From 1968 to 2000, for example, 6 yrs had wet-season rainfall that was >90% of that in 1989, and 14 yrs had >80% of that rainfall (Laurance et al. 2005). Thus, many forest microhabitats that flooded in 1989 would also have flooded in preceding years, greatly reducing the likelihood that a single, marginally wetter year would have had exceptional effects on tree communities. Moreover, the tree genera that declined significantly during our study (Table 9.1) did not exhibit larger population declines in plots with flood-prone microhabitats (gullies and plateau depressions) than in plots that lacked flood-prone areas (Laurance et al., in press). Finally, effects of flooding also would not explain accelerated tree growth.

Sampling artefacts

Nelson (in press) suggested that physical damage to trees incurred during the collection of herbarium specimens might increase tree mortality, or render them more vulnerable to disease. Old-growth subcanopy trees, which are strongly energy limited, might be especially vulnerable to such disturbances. If this were the case, this might explain the decline of smaller, slower growing trees observed in our study (e.g. Fig. 9.2).

Notably, however, a previous study (Phillips *et al.* 1998a) concluded that collecting vouchers in tropical forests (including tree-climbing with spiked ascenders that can cause >400 small wounds to the tree trunk) did not increase overall tree mortality (although the authors did not explicitly assess mortality among different size classes of trees). Moreover, our field-sampling methods were less damaging to trees than Nelson (in press) implied: (1) for most trees, only three small leaf samples were collected, usually taken from a single branchlet (flowers or fruits were collected from just 1–2 individuals of each species); (2) slashes on the lower trunks of trees were usually small (<15 cm²) and superficial (<1 cm deep); and (3) trees were climbed only with cloth ankle bands and rubber-soled shoes, not with spiked ascenders.

If botanical collecting had a significant impact on tree composition, then tree-mortality rates should have peaked soon after the initial census of each plot, and then declined afterwards. In fact, we observed the opposite trend—mortality rates increased over time in our plots (Fig. 9.3), a pattern seen at many other sites in Amazonia (Phillips and Gentry 1994; Phillips *et al.* 2004). In fact, old-growth subcanopy trees, which generally have dense, strong wood to withstand recurring damage from litterfall (Thomas 1996; Laurance *et al.* 2004b), may actually be relatively robust to minor physical damage. If they are not, then the enhanced-mortality effect that Nelson proposes should plague many permanent-plot studies, not just ours. We are aware of no evidence to this effect.

Effects of droughts

Another possibility is that the observed changes in our study might reflect differential vulnerability of trees to El Niño-related droughts (e.g. Condit *et al.* 1996a,b; Tian *et al.* 1998). Our study area experienced major droughts in 1983 and 1997, and a smaller drought in 1992; such events have increased in frequency this century (Dunbar 2000), possibly because of global warming (Timmerman *et al.* 1999).

We found little direct support for the drought hypothesis. First, we contrasted the geographic distributions of locally occurring species within the increasing and decreasing genera across the Amazon basin. The former did not show stronger associations with drier forest types in the Amazon Basin, as might be expected if the increasing genera were more drought tolerant (Laurance *et al.* 2004a). Second, we tested whether more drought-tolerant tree species had increased in density during our study, in response to the strong droughts in 1983 and 1997. Our index of drought tolerance was generated by dividing the mortality rate of each species during the 1997 drought year, by the baseline mortality rate in years preceding the drought (Williamson *et al.* 2000; Laurance *et al.* 2001c). For the 30 most abundant species in our plots, there was no relationship between the drought-tolerance index and its percentage change in population density during our study (Laurance *et al.* 2004a). Nonetheless, in the only other long-term study of floristic change in mature tropical forest, strong droughts evidently caused a shift in tree-community composition in Panama (Condit *et al.* 1996a,b), so the drought hypothesis requires further examination.

Multi-decadal changes in rainfall

Yet another possibility is that our forests might be responding to multi-decadal changes in rainfall that affect forest productivity and species composition. Drier conditions in rainforests may increase tree growth and reproduction (Clark and Clark 1994; Wright *et al.* 1999), possibly because cloud cover is reduced, increasing available sunlight for light-limited trees.

To test this hypothesis we assessed rainfall data collected near our study area (Manaus, Brazil), contrasting the first (1984–91) and second (1992–9) halves of our study. There was no significant difference between the two intervals for dry-season (June–October) rainfall, wet-season (November–May) rainfall, total annual rainfall, and the number of dry (<100 mm rain) months per year ($t < 1.5$, d.f. = 14, $P > 0.15$ in all cases; t-tests), nor did any rainfall variable change significantly with calendar year ($r < 0.30$, d.f. = 14, $P > 0.25$; Pearson correlations; Laurance et al. 2004a). In addition, a study of tropical climates in the twentieth century (Malhi and Wright 2004) revealed no obvious trend in rainfall at Manaus, with the exception of higher precipitation in the first quarter of the century. Thus, at least in recent decades, it appears unlikely that these forests have been markedly affected by changing rainfall patterns.

Increasing forest productivity

Finally, the observed changes in floristic composition, tree growth, and forest dynamics could be driven by accelerated forest productivity. We believe the most likely cause of higher productivity is rising atmospheric CO_2 levels (cf. Reekie and Bazzaz 1989; Grace et al. 1994; Phillips and Gentry 1994; Phillips et al. 1998b, 2002, 2004; Winter and Lovelock 1999; Baker et al. 2004b; Lewis et al. 2004a,b). However, other agents, such as higher airborne nutrient deposition (Artaxo et al. 2003) from increasing forest fires, and possible increases in solar radiation from reduced tropical cloudiness (see Wielicki et al. 2002; Lewis et al. 2004b), are also plausible causes of rising productivity.

Of all the hypothesized factors, rising productivity best explains key observations of this study: (1) that tree growth, mortality, and recruitment have increased markedly, all of which could result from greater productivity (Phillips and Gentry 1994; Lewis et al. 2004b; Phillips et al. 2004); (2) that many faster growing genera are increasing in basal area, possibly because fast-growing trees show stronger growth enhancement under elevated CO_2 (Reekie and Bazzaz 1989; Körner 1998; Winter and Lovelock 1999); and (3) that forests are experiencing non-random changes in species composition, with fast-growing canopy and emergent genera evidently gaining a competitive advantage over smaller, slower growing genera. That rapidly growing pioneers have not increased in abundance is surprising, but these species usually establish in large treefall gaps, which may be uncommon in our

study area because mortality is greatest among small trees. The group most likely to decline further, we suggest, is old-growth subcanopy species, a highly diverse assemblage that are notable for their slow growth, dense wood, and ability to reproduce in full shade (Thomas 1996; Laurance et al. 2004b).

Conclusions and implications

The suite of changes observed in this study appears to be most consistent with those expected from increasing forest productivity, possibly in response to rising atmospheric CO_2 concentrations. This conclusion is bolstered by other studies that also suggest that forest productivity in neotropical forests has generally increased in recent decades (e.g. Phillips and Gentry 1994; Phillips et al. 1998b, 2002, 2004; Baker et al. 2004b; Lewis et al. 2004b; Paruelo et al. 2004, Chapter 5 this volume; but see Clark 2002a, 2004 for a different perspective). Regardless of the underlying mechanisms involved, the fact that changes in tree-community composition were positively correlated between two independent studies in central Amazonia (Fig. 9.1) suggests that these trends are real, and not the result of sampling or plant-identification errors.

If Amazonian forests are truly experiencing shifts in tree-community composition and forest dynamics, then these changes could potentially have important consequences. For example, undisturbed Amazonian forests appear to be functioning as a significant carbon sink (Grace et al. 1995a; Malhi et al. 1998; Phillips et al. 1998b; Baker et al. 2004b), helping to slow down global warming, but pervasive changes in tree communities could modify this effect (Körner 1998, 2004, Chapter 6). In particular, increases in forest carbon storage might be slowed down by the tendency of canopy and emergent trees to produce wood of reduced density as their size and growth rate increases (Thomas 1996), and by the decline of densely wooded subcanopy species. Forest-wide changes in tree communities, which sustain assemblages of often-specialized pollinators, herbivores, symbiotic fungi, and other species (Bazzaz 1998), may also have serious ecological repercussions for the diverse Amazonian biota. Further studies are urgently needed to determine whether comparably large shifts in tree communities are occurring throughout the tropics—in concert with widespread increases in forest growth and turnover (Phillips and Gentry 1994; Phillips *et al.* 1998b, 2004; Baker *et al.* 2004b; Lewis *et al.* 2004b)—and to identify the environmental agents driving these changes.

Acknowledgements

See Laurance *et al.* (2004a) for a list of project sponsors. Comments from Oliver Phillips improved the chapter. This is publication number 431 in the BDFFP technical series.

CHAPTER 10

Late twentieth-century patterns and trends in Amazon tree turnover

Oliver L. Phillips, Timothy R. Baker, Luzmila Arroyo, Niro Higuchi, Timothy Killeen, William F. Laurance, Simon L. Lewis, Jon Lloyd, Yadvinder Malhi, Abel Monteagudo, David A. Neill, Percy Nuñez Vargas, J. Natalino N. Silva, John Terborgh, Rodolfo Vásquez Martinez, Miguel Alexiades, Samuel Almeida, Sandra Brown, Jerome Chave, James A. Comiskey, Claudia I. Czimczik, Anthony Di Fiore, Terry Erwin, Caroline Kuebler, Susan G. Laurance, Henrique E. M. Nascimento, Jean Olivier, Walter Palacios, Sandra Patiño, Nigel Pitman, Carlos A. Quesada, Mario Saldias, Armando Torres Lezama, and Barbara Vinceti

Previous work has shown that tree turnover, tree biomass, and large liana densities have increased in mature tropical forest plots in the late twentieth century. These results point to a concerted shift in forest ecological processes that may already be having significant impacts on terrestrial carbon stocks, fluxes, and biodiversity. However, the findings have proved controversial. Here we characterize regional-scale patterns of 'tree turnover'—the rate with which trees die and recruit into a population—using improved datasets now available for Amazonia that span the last 25 yr. Specifically, we assess whether concerted changes in turnover are occurring, and if so whether they are general throughout the Amazon or restricted to one region or environmental zone. In addition, we ask whether they are driven by changes in recruitment, mortality, or both. We find that: (1) trees ≥ 10 cm diameter recruit and die twice as fast on the richer soils of western Amazonia compared to trees on the poorer soils of eastern Amazonia, (2) turnover rates have increased throughout Amazonia over the last two decades, (3) mortality and recruitment rates have both increased significantly in every region and environmental zone, with the exception of mortality in eastern Amazonia, (4) recruitment rates have consistently exceeded mortality rates, (5) absolute increases in recruitment and mortality rates are greatest in western Amazonian sites, (6) mortality appears to be lagging recruitment at regional scales. These spatial patterns and temporal trends are not caused by obvious artefacts in the data or the analyses. The trends cannot be directly driven by a mortality driver (such as increased drought or fragmentation-related death) because the biomass in these forests has simultaneously increased. Our findings therefore indicate that long-acting and widespread environmental changes are stimulating the growth and productivity of Amazon forests.

Introduction

Ecosystems are changing worldwide as a result of myriad anthropogenic processes. Some processes are physically obvious (e.g. deforestation), others may be less so but also affect biodiversity (e.g. fragmentation). Atmospheric changes such as increasing carbon dioxide (CO_2) concentrations, increasing temperatures, and increasing nitrogen deposition are altering the environment of even remote regions. Anthropogenic atmospheric change will certainly become more significant through the century as CO_2 concentrations will reach values unprecedented for at least 20 million yrs (Retallack 2001a). Climates and nitrogen-deposition will move far beyond Quaternary envelopes (Prentice et al. 2001; Galloway and Cowling 2002).

While we can measure most physical and chemical drivers with reasonable accuracy and precision, quantifying ecological responses to atmospheric change is extremely difficult. The task is particularly urgent for tropical forests as much of earth's biodiversity, carbon stocks, and productivity is centred within this biome (Malhi and Grace 2000). Assessment and monitoring sites have traditionally been few and poorly integrated but over the last decade we have sought to overcome these limitations by developing collaborative networks of researchers—recognizing that by pooling local efforts and small-scale datasets we can tackle large-scale questions. In particular, the Amazon Forest Inventory Network (RAINFOR, www.geog.leeds.ac.uk/projects/rainfor/), seeks to understand patterns and changes in mature Amazon forests (Malhi et al. 2002a).

Earlier analyses suggested that significant changes occurred in the structure and function of mature tropical forests by the close of the twentieth century. For example, turnover rates of trees in mature tropical forest plots increased throughout the 1980s and early 1990s (Phillips and Gentry 1994; Phillips 1996). The trend was demonstrated for both the neotropics and the paleotropics, with the changes robust to concerns such as the effects of individual El Niño-Southern Oscillation (ENSO) cycles (Phillips 1995), site-specific bias when establishing plots (Phillips et al. 1997), damage by botanical collectors (Phillips et al. 1998a), and census-interval artefacts (Lewis et al. 2004c). We have also shown that the structure and composition of non-fragmented Amazonian forests are changing, with an increase in tree biomass (Phillips et al. 1998b; Baker et al., Chapter 11, this volume) and large liana dominance (Phillips et al. 2002b). Together these results suggest that changes in structure, composition, and dynamics are common manifestations of a profound shift in the whole ecology of tropical forests. However, to fully test the proposition that ecological processes in mature tropical forests are changing systematically, additional evidence needs evaluating against these criteria:

1. *Are the changes observed so far concerted across space and time?* Are they geographically coincident (occurring together in the same forest region and sites), geographically widespread (occurring across spatial and environmental gradients), and temporally robust (relatively insensitive to short-term climatic fluctuations)?

2. *Can the phenomena be explained in terms of underlying ecological processes, such as growth, mortality, and recruitment?* Specifically, is the turnover increase driven by changes in only recruitment, or only mortality, or both? Is the biomass increase driven by greater basal area growth or reduced basal area mortality? Are these changes consistent with one another and with possible mechanistic drivers?

Based on Phillips et al. (2004), we provide here a description of the patterns of tree turnover in Amazonia, which comprises more than half of the world's humid tropical forests. We explore the aspects of the two sets of questions described above, and show results accounting for potential sources of error. Elsewhere (Lewis, et al., Chapter 4, this volume) we have developed a conceptual framework that links plausible physical and chemical mechanistic drivers to predicted changes, and suggest some tests to fingerprint potential drivers. We define 'turnover' as the rate at which trees move through a population (the flux) in relation to the number of trees in the population (the pool), and estimate this flux by the mean rate with which they recruit and die.

Our specific objectives are to determine:

1. The extent to which turnover rates have changed (or not) throughout the Amazon basin (note that the turnover increase has so far only been shown for the neotropics and paleotropics as a whole).
2. Any patterns in the changes in turnover rates with respect to the different climatic, edaphic, and geographic regions within Amazonia (Amazon forests vary greatly, so it is important to know if the patterns of change vary too).
3. Whether these changes are driven by increases in recruitment rates, mortality rates, or both (turnover changes in neotropics and paleotropics have only been shown so far in aggregate, and have not been deconstructed into component processes).

Addressing these questions requires careful consideration of possible sources of error, correcting for these where possible. Potential sources of error include the differing census intervals with which plots are monitored, the timing of censuses, the possible tendency of foresters and ecologists to select 'good-looking', mature-phase patches for plots ('majestic forest'), and 'site-switching' changes through time in the spatial and environmental distribution of available datasets (see Phillips *et al.* 2004 and literature cited therein). Below we describe how we address these problems.

Methods

Site selection

The region considered is the Amazon River Basin and contiguous forested areas, including all mature forest except for areas with obvious anthropogenic disturbances (logging, fragmentation, and fires) and excluding small forest patches in forest/savanna mosaics. Data were obtained from published sources where available, but most analysed data come from unpublished permanent plots maintained by the authors in Bolivia, Brazil, Ecuador, French Guiana, Peru, and Venezuela, comprising much of the RAINFOR network (Malhi *et al.* 2002a). The criteria used for selecting tree turnover data include a minimum initial population of ≥ 200 trees ≥ 10 cm diameter, a minimum area of 0.25 ha, and a minimum monitoring period of 2 yr. Mean (and median)

values of the initial population are 954 (572) trees, of the area monitored 1.7 (1.0) ha, and of the monitoring period 10.1 (9.6) yr.

Turnover rate calculations

Annual mortality and recruitment rates were separately estimated using standard logarithmic models which assume a constant probability of mortality and recruitment through each inventory period (Phillips *et al.* 1994). To reduce noise, turnover rates for each period are represented by the mean of recruitment and mortality (91 sites), or as mortality rates alone when recruitment data were unavailable (Phillips *et al.* 2004: table 1).

Analytical approach

Change in a rate process can be evaluated in many ways, depending on the exact hypothesis being tested and the data quality (Phillips 1996). Some sites have only one measurement interval, while others have turnover rates reported for multiple intervals. To use the greatest information content possible we have used several different approaches. The core approach used in *this* chapter involves calculating mortality and recruitment rates for all sites for all years in which they were monitored, and plotting these rates as a function of calendar year (in Lewis *et al.*, Chapter 12, (this volume) changes *within* plots are evaluated). The null hypothesis being tested is not that 'tree turnover rates have not increased *within a specific, individual site*', but rather that 'tree turnover rates have not systematically increased *across all sites in a region*'. We test for change by comparing rates in the last year in which at least 10 sites were monitored with the rates in the first year in which at least 10 sites were monitored. This typically allows comparisons across two decades from the early 1980s to 2001.

This method has the advantages of using all available turnover data and of being able to show graphically the statistical range of site values within each calendar year and across all calendar years. However, a potential concern is that the results may be skewed by using short or varying census intervals through time as it is not possible to co-ordinate censuses at the Amazonian scale, nor is it even possible to select censuses retrospectively so that they are simultaneous and equally frequent at all sites. We take a pragmatic approach to minimize the impact of this concern. Thus, all rates are calculated for each site over intervals of as close to 5 yr as practical, so that short intervals are collapsed together where possible (Phillips *et al.* 2004: appendix 1). Adjacent intervals less than 5 yr are combined when the difference between the combined period and 5 yr is less than the summed difference between each of the constituent intervals and 5 yr. To account for any residual census interval effect we present results after an empirical correction for the census interval effect derived from 10 long-term sites from Latin America, Africa, Asia, and Australia (Lewis *et al.* 2004c).

We also need to identify those plots potentially affected by a 'majestic forest' bias, as a breakdown of mature phase forest may lead to locally accelerated dynamics. For most plots we can rule out the possibility that a majestic forest effect could be artificially accelerating dynamics, based on either the sample unit shape and size, and/or the site selection procedures used, and/or the fact that the stand has gained basal area through the monitoring period as rate processes are unlikely to be driven by locally accelerated dynamics resulting from death of large trees (Table 10.1). The remaining seven plots potentially most susceptible to majestic forest bias were excluded from analyses.

A further concern with our approach is that a calendar year signal confounds within-site change with among-site change, so aggregated results could be influenced by biases that could arise through unequal sampling of forest types across time ('site-switching'). Therefore, we present results in a way that eliminates site-switching, to show only the aggregate of *within-site* changes. This is achieved by 'stretching' all multi-interval data backward and forward. We do this by applying the rate actually recorded in the first interval rate for each year before the first census back to 1976 (for each site initiated after 1976), and applying the rate actually recorded in the last interval forward to 2001 (for each site last censused before 2001). This should be a conservative procedure with respect to null hypothesis since we are assuming no change in rates for

all years in which a site was not monitored. Most plots have been monitored for less than 25 yr and so stretching always flattens the average gradient of any trend in rates. The main analyses—correcting for site-switching, census-interval, and majestic-forest effects—are shown graphically (Fig. 10.3) and in Table 10.2. Results using the raw uncorrected data are shown in tabular form.

To be able to test whether patterns are widespread or driven by change in only one region, we arbitrarily divided Amazonia into two roughly equal areas with roughly equal sample sizes: western and southern Amazonia ('west'), and eastern and central Amazonia ('east') (Fig. 10.1). Most eastern forests are on the actively weathering Guianan or Brazilian shield or associated Tertiary planation surfaces, whereas most west Amazon forests are located on Quaternary or Holocene Andean sediment (Irion 1978; Sombroek 1984). Our geographical division is also consistent with what we know about the floristic make-up of Amazon forests, lying roughly perpendicular to the main south-west–north-east gradient in composition. In separate disaggregations we divided Amazonia in a climatic sense ('aseasonal' versus 'seasonal', using the criterion of ≥ 1 month receiving <100 mm rain to define seasonality), and in an edaphic sense (poor soil versus richer soils, with oxisols, oligotrophic histosols, and spodosols and other white sands defined as 'poor', and alfisols, eutrophic histosols, ultisols, clay-rich entisols, and alluvial and basaltic inceptisols defined as 'richer'). Climate data come from local meteorological stations where possible, else from the University of East Anglia global climatology for the period 1960–98 (http://ipcc-ddc.cru.uea.ac.uk). Soil classifications come from published profiles where possible, and otherwise are based on our own analyses. These categories represent an advance on previous approaches that lumped the neotropics into a single category (e.g. Phillips 1996) and allow us to maintain reasonable sample sizes in each through the late twentieth century.

For data that were not corrected for site-switching we used simple two-sample *t*-tests or the non-parametric equivalent (Mann–Whitney *U*-test), comparing values recorded at all sites monitored at the start of the period (e.g. 1976) with

Table 10.1 Site-by-site summary of structural and dynamic properties, all sites

Site name	Country	Site code	First census (n censuses)	'Majestic forest' bias possible?[a]	Monitoring period (years)	Basal area start, (m²ha⁻¹)	Stems start (ha⁻¹)	Recruitment[b] (% yr⁻¹)	Mortality[b] (% yr⁻¹)	Turnover[b] (% yr⁻¹)	Recruitment, interval corrected[b], (% yr⁻¹)	Mortality, interval corrected[b], (% yr⁻¹)	Turnover, interval corrected[b], (% yr⁻¹)
Cerro Pelao 1	Bolivia	CRP-01	1994.21 (3)	Y? d?	7.25	20.30	552	1.87	3.32	2.59	2.19	3.89	3.03
Cerro Pelao 2	Bolivia	CRP-02	1994.27 (3)	N, df	7.19	24.09	472	3.05	2.30	2.67	3.57	2.69	3.13
Chore 1	Bolivia	CHO-01	1996.53 (2)	N, cf	4.91	14.08	565	2.39	2.61	2.50	2.71	2.96	2.84
Huanchaca Dos, plot1	Bolivia	HCC-21	1996.52 (2)	N, cef	4.91	24.69	529	2.38	2.85	2.61	2.70	3.24	2.96
Huanchaca Dos, plot2	Bolivia	HCC-22	1996.54 (2)	N, cef	4.89	26.66	644	1.18	1.79	1.48	1.34	2.03	1.68
Las Londras, plot 1	Bolivia	LSL-01	1996.53 (2)	N, cf	4.95	17.52	560	1.56	2.86	2.21	1.77	3.25	2.51
Las Londras, plot 2	Bolivia	LSL-02	1996.53 (2)	N, cf	4.95	20.45	630	1.25	1.19	1.22	1.42	1.35	1.39
Los Fierros Bosque I	Bolivia	LFB-01	1993.62 (3)	N, cf	7.78	23.57	557	2.86	3.36	3.11	3.37	3.96	3.66
Los Fierros Bosque II	Bolivia	LFB-02	1993.65 (3)	N, cf	7.76	28.03	540	2.82	2.73	2.78	3.32	3.22	3.28
BDFFP, 1101 Gaviao	Brazil	BDF-03	1981.13 (4)	N, fg	18.17	28.39	593	1.10	1.21	1.15	1.39	1.53	1.45
BDFFP, 1102 Gaviao	Brazil	BDF-04	1981.13 (4)	Y? g?	18.17	28.39	590	2.49	2.81	2.65	3.14	3.54	3.34
BDFFP, 1103 Gaviao	Brazil	BDF-05	1981.21 (4)	N, fg	18.08	25.28	650	0.93	1.45	1.19	1.19	1.84	1.47
BDFFP, 1113 Florestal	Brazil	BDF-09	1987.04 (3)	N, fg	10.25	29.49	571	0.67	1.25	0.96	0.81	1.51	1.16
BDFFP, 1201 Gaviao	Brazil	BDF-06	1981.29 (4)	N, fg	18.00	25.48	632	1.17	1.48	1.33	1.47	1.87	1.68
BDFFP, 1301 Florestal 1	Brazil	BDF-10	1983.46 (3)	N, fg	13.67	27.47	632	1.49	1.40	1.44	1.84	1.73	1.78
BDFFP, 1301 Florestal 2	Brazil	BDF-11	1983.46 (3)	N, fg	13.67	28.85	629	0.61	0.68	0.65	0.75	0.84	0.80
BDFFP, 1301 Florestal 3	Brazil	BDF-12	1983.46 (3)	N, fg	13.67	28.45	617	0.60	0.61	0.60	0.74	0.75	0.74
BDFFP, 2303 Faz. Dimona 4-6	Brazil	BDF-01	1985.29 (4)	N, dfg	12.42	30.15	688	1.23	1.18	1.20	1.50	1.44	1.47
BDFFP, 3304 Porto Alegre	Brazil	BDF-14	1984.21 (5)	N, beg	14.17	32.03	651	1.25	1.29	1.27	1.55	1.59	1.57
BDFFP, 3402 Cabo Frio	Brazil	BDF-13	1985.86 (4)	N, defg	13.02	26.52	565	1.31	0.98	1.14	1.61	1.20	1.40
Bionte 1	Brazil	BNT-01	1986.50 (11)	N, af	12.70	28.04	561	1.07	0.91	0.99	1.31	1.12	1.21
Bionte 2	Brazil	BNT-02	1986.50 (11)	N, af	12.70	30.14	692	0.64	0.64	0.64	0.78	0.78	0.78
Bionte 4	Brazil	BNT-04	1986.50 (10)	N, af	12.70	27.76	608	1.07	1.22	1.14	1.31	1.50	1.40
Bionte T4 B1 SB3	Brazil	BNT-06	1986.50 (5)	N, a	7.00	32.33	576	1.21	1.43	1.32	1.41	1.67	1.54
Bionte T4 B2 SB1	Brazil	BNT-05	1986.50 (5)	N, af	7.00	26.05	565	1.72	1.52	1.62	2.01	1.78	1.89
Bionte T4 B4 SB4	Brazil	BNT-07	1986.50 (5)	N, af	7.00	30.59	643	1.23	1.01	1.12	1.44	1.18	1.31
Caxiuana 1	Brazil	CAX-01	1994.50 (3)	N, cfg	8.38	30.07	524	0.78	0.90	0.84	0.93	1.06	1.00
Caxiuana 2	Brazil	CAX-02	1995.50 (2)	N, cg	4.00	33.11	508	1.56	1.31	1.44	1.74	1.46	1.61
Jacaranda, plots 1–5	Brazil	JAC-01	1996.50 (2)	N, cg	6.00	27.51	593	1.61	1.03	1.32	1.86	1.19	1.53
Jacaranda, plots 6–10	Brazil	JAC-02	1996.50 (3)	N, cg	6.00	26.60	573	1.34	1.11	1.23	1.55	1.28	1.41
Jari 1	Brazil	JRI-01	1985.50 (6)	N, bfg	10.50	32.99	572	1.59	1.17	1.38	1.92	1.41	1.67
Mocambo	Brazil	MBO-01	1956.50 (2)	N, deg	15.00	27.70	453	0.93	1.37	1.15	1.15	1.70	1.43
Tapajos, RP014, 1–4[c]	Brazil	TAP-01	1983.50 (4)	N, af	12.00	23.61	527	1.56	0.69	1.13	1.90	0.84	1.38

Table 10.1 (*Continued*)

Site name	Country	Site code	First census (n censuses)	'Majestic forest' bias possible?[a]	Monitoring period (years)	Basal area start, (m²ha⁻¹)	Stems start (ha⁻¹)	Recruitment[b] (% yr⁻¹)	Mortality[b] (% yr⁻¹)	Turnover[b] (% yr⁻¹)	Recruitment, interval corrected[b] (% yr⁻¹)	Mortality, interval corrected[b] (% yr⁻¹)	Turnover, interval corrected[b] (% yr⁻¹)
Tapajós, RP014, 5–8[c]	Brazil	TAP-02	1983.50 (4)	N, af	12.00	27.82	479	1.63	0.61	1.12	1.99	0.74	1.37
Tapajós, RP014, 9–12[c]	Brazil	TAP-03	1983.50 (4)	N, af	12.00	31.25	491	1.50	0.82	1.16	1.83	1.00	1.42
Añangu, A1	Ecuador	ANN-01	1982.48 (2)	N, cg	8.50	36.80	417	N/A	3.08	3.08	N/A	3.66	3.66
Añangu, A2	Ecuador	ANN-02	1982.48 (2)	N, cg	8.50	33.82	728	N/A	1.88	1.88	N/A	2.23	2.23
Añangu, A3	Ecuador	ANN-03	1986.04 (2)	N, fg	4.92	22.20	734	1.80	1.89	1.84	2.04	2.15	2.09
Bogi 1	Ecuador	BOG-01	1996.29 (2)	N, cfg	5.83	28.40	544	2.88	2.08	2.48	3.32	2.40	2.86
Bogi 2	Ecuador	BOG-02	1996.29 (2)	N, cfg	5.83	25.30	611	4.05	2.96	3.51	4.66	3.41	4.04
Cuyabeno	Ecuador	CYB-01	1988.40 (2)	N, fg	2.54	27.20	697	3.05	1.03	2.04	3.29	1.11	2.20
Jatun Sacha 2	Ecuador	JAS-02	1987.63 (4)	Y? g?	14.42	30.18	724	1.94	1.98	1.96	2.40	2.45	2.43
Jatun Sacha 3	Ecuador	JAS-03	1988.88 (4)	N, fg	13.17	27.96	648	2.09	1.92	2.00	2.57	2.36	2.46
Jatun Sacha 4	Ecuador	JAS-04	1990.45 (3)	N, ef	11.55	32.47	720	3.01	1.22	2.12	3.54	1.43	2.49
Jatun Sacha 5	Ecuador	JAS-05	1989.38 (4)	N, ef	12.67	30.90	536	2.53	2.10	2.31	3.10	2.57	2.83
Tiputini 2	Ecuador	TIP-02	1997.71 (2)	N, acf	4.42	27.18	626	2.37	2.04	2.20	2.67	2.30	2.48
Tiputini 3	Ecuador	TIP-03	1998.13 (2)	N, ef	4.00	23.77	444	2.77	2.55	2.66	3.09	2.85	2.97
Nouragues GP	French Guiana	NOR-02	1993 and 1994 (2)	N, cf	7.52	28.13	493	1.23	2.07	1.65	1.45	2.43	1.94
Nouragues PP	French Guiana	NOR-01	1992.50 (2)	N, cf	9.55	30.28	524	1.13	1.51	1.32	1.35	1.81	1.58
Paracou	French Guiana	PAR	1984.50 (7)	N, cf	11.00	30.60	625	0.83	1.05	0.94	1.01	1.27	1.14
Saint Elie Transect 1	French Guiana	ELI-01	1981.50 (2)	N, c	10.00	35.83	615	0.82	0.85	0.83	0.99	1.02	1.00
Saint Elie Transect 2	French Guiana	ELI-02	1981.50 (2)	N, cf	10.00	37.94	609	0.95	1.02	0.98	1.14	1.23	1.18
Allpahuayo A poorly drained	Peru	ALP-11	1990.87 (3)	N, acf	10.15	27.36	580	2.26	2.69	2.48	2.72	3.24	2.99
Allpahuayo A, well drained	Peru	ALP-12	1990.87 (3)	N, ac	10.15	25.19	570	1.68	2.44	2.06	2.02	2.94	2.48
Allpahuayo B, clayey	Peru	ALP-22	1990.87 (3)	N, acf	10.16	25.49	614	2.32	1.93	2.12	2.79	2.32	2.55
Allpahuayo B, sandy	Peru	ALP-21	1990.87(3)	N, acf	10.16	26.88	575	2.47	2.05	2.26	2.97	2.47	2.72
Altos de Maizal	Peru	ALM-01	1994.75 (2)	N, df	5.00	30.99	672	1.04	1.68	1.36	1.18	1.91	1.55
Cocha Salvador Manu	Peru	MNU-08	1991.75 (3)	N, cf	10.07	36.81	563	1.52	1.33	1.43	1.83	1.60	1.71
Cuzco Amazonico, 1E	Peru	CUZ-01	1989.39 (3)	N, acf	9.38	25.41	489	2.56	1.70	2.13	3.06	2.03	2.55
Cuzco Amazonico, 1U	Peru	CUZ-02	1989.42 (3)	N, acf	9.35	25.27	509	2.08	1.51	1.8	2.49	1.81	2.15
Cuzco Amazonico, 2E	Peru	CUZ-03	1989.40 (3)	N, acf	9.37	21.69	470	2.72	2.13	2.43	3.25	2.55	2.91
Cuzco Amazonico, 2U	Peru	CUZ-04	1989.44 (3)	N, acf	9.34	27.26	571	2.57	2.13	2.35	3.07	2.55	2.81
Infierno	Peru	INF-01	1988.88 (2)	N, cg	7.08	N/A	809	1.94	2.08	2.01	2.27	2.43	2.35
Jenaro Herrera: Spichiger	Peru	JEN-10	1976.50 (2)	N, a	5.00	23.60	504	N/A	1.14	1.14	N/A	1.30	1.30
Jenaro Restinga 3	Peru	JEN-03	1993.71 (4)	N, f	4.04	25.19	452	4.57	3.16	3.87	5.11	3.53	4.33
Jenaro Restinga 6	Peru	JEN-06	1993.71 (4)	N, f	3.87	24.32	569	3.56	2.20	2.88	3.97	2.45	3.21
Jenaro Tahuampa 9	Peru	JEN-09	1993.71 (4)	N, f	4.04	26.91	532	2.99	2.49	2.74	3.34	2.78	3.06

Manu, Cocha Cashu Trail 12	Peru	MNU-05	1989.99 (3)	N, df	10.00	33.59	599	1.84	1.63	1.73	2.21	1.96	2.08
Manu, Cocha Cashu Trail 2 and 31	Peru	MNU-06	1989.80 (3)	N, df	10.00	32.21	511	1.96	1.83	1.89	2.35	2.20	2.27
Manu, Cocha Cashu Trail 3	Peru	MNU-01	1975.00 (6)	N, f	25.75	28.56	549	2.31	2.31	2.31	3.00	2.99	2.99
Manu, clay	Peru	MNU-02	1974.50 (3)	N, f	15.00	30.61	610	2.32	2.79	2.56	2.88	3.46	3.18
Manu, trans-Manu upland	Peru	MNU-07	1986.60 (2)	Y? g?	3.00	N/A	617	3.53	2.99	3.26	3.85	3.26	3.56
Manu, terra firme ravine	Peru	MNU-04	1991.75 (3)	N, df	10.00	27.12	587	2.30	2.13	2.20	2.77	2.56	2.64
Manu, terra firme terrace	Peru	MNU-03	1991.75 (3)	N, df	10.00	25.90	578	3.34	3.13	3.24	4.02	3.76	3.89
Mishana	Peru	MSH-01	1983.04 (2)	N, f	7.67	28.66	829	1.39	1.58	1.49	1.64	1.86	1.75
Pakitza, Manu River, plot2	Peru	PAK-02	1987.50 (2)	Y? g?	4.00	37.20	610	1.14	2.27	1.71	1.27	2.54	1.91
Pakitza, Manu River, plot1	Peru	PAK-01	1987.50 (2)	Y? g?	4.00	27.11	550	1.59	2.66	2.13	1.78	2.97	2.38
Pakitza, Manu River, swamp	Peru	PAK-03	1987.50 (2)	Y? e?g?	4.00	29.98	714	3.94	2.01	2.97	4.40	2.25	3.32
Sucusari A	Peru	SUC-01	1992.13 (3)	N, ac	8.93	28.25	612	1.76	2.02	1.89	2.10	2.41	2.25
Sucusari B	Peru	SUC-02	1992.13 (3)	N, ac	8.93	29.46	607	2.25	2.53	2.39	2.68	3.01	2.85
Tambopata plot four	Peru	TAM-06	1983.71 (5)	N, efg	16.84	30.54	520	2.94	1.54	2.24	3.69	1.93	2.81
Tambopata plot one	Peru	TAM-02	1979.87 (8)	N, fg	20.71	27.44	576	2.12	1.50	1.81	2.70	1.91	2.31
Tambopata plot six	Peru	TAM-07	1983.76 (5)	N, fg	14.97	27.36	548	2.55	2.55	2.55	3.17	3.17	3.17
Tambopata plot three	Peru	TAM-05	1983.70 (6)	N, fg	16.86	24.27	548	2.59	2.33	2.46	3.25	2.92	3.08
Tambopata plot two swamp	Peru	TAM-03	1983.79 (4)	N, eg	15.00	N/A	617	0.81	1.09	0.95	1.01	1.35	1.18
Tambopata plot two swamp edge	Peru	TAM-04	1983.79 (4)	N, af	14.96	28.56	705	2.26	2.42	2.34	2.81	3.00	2.91
Tambopata plot zero	Peru	TAM-01	1983.78 (6)	N, fg	16.81	26.91	555	2.49	2.18	2.33	3.12	2.73	2.92
Yanamono A	Peru	YAN-01	1983.46 (6)	N, efg	17.59	30.95	570	2.44	2.47	2.46	3.07	3.11	3.09
CELOS 67/9A plot8	Suriname	CEL-08	1967.50 (?)	N, a	18.00	N/A	N/A	N/A	1.70	1.70	N/A	1.70	1.70
Cano Rosalba, Z1, CR1	Venezuela	CRS-01	1970.68 (2)	N, f	2.02	17.29	532	2.21	0.66	1.43	2.34	0.70	1.51
Cano Rosalba, Z2, CR2	Venezuela	CRS-02	1970.71 (2)	N, f	1.99	28.61	266	1.54	1.73	1.64	1.63	1.83	1.73
El Dorado km 91	Venezuela	ELD-12	1971.55 (15)	N, f	22.89	27.02	468	1.11	0.82	0.96	1.43	1.05	1.23
El Dorado km 98	Venezuela	ELD-34	1971.56 (11)	N, f	9.63	23.74	492	2.89	1.54	2.22	3.46	1.85	2.66
Rio Grande	Venezuela	RIO-12	1971.58 (16)	N, f	22.88	28.17	540	1.20	0.86	1.03	1.54	1.10	1.32
San Carlos de Rio Negro, SC1	Venezuela	SCR-01	1975.71 (2)	N, af?	10.71	27.80	786	1.43	1.14	1.29	1.73	1.38	1.56
San Carlos de Rio Negro, SC2	Venezuela	SCR-02	1976.14 (2)	N, af	4.01	31.89	680	0.74	0.44	0.59	0.83	0.49	0.66
San Carlos de Rio Negro, SC3	Venezuela	SCR-03	1975.50 (2)	N, d	4.00	33.05	964	1.54	1.63	1.58	1.72	1.82	1.77

[a] a, Pre-selected randomly or systematically on a larger grid; b, randomized with respect to the forest growth phase; c, ≥ 300 m long; d, ≥ 2 ha, much larger than the typical scale of gap-phase dynamics; e, sampling most of the total area of the target stratum; f, gained basal area in the monitoring period; g, consciously selected as unbiased with respect to topography and other microsite factors. In remaining plots there is a possibility that stem dynamic patterns might have been affected by unconscious 'majestic forest' bias when the plot was located. Note that these descriptions may be incomplete; they represent our best knowledge and some are subject to revision if more information comes to light.

[b] rates are calculated over the total observation period for the plot, treating it as a single interval.

[c] Tapajós: these are 12 × 0.25 ha plots laid out in a randomized fashion over an area of 300 m × 1200 m; at the time of analysis treated as 3 × 1 ha units.

Note. Data are the best available at the time of final analyses (1 March 2003).

Table 10.2 Tests of increase in dynamic parameters. Results are given for t-tests without assuming equal variance, or for non-parametric equivalents when assumptions of normality are clearly violated

	First interval rates		Final interval rates		Final – initial difference
	Mean, s.e. of mean (% yr^{-1})	n, Mean mid-year of monitoring	Mean, s.e. of mean (% yr^{-1})	n, Mean mid-year of monitoring	Mean ± 95% CI, % yr^{-1} (median for non-parametric comparisons)
A. Turnover					
Pan-Amazon					
Raw data: 1976–2001	1.22, 0.13	10, 1977	2.34, 0.11	31, 1999	1.12 ± 0.34 T=6.75, P=0.000, d.f.= 23
Corrected for census-interval only: 1976–2001	1.30, 0.21	10, 1977	2.67, 0.13	32, 1999	1.24 ± 0.39 T=5.58, P=0.000, d.f.= 14
Corrected for majestic forest effects only: 1976–2001	1.24, 0.13	10, 1977	2.35, 0.12	29, 1999	1.23 ± 0.33 T=6.46, P=0.000, d.f.= 25
Corrected for site-switching only	1.42, 0.08	56, 1987	1.91, 0.11	56, 1997	0.49 ± 0.18 T=5.38, P=0.000, n= 56
Corrected for census-interval, site-switching, and majestic forest effects	1.65, 0.09	55, 1987	2.11, 0.12	55, 1997	0.46 ± 0.28 T=5.03, P=0.000, n= 55
West Amazonia					
Raw data: 1983–2001	1.94, 0.18	13, 1984	2.48, 0.11	27, 1999	0.54 ± 0.43 T= 2.61, P=0.017, d.f.= 20[**]
Corrected for census-interval, site-switching, and majestic forest effects	2.18, 0.10	27, 1987	2.80, 0.13	27, 1997	0.62 ± 0.32 T=4.12, P=0.000, n=27
East Amazonia					
Raw data: 1981–99	1.01, 0.10	11, 1984	1.38, 0.19	14, 1996	+ 0.38 ± 0.43 W=112, P=0.08, n= 25
Corrected for census-interval, site-switching, and majestic forest effects	1.20 ± 0.07	28, 1986	1.43 ± 0.10	28, 1996	+ 0.24 ± 0.20 T= 2.50, P=0.02, n= 28
Richer soils					
Raw data: 1983–2001	2.01 ± 0.16	11, 1986	2.44 ± 0.13	20, 1999	+ 0.42 ± 0.42 T= 2.05, P=0.05, d.f.= 21
Corrected for census-interval, site-switching, and majestic forest effects	2.20 ± 0.11	24, 1989	2.83 ± 0.13	24, 1998	+ 0.66 ± 0.28 T= 4.15, P=0.000, n= 44
Poorer soils					
Raw data: 1981–2001	0.97 ± 0.08	11, 1983	2.17 ± 0.20	12, 1999	+ 1.20 ± 0.46 T= 5.73, P=0.000, d.f.= 13
Corrected for census-interval, site-switching, and majestic forest effects	1.22 ± 0.08	31, 1986	1.54 ± 0.12	31, 1996	+ 0.32 ± 0.17 T= 2.99, P=0.003, n= 31
Aseasonal climate					
Raw data: 1990–2001	1.94 ± 0.16	13, 1990	2.58 ± 0.11	15, 1999	+ 0.64 ± 0.40 T= 3.30, P=0.003, d.f.= 24[*]

Corrected for census-interval, site-switching, and majestic forest effects	2.08 ± 0.13	10, 1991	2.78 ± 0.19	10, 1998	+ 0.69 ± 0.49 $T = 3.21$, $P = 0.011$, $n = 10$
Seasonal climate					
Raw data: 1981–2001	1.17 ± 0.14	13, 1982	2.19 ± 0.18	16, 1999	+ 0.91 ± 0.51 $T = 4.42$, $P = 0.002$, d.f. = 25
Corrected for census-interval, site-switching, and majestic forest effects	1.55 ± 0.10	45, 1985	1.96 ± 0.13	45, 1996	+ 0.41 ± 0.20 $T = 4.15$, $P = 0.000$, $n = 45$
B. Recruitment					
Pan-Amazon					
Raw data: 1979–2001	1.35 ± 0.15	10, 1981	2.33 ± 0.13	32, 1999	+ 0.98 ± 0.43 $T = 4.86$, $P = 0.000$, d.f. = 21
Corrected for census-interval, site-switching, and majestic forest effects	1.70 ± 0.11	57, 1987	2.34 ± 0.15	57, 1997	+ 0.64 ± 0.28 $T = 4.45$, $P = 0.000$, $n = 57$
West Amazonia					
Raw data: 1983–2001	1.82 ± 0.22	10, 1985	2.45 ± 0.14	27, 1999	+ 0.64 ± 0.57, $T = 2.39$, $P = 0.03$, d.f. = 15[**]
Corrected for census-interval, site-switching, and majestic forest effects	2.24 ± 0.14	27, 1988	2.86 ± 0.18	27, 1997	+ 0.62 ± 0.39 $T = 3.30$, $P = 0.003$, $n = 27$
East Amazonia					
Raw data: 1981, 1999	0.87 ± 0.10	10, 1984	1.43 ± 0.28	14, 1996	+ 0.55 ± 0.60 $T = 1.91$, $P = 0.07$, d.f. = 16[**]
Corrected for census-interval, site-switching, and majestic forest effects	1.23 ± 0.10	27, 1986	1.60 ± 0.11	27, 1996	+ 0.38 ± 0.30 $T = 2.64$, $P = 0.014$, $n = 27$
Richer soils					
Raw data: 1983–2001	1.92 ± 0.23	10, 1986	2.50 ± 0.16	20, 1999	+ 0.55 ± 0.59 $T = 1.95$, $P = 0.06$, d.f. = 17
Corrected for census-interval, site-switching, and majestic forest effects	2.30 ± 0.14	25, 1989	3.01 ± 0.18	25, 1999	+ 0.69 ± 0.44 $T = 3.09$, $P = 0.003$, $n = 25$
Poorer soils					
Raw data: 1981–2001	0.87 ± 0.10	10, 1983	2.10 ± 0.22	12, 1999	+ 1.22 ± 0.51 $T = 5.12$, $P = 0.000$, d.f. = 15
Corrected for census-interval, site-switching, and majestic forest effects	1.26 ± 0.10	29, 1986	1.67 ± 0.12	29, 1996	+ 0.41 ± 0.27 $T = 3.03$, $P = 0.005$, $n = 29$
Aseasonal climate					
Raw data: 1990–2001	1.96 ± 0.20	11, 1991	2.66 ± 0.15	15, 1999	+ 0.70 ± 0.31 $W = 101.0$, $P = 0.006$, $n = 20$[**]
Corrected for census-interval, site-switching, and majestic forest effects	2.12 ± 0.10	10, 1990	2.69 ± 0.21	10, 1997	+ 0.57 ± 0.48 $T = 2.67$, $P = 0.022$, $n = 10$

Table 10.2 (*Continued*)

	First interval rates		Final interval rates		Final − initial difference
	Mean, s.e. of mean (% yr^{-1})	n, Mean mid-year of monitoring	Mean, s.e. of mean (% yr^{-1})	n, Mean mid-year of monitoring	Mean ± 95% CI, % yr^{-1} (median for non-parametric comparisons)
Seasonal climate					
Raw data: 1981–2001	1.03 ± 0.20	11, 1983	2.04 ± 0.19	17, 1999	+1.01 ± 0.55 $T=3.61$, $P=0.001$, d.f. = 23
Corrected for census-interval, site-switching, and majestic forest effects	1.60 ± 0.13	46, 1986	2.25 ± 0.17	41, 1996	+0.65 ± 0.34 $T=3.75$, $P=0.001$, $n=46$
C. Mortality					
Pan-Amazon					
Raw data: 1976–2001	1.16 ± 0.14	10, 1981	2.30 ± 0.14	31, 1999	+1.14 ± 0.39 $T=5.97$, $P=0.000$, d.f. = 25
Corrected for census-interval, site-switching, and majestic forest effects	1.58 ± 0.10	52, 1987	1.91 ± 0.13	52, 1997	+0.32 ± 0.19 $Z=360$, $P=0.001$, $n=55$
West Amazonia					
Raw data: 1983–2001	1.85 ± 0.18	12, 1985	2.50 ± 0.13	27, 1999	+0.65 ± 0.46 $T=2.94$, $P=0.007$, d.f. = 23
Corrected for census-interval, site-switching, and majestic forest effects	2.03 ± 0.13	27, 1988	2.59 ± 0.14	24, 1997	+0.56 ± 0.35 $T=3.31$, $P=0.003$, $n=27$
East Amazonia					
Raw data: 1981–99	1.12 ± 0.11	12, 1984	1.33 ± 0.13	14, 1996	+0.21 ± 0.33 $T=1.24$, $P=0.23$, d.f. = 23
Corrected for census-interval, site-switching, and majestic forest effects	1.16 ± 0.08	28, 1986	1.27 ± 0.09	28, 1996	+0.11 ± 0.20 $T=1.01$, $P=0.33$, $n=28$

Richer soils					
Raw data: 1983–2001	1.98 ± 0.15	11, 1986	2.40 ± 0.15	20, 1999	+0.42 ± 0.43 $T = 2.01$, $P = 0.056$, d.f. = 25
Corrected for census-interval, site-switching, and majestic forest effects	2.12 ± 0.12	24, 1990	2.55 ± 0.12	21, 1999	+0.42 ± 0.30 $T = 2.89$, $P = 0.009$, $n = 24$
Poorer soils					
Raw data: 1981–2001	1.06 ± 0.10	12, 1983	2.24 ± 0.26	12, 1999	+1.18 ± 0.60 $T = 4.22$, $P = 0.001$, d.f. = 14
Corrected for census-interval, site-switching, and majestic forest effects	1.16 ± 0.07	31, 1986	1.41 ± 0.14	31, 1996	+0.32 ± 0.19 $Z = 365$, $P = 0.023$, $n = 31$
Aseasonal climate					
Raw data: 1990–2001	1.84 ± 0.18	15, 1991	2.49 ± 0.17	13, 1999	+0.66 ± 0.51 $T = 2.66$, $P = 0.013$, d.f. = 25[*]
Corrected for census-interval, site-switching, and majestic forest effects	2.04 ± 0.18	10, 1990	2.86 ± 0.29	10, 1997	+0.82 ± 0.81 $T = 2.29$, $P = 0.048$, $n = 10$
Seasonal climate					
Raw data: 1981, 2001	1.26 ± 0.11	13, 1983	2.21 ± 0.21	17, 1999	+0.94 ± 0.46 $T = 4.26$, $P = 0.000$, d.f. = 23
Corrected for census-interval, site-switching, and majestic forest effects	1.48 ± 0.10	45, 1985	1.70 ± 0.11	45, 1996	+0.22 ± 0.18 $Z = 763$, $P = 0.006$, $n = 45$

[*] P-values should be interpreted with caution because seven sites monitored in 1990 were also monitored in 2001.

[**] P-values should be interpreted with caution because four sites monitored in the start year were also monitored in the end year.

For raw data, census-interval corrected data, and majestic-forest corrected data, we compare the first year in which ≥10 sites monitored with the last, using two sample t-tests or Mann–Whitney U-tests. For all data corrected for site-switching, we compare the end interval with the start interval for all multi-interval sites except those with the end interval starting in 1976 or earlier, using paired t-tests or Wilcoxon signed-rank tests. See text for further details.

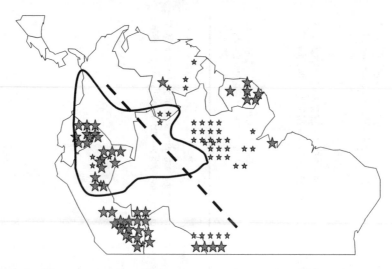

Figure 10.1 Map of site locations, showing approximate boundary between 'aseasonal' (rainfall averages <1 month with <100 mm) and 'seasonal' Amazonia. Dotted line divides 'east' and 'west' sites. Poor soil sites (oxisol, spodosol, and oligotrophic histosol) represented by small stars, richer soil sites (ultisol, inceptisol, entisol, histosol) by large stars.

values at all sites monitored at the end of the period (2001). For data corrected for site-switching we used paired *t*-tests or the non-parametric equivalent (Wilcoxon tests), evaluating change across all sites monitored for at least two intervals by comparing the final interval rate with the first interval rate for the same site. These statistical tests supplement graphical display of time-dependent patterns for each major pan-Amazon and regional analysis. The focus here is on detecting broad spatial and temporal pattern, rather than determining causes—the data are not yet of sufficient quality to disaggregate the potential environmental and spatial drivers of turnover processes nor to pinpoint annual fluctuations, but they are sufficient to test whether change is confined to specific Amazonian environments or if it is a general phenomenon, and whether process rates are changing at different rates.

Results

A total of 97 sites met our criteria for inclusion, of which 61 have at least two intervals (see Phillips *et al.* 2004, table 1). Sites are distributed across the region (Fig. 10.1). The data represent 1640 ha-yr of monitoring by more than 20 research groups. Across all

97 sites recruitment and mortality rates average ca. 2% yr^{-1} (Table 10.1), but recruitment marginally exceeds mortality, using only sites with both mortality and recruitment values (Wilcoxon signed-rank test, $Z = 2359$, $P < 0.05$, $n = 93$, for both uncorrected and census-interval corrected values).

When results are plotted from individual sites, turnover varies substantially from site-to-site and interval-to-interval (Fig. 10.2), suggesting that many sites may be needed to statistically distinguish large-scale patterns in time and space. However, in spite of the noise, these data show that turnover rates have increased substantially across all Amazonian sites regardless of the method of data treatment (Fig. 10.3, Table 10.2). Each correction produces different patterns in terms of magnitude of overall change and inter-annual fluctuations. Nevertheless, irrespective of whether the procedures are applied singly or in combination, the overall result of turnover increase is always highly significant ($P < 0.001$).

The remaining results—broken down by process, spatial region, and environmental attributes—are given after correcting for all three potential artefacts.

Both recruitment and mortality have increased across all sites (Fig. 10.4), with mean recruitment

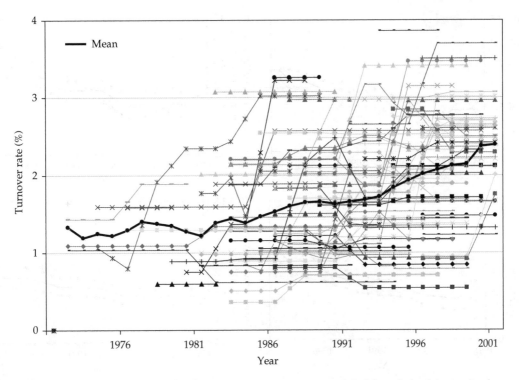

Figure 10.2 Interval-by-interval turnover rates for all sites. Average turnover rates were derived from all plots being monitored simultaneously (See Plate 3).

rates exceeding mean mortality rates throughout the period. This difference is not significant initially but it becomes so by the end of the period (paired t-test for all 55 multi-census sites: for first interval rates, $t = 1.51$, $P < 0.15$; for final interval rates, $t = 2.90$, $P < 0.01$). Elsewhere (Lewis *et al.*, Chapter 12, this volume) we used within-plot analyses to show that a logical corollary of this—increased stem density—is also apparent.

Turnover is twice as high in the west as it is in the east (median values 2.60, 1.35% yr^{-1}, respectively; $t = 7.94$, $P < 0.001$, d.f. $= 43$). Turnover rates have increased significantly in both regions (Fig. 10.8, Table 10.2). The absolute rate of change is greater in the west (Fig. 10.5, Mann–Whitney U-test, $W = 657$, $P < 0.03$, $n = 55$; test compares regions using the increase in turnover rate between consecutive intervals, correcting for variable census interval length). In the west, mortality and recruitment have both increased significantly

(Fig. 10.6(a)). In east Amazonia mortality and recruitment trends are positive but only significantly so for recruitment (Fig. 10.6(b)).

The east/west differences and the within-region trends in turnover, recruitment, and mortality are largely mirrored by patterns among and within the soil-based categories (Figs 10.7 and 10.8), because poor soils dominate in the east and richer soils are more common in the west. Richer soil forests are twice as dynamic as poor soil forests (median turnover rates 2.72, 1.37% yr^{-1}, respectively, $t = 9.23$, $P < 0.001$, d.f. $= 39$; test includes all census-corrected sites monitored in 1995 except those with potential majestic forest effects). Recruitment and mortality have tended to increase on both substrates but with the largest absolute increases on richer soils and in recruitment rates (Table 10.2).

Only the north-western quadrant of Amazonia is generally aseasonal, and so our aseasonal dataset

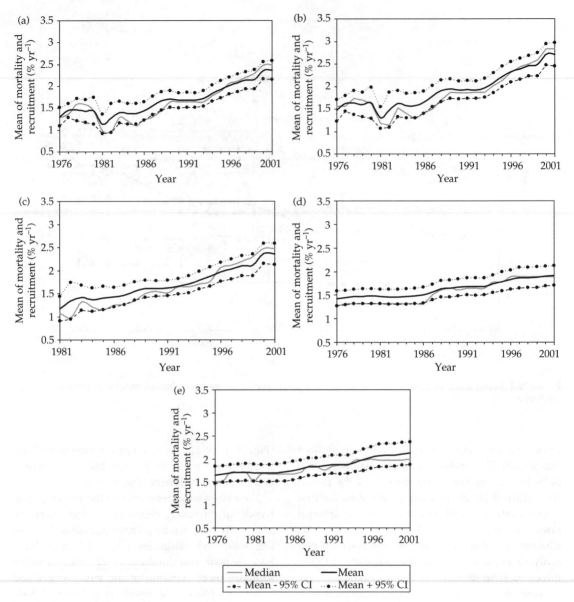

Figure 10.3 Pan-Amazon turnover rates. (a) all sites; no census-interval correction or smoothing of site-switching, (b) all sites; census-interval corrected, (c) potential majestic forest sites removed, (d) sites with single interval removed; all multi-census sites corrected for site-switching, and (e) census-interval, site-switching, and majestic-forest corrected.

is smaller than the seasonal one. Aseasonal Amazon forests are more dynamic than seasonal Amazon forests but not significantly so (mean turnover rates 2.64, 2.12% yr^{-1}, respectively; $t = 1.88$, $P < 0.08$, d.f. = 18). Regardless, forests in

both climate regimes have become significantly more dynamic (Fig. 10.9, Table 10.2). In both seasonal and aseasonal Amazonia both recruitment and mortality have increased significantly (Fig. 10.10(a) and (b)).

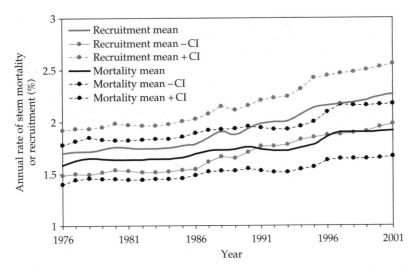

Figure 10.4 Recruitment and mortality rates, Amazonia 1976–2001. Both have increased. Census-interval, site-switching, and majestic-forest corrected.

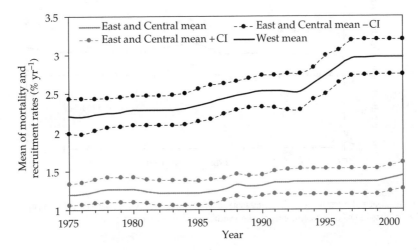

Figure 10.5 Turnover trends in west and east Amazonia. Turnover increased significantly in both regions, but was higher in the west than in the east throughout. Census-interval, site-switching, and majestic-forest corrected.

Discussion

Mature forests of Amazonia have experienced accelerated tree turnover in the latter part of the twentieth century. This finding is consistent with earlier reports at larger temporal and spatial scales—tropical forest plots were on average twice as dynamic in the 1990s as in the 1950s, and increases occurred in both the Old and New World tropics (Phillips and Gentry 1994; Phillips 1996). The current analysis also expands upon these findings in several important ways.

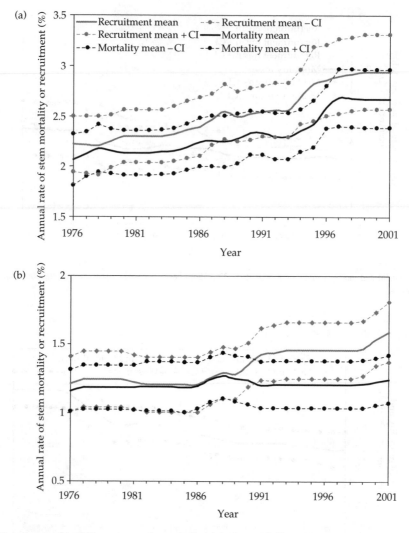

Figure 10.6 (a) Recruitment and mortality rates, west Amazonia. Both have increased, (b) recruitment and mortality rates, east Amazonia. Only recruitment has increased significally. Census-interval, site-switching, and majestic-forest corrected.

First, the consistent patterns observed here suggest that the increases in tropical tree turnover rates are not driven by artefactual concerns. Neither the tendency for turnover rates to appear greater when measured over shorter intervals (Sheil and May 1996), nor the possible preference of ecologists to select high-biomass 'majestic' forest that subsequently develops gaps and thus locally accelerated mortality and recruitment (Condit 1997; Phillips *et al.* 1997, 2002a), nor possible

'switching' of monitoring effort to intrinsically more dynamic forests (Condit 1997) can explain the result.

Second, the increasing turnover result sheds light on the parallel observation of increasing biomass in long-term forest plots (Phillips *et al.* 1998b; Baker *et al.*, Chapter 11, this volume) and vice versa. Thus, the net increase in biomass in Amazon plots cannot reflect widespread natural recovery from earlier catastrophic disturbance,

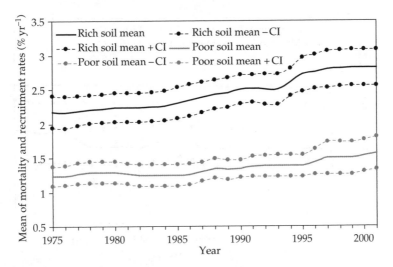

Figure 10.7 Turnover through time in poor soil and richer soil Amazonia. Census-interval, site-switching, and majestic-forest corrected.

since succession involves *reduced* recruitment rates of small trees as maturing forests thin. Conversely, progressive fragmentation and advancing edge effects—changes that accelerate turnover by killing large trees (Laurance *et al.* 2000)—cannot be responsible for the turnover increases because most plots with increasing turnover are also gaining biomass (Lewis *et al.*, Chapter 12, this volume). In sum, the coincidence of increasing turnover with increasing biomass makes it difficult to explain either as artefacts of sampling bias or landscape processes.

Third, we have shown that the increase in turnover is not simply an outcome of an increase in mortality, or an increase in recruitment. For Amazonia at least it is both. Forest dynamic processes have accelerated in a concerted manner.

Fourth, we have shown how tropical tree turnover rates also vary with geographic factors. Turnover is highest on richer soils, in aseasonal forests, and in western Amazonia. Long-term process rates vary by a factor of 2 (Table 10.2). Moreover, despite these systematic differences in Amazon tree dynamics, change occurred simultaneously in a consistent direction.

Finally, these findings show the value of large-scale, long-term research programmes in tropical ecology. No single region, soil class, or climate regime can represent 'typical' conditions for the lowland Amazon. Landscape-scale studies alone cannot be used to test hypotheses of regional- and continental-scale change.

Mechanisms of forest dynamics

The data can also provide insight into the mechanisms of Amazon forest dynamics. Tree turnover is an emergent property of underlying structural, floristic, and dynamic processes. At its simplest we can envision two extreme situations: (1) a system driven entirely by catastrophic mortality, in which external disturbance events such as fire, drought, flood, and storm determine forest structure and dynamics (cf. Connell 1978), or (2) a system driven entirely by internal growth and recruitment processes, in which resource supply provides the ultimate driver for forest ecology so that trees mostly die competing for these resources (cf. Enquist and Niklas 2001). Which model best reflects reality in the Amazon? We know of course that both processes operate—weather extremes kill trees but so does competition for resources—but it is possible to test which mode is dominant at the regional scale. One approach is to assess temporal lags between mortality and recruitment within plots and within regions. Another is to ask

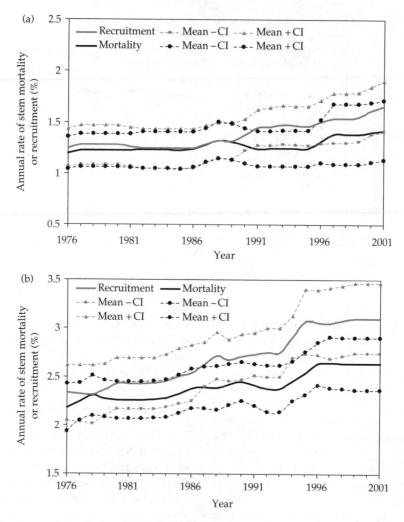

Figure 10.8 (a) Recruitment and mortality through time, poor soil Amazonia, (b) recruitment and mortality through time, richer soil Amazonia. Census-interval, site-switching, and majestic-forest corrected.

whether catastrophic disturbances occur frequently enough and synchronously enough to generate large-scale lags of recruitment following mortality? Or, are they rare and random, so that instead pulses of recruitment lead pulses of mortality?

The pan-Amazon and regional datasets show mean mortality rates lagging mean recruitment rates, which implies that recruitment and growth increases are driving the turnover increase. Still, mortality-led dynamics certainly do occur in the Amazon. How frequent are catastrophic disturbances? Long-term plots should provide better estimates of their frequency and impact than anecdotal reports of individual events. In central Amazonia, intense rainfall and wind-storms associated with La Niña brought increased risk of death by flooding and wind-throw (N. Higuchi, personal observation). Likewise, in south-western Amazonia storms can topple emergent trees (Foster and Terborgh 1998). But in 1640 ha-yr of monitoring we have not yet observed a single really catastrophic disturbance in any one of our

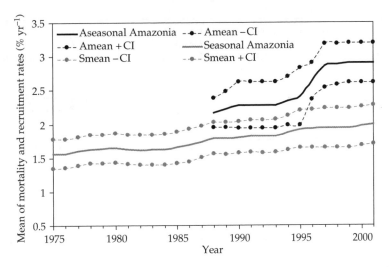

Figure 10.9 Turnover through time, aseasonal versus seasonal Amazonia. Census-interval, site-switching, and majestic-forest corrected.

plots. Although space and time are not perfectly substitutable, clearly such events must have been very rare and localized since at least 1980.

However, late twentieth-century Amazonia is perhaps not an ideal setting for testing equilibrium or stochastic models of forest behaviour, because the system is shifting as turnover rates accelerate and forest basal area increases. The shift is ubiquitous but asymmetric—turnover has risen most in absolute terms in the already-dynamic forests of the west, and is being led by recruitment changes, with recruitment exceeding mortality in most forests for most of the time. Growth and mortality rates are higher on more productive soils (Phillips *et al.* 1994, 2004; Malhi *et al.* 2004), so spatial variation in growth rates is primarily due to factors which influence growth of plants, and therefore the temporal variation may be too. The patterns of change in dynamics and stand structure also conform to common-sense predictions for a growth driver (Lewis *et al.*, Chapter 4, this volume). Here, growth rates across all size-classes including recruitment rates into the 10-cm size-class respond instantaneously to an increase in resource provision, with adult mortality lagging as the system approaches, perhaps, a new equilibrium at higher biomass and turnover (Lloyd and Farquhar 1996). Given an equal *proportional* effect in all forests the

absolute effect should be greater in faster forests and therefore the signal easier to detect, which is what we observe (cf. for example, western versus eastern Amazon significance levels for the final-interval versus first-interval change in recruitment and mortality rates, Table 10.2). Similarly, faster systems should respond to a stimulating effect in a more synchronized manner than slower systems. Mortality and recruitment curves do appear to be more closely synchronized with one another in the faster forests.

Causes of changes in forest dynamics

What might the environmental parameter(s) driving these changes be? We have two sets of circumstantial evidence to guide us. First, a priori knowledge of changes in drivers and their likely ecophysiological effects (Lewis, *et al.*, Chapter 4, this volume) allows us to estimate the potential impact of any given process. Second, the geographical and temporal pattern of response provides further clues. Change has occurred over large areas (different regions of Amazonia and beyond) for at least two decades. We note that long-term satellite studies also report a widespread, simultaneous increase in tropical terrestrial net primary productivity (NPP) (Cao *et al.* 2004),

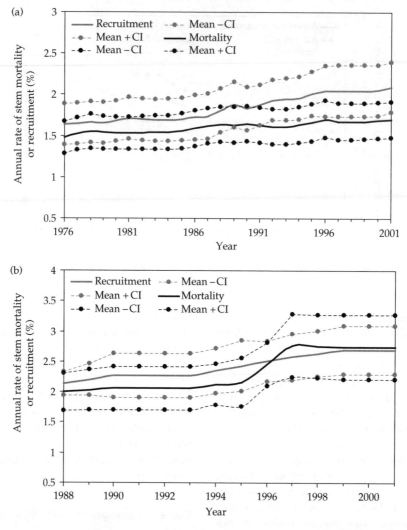

Figure 10.10 (a) Recruitment and mortality through time, seasonal Amazonia, (b) recruitment and mortality through time, aseasonal Amazonia. Census-interval, site-switching, and majestic-forest corrected.

although at a slower rate (0.19–0.66% yr^{-1}) than the turnover increase. The driver must be either a set of coincident yet independent local changes at dozens of sites, or more parsimoniously a single 'global' environmental change. Elsewhere, Laurance *et al.* (2004a), Malhi and Phillips (2004), and Phillips *et al.* (2004) argue that rainfall increases are not a plausible dominant driver of recent tropical ecological change. Other candidate growth drivers such as nitrogen deposition via biomass burning are too poorly characterized,

ecophysiologically uncertain, and too localized to be able to make a coherent case for, although contributory effects cannot be ruled out. Two growth drivers remain as serious candidates (discussed in Malhi and Phillips (2004) and Phillips *et al.* (2004))—the ubiquitous long-term increase in atmospheric CO_2 concentrations, and a possible recent increase in insolation in western Amazonia (Wielicki *et al.* 2002; Nemani *et al.* 2003).

The CO_2 and sunlight explanations are not mutually exclusive (growth responses to CO_2

could improve synergistically with increased radiation), but because the first is universal and the second has a strong spatial pattern, we can make distinct predictions that allow us to eventually discriminate their ecological footprints. Thus: if a CO_2 effect is dominant we should see growth and dynamics responses everywhere we look in the tropics (except where constrained by large climate change); if a radiation effect is dominant we expect to see growth and dynamics responses proportionate to simultaneous local radiation trends. To test this requires estimating growth rates and trends for permanent plots across the biome, building on the painstaking, cumulative, and collaborative work by field biologists which has been synthesized here. The need for plot-based global vegetation monitoring has never been greater.

Acknowledgements

Invaluable contributions of colleagues, assistants, research agencies, and institutes are gratefully acknowledged in Phillips *et al.* (2004).

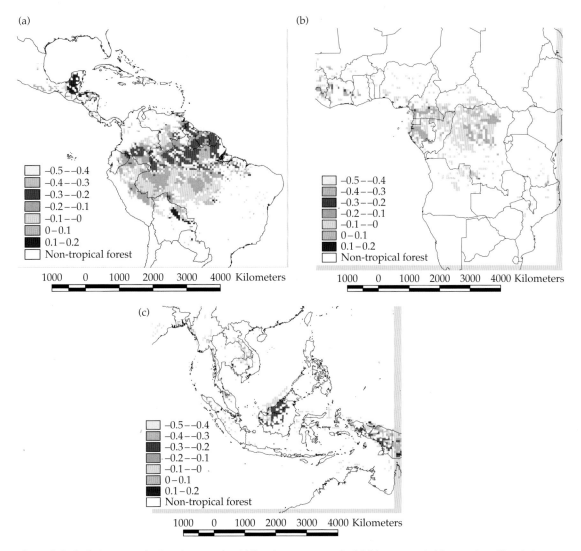

Plate 1(a,b,c) El Niño events dominate interannual variability of temperature and rainfall in most tropical forest regions. Plate 1 shows the cross-correlation between rainfall and the multivariate ENSO index for the period 1960–98, for all three major tropical forest regions. Areas with high negative values (red and yellow) experience strong reductions in rainfall during El Niño events. For more details see Chapter 1 by Malhi and Wright.

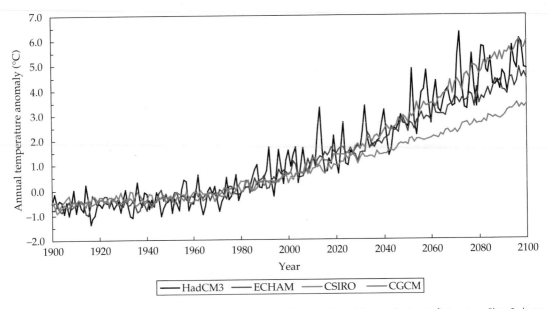

Plate 2 Global climate models suggest that tropical rainforest regions will warm substantially over the twenty-first century. Plate 2 shows simulated change in the temperature of the tropical Americas (30° S–30° N) in the period 1901–2100, based on four different general circulation models. The anomalies are relative to the mean over 1969–98. For more details, see Chapter 2 by Cramer et al.

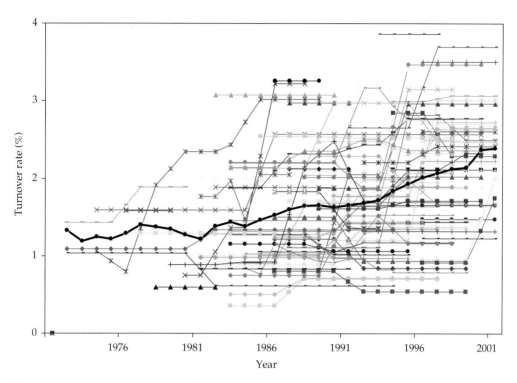

Plate 3 Forest dynamism appears to have accelerated in recent decades. Interval-by-interval turnover rates for individual sites monitored in Amazonia since the 1970s. Each line represents a different forest plot, and the measured turnover rate between successive censuses is applied to all years in the census interval. Average turnover rates were derived from all plots being monitored simultaneously. Thick black line = mean. For more details see Chapter 10 by Phillips et al.

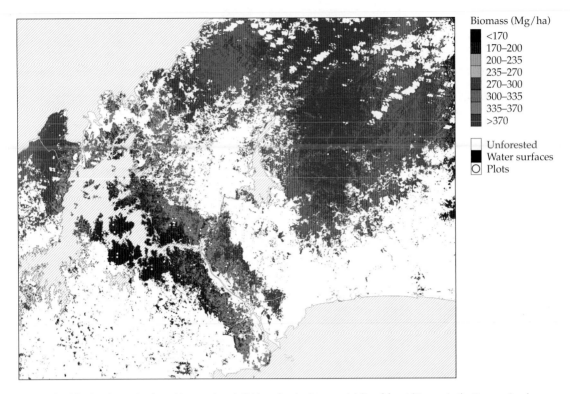

Biomass (Mg/ha)
- ■ <170
- 170–200
- 200–235
- 235–270
- 270–300
- 300–335
- 335–370
- >370

- □ Unforested
- ■ Water surfaces
- ◯ Plots

Plate 4 The difficulty of assessing forest biomass through field studies: landscape variability of forest biomass in the Panama Canal Watershed based on a generalized linear model. The model is extrapolated to the rainforest (young secondary forest, old secondary forest and old-growth forest) which was previously classified with a maximum likelihood classification algorithm (classification error: 1%). In white, non-forested land: agricultural and urban land, mangroves, shrubland, and clouds. For more details see Chapter 13 by Chave et al.

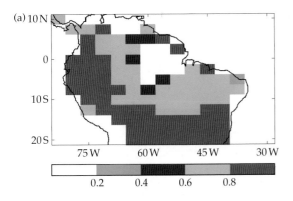

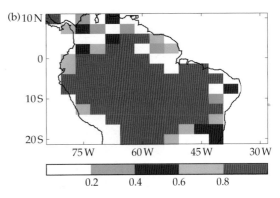

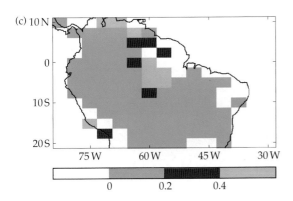

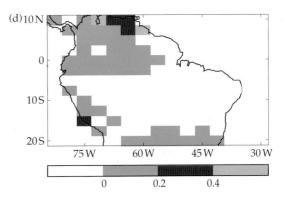

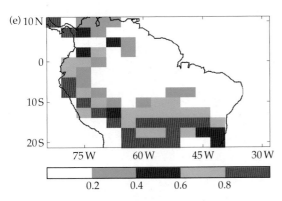

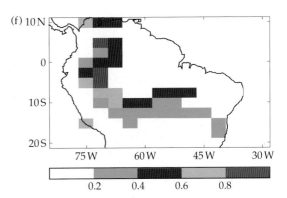

Plate 5 Model simulations of vegetation cover in tropical South America under past and future climates. Fraction of broadleaf vegetation simulated for (5a) the last glacial maximum (LGM), and (5b) Younger Dryas (YD)-like palaeoclimate. Difference in the fraction of C4-grass simulated for (5c) the last glacial maximum (LGM), and (5d) Younger Dryas (YD)-like palaeoclimate. Positive numbers represent an increase in grass percentage in the LGM and YD climate scenarios relative to pre-1850 late Holocene conditions. Simulations of (5e) percent broadleaf cover and (5f) C4-grass under a mid-range scenario of future climate change (IS92a). For more details see Chapter 16 by Cowling et al.

Plate 6 Seasonally dry tropical forests may increase in importance in rainforest areas experiencing increased water stress under climate change. The example in this photo is from Mexico, with abundant cacti (photo T. Pennington). See Chapter 18 by Phillips and Malhi.

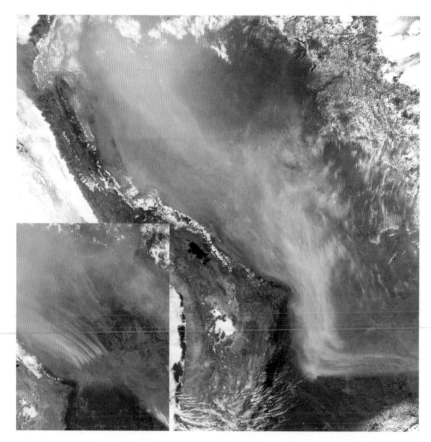

Plate 7 Forest fires can modify climate for substantial areas downwind. Fires and associated smoke plumes in a forest clearance frontier in Santa Cruz, Bolivia (inset), generating a smoke haze that extends thousands of kilometres across Peru and Brazil to Colombia and Venezuela. This image was taken by the Moderate Resolution Imaging Spectroradiometers (MODIS) on NASA's Terra and Aqua satellites in September 2004. See Chapter 18 by Phillips and Malhi.

Plate 8 Large canopy trees account for a substantial fraction of the biomass carbon dynamics of a forest, but can be a challenge to measure with sufficient accuracy. There is some evidence that they may be increasing disproportionately in growth rates: this example is from Noel Kempff National Park, Bolivia (photo Y. Malhi). See Chapter 18 by Phillips and Malhi.

Plate 9 Atmospheric changes may bring dramatic changes in the relative abundance of different plant functional groups, and hence to the ecological dynamics of tropical forests. There is already evidence that lianas may be increasing in abundance. This example is from Barro Colorado Island, Panama, where liana leaf productivity has increased dramatically over the last two decades (photo Y. Malhi). See Chapter 18 by Phillips and Malhi.

Plate 10 Atmospheric change may have different effects on canopy trees versus gap trees versus understorey trees, altering the competitive balance between these functional groups. Forest understoreys vary substantially with soil and climate conditions: this example is from Sucusari near Iquitos, Peru (photo Y. Malhi). See Chapter 18 by Phillips and Malhi.

CHAPTER 11

Late twentieth-century trends in the biomass of Amazonian forest plots

**Timothy R. Baker, Oliver L. Phillips, Yadvinder Malhi,
Samuel Almeida, Luzmila Arroyo, Anthony Di Fiore,
Terry Erwin, Niro Higuchi, Timothy J. Killeen, Susan G. Laurance,
William F. Laurance, Simon L. Lewis, Abel Monteagudo,
David A. Neill, Percy Núnez Vargas, Nigel C. A. Pitman,
J. Natalino M. Silva, and Rodolfo Vásquez Martínez**

There has been substantial debate about whether the old-growth forests of Amazonia are increasing in biomass and therefore acting as a sink for atmospheric CO_2. We present a new analysis of biomass change in old-growth Amazonian forest plots using updated inventory data. We find that across 59 sites, the above-ground dry biomass in trees that are more than 10 cm in diameter (AGB) has increased since plot establishment by 1.22 ± 0.43 Mg per hectare per year (ha^{-1} yr^{-1}, where 1 ha $= 10^4$ m^2), or 0.98 ± 0.38 Mg ha^{-1} yr^{-1} if individual plot values are weighted by the number of hectare years of monitoring. This significant increase is neither confounded by spatial or temporal variation in wood specific gravity, nor dependent on the allometric equation used to estimate AGB. The conclusion is also robust to uncertainty about diameter measurements for problematic trees: for 34 plots in western Amazon forests a significant increase in AGB is found even with a conservative assumption of zero growth for all trees where diameter measurements were made using optical methods and/or growth rates needed to be estimated following fieldwork. Overall, our results suggest a slightly greater rate of net stand-level change than was reported by Phillips *et al.* (1998b). Considering the spatial and temporal scale of sampling and associated studies showing increases in forest growth and stem turnover, the results presented here suggest that the total biomass of these plots has on average increased and that there has been a regional-scale carbon sink in old-growth Amazonian forests during the previous two decades.

Introduction

Quantifying changes over time in the carbon storage of Amazonian forests is extremely important for understanding current and future trends in the global carbon cycle (Prentice *et al.* 2001). Variation occurs over a range of timescales and monitoring these patterns remains a considerable challenge. Over short timescales, at a number of Amazonian sites, measurements of carbon dioxide fluxes between the forest and atmosphere have been made by eddy covariance systems to estimate forest carbon balance (e.g. Grace *et al.* 1995a) but it is difficult to extend these measurements over many years, or many sites. Inversion models, which combine data on the concentrations of carbon dioxide, oxygen, and their isotopes with atmospheric circulation models to predict patterns of carbon dioxide sources and sinks, can be used at large scales (e.g. Gurney *et al.* 2002), but are poorly constrained in tropical regions. By contrast, repeated measurements of permanent sample plots can potentially provide direct estimates of changes in tropical forest biomass with the requisite spatial and temporal coverage from a wide variety of sites.

The potential value of using long-term data from tropical forest plots for studying changes in biomass was highlighted by a study of 68 pantropical sites (Phillips *et al.* 1998b). Over the period 1975–96, in 40 sites across Amazonia, total above-ground dry biomass increased by 0.97 ± 0.58 Mg ha^{-1} yr^{-1}, which is equivalent to 0.88 ± 0.53 Mg ha^{-1} yr^{-1} for trees ≥ 10 cm diameter. This value was used to estimate a total carbon sink across Amazonia of 0.44 ± 0.26 Gt C yr^{-1}. However, the result generated a vigorous debate about the methodology that should be used to estimate changes in forest biomass from permanent plot measurements. For example, it was suggested that it could be explained by a potential sampling bias towards successional forests on floodplain sites or by poor tree measurement techniques (Clark 2002a, but see also Phillips *et al.* 2002a). In addition, the problems inherent in including small plot sizes, where the biomass of all trees ≥ 10 cm diameter above-ground biomass (AGB) is not normally distributed, and the potential importance of changes in the carbon stocks of other compartments, such as coarse woody debris, have also been noted (Chave *et al.* 2003, 2004; Rice *et al.* 2004).

The method of AGB estimation used by Phillips *et al.* (1998b), on a stand-level basis using plot basal area values is also open to criticism. It is well known that the large number of published biomass equations can give substantial variation in stand-level AGB estimates (e.g. Chambers *et al.* 2001b; Baker *et al.* 2004a). However, it is not known whether the observed patterns of net biomass *change* are sensitive to the equation used to estimate AGB. In addition, the method of Phillips *et al.* (1998b) does not explicitly account for spatial or temporal variation in tree size–frequency distributions or variation in wood specific gravity. As mean tree size and wood specific gravity vary at a regional scale across the Amazon basin (Malhi *et al.* 2002a; Baker *et al.* 2004a), estimates of AGB change across all sites should ideally include these factors. Also, given the substantial changes over time in Amazon forest dynamics (Phillips and Gentry 1994; Lewis *et al.* 2004b; Phillips *et al.* 2004), estimates of AGB change

need to incorporate any potential changes in forest structure or functional composition.

A re-examination of pan-Amazonian forest plot data is therefore needed to directly address these issues, and provide improved estimates of AGB change. Using old-growth forest plot data, we ask the following questions:

1. Do the patterns of AGB change depend on the allometric equation used to calculate biomass?
2. Are the patterns of change sensitive to spatial or temporal variation in tree size-frequency distributions, or wood specific gravity?
3. Is there any consistent regional-scale change in AGB?
4. Are conclusions about the direction of change influenced by uncertainty concerning problematic tree records?

Methods

Inventory data were used from 59 forest sites from across the range of local and regional environmental gradients that occur in Amazonia, including *terra firme* forests on both clay-rich and white sand substrates, and seasonally flooded forests (Fig. 11.1, Table 11.1). All sites examined were in lowland forest (<500 m a.m.s.l.) consisting of an apparently mature forest with natural gap-phase dynamics and a canopy dominated by non-pioneer species. None of the plots are believed to have experienced any recent, major, direct human impact. The individual plots range in size from 0.4 to 9.0 ha (median 1.0, mean 1.3 ha), and in total encompass 78.9 ha of forest (Table 11.1). Initial measurement dates vary from 1979 to 1998, and census intervals from 4.0 to 21.7 yr (median 10.2, mean 10.9 yr) (Table 11.1). Overall, the results are based on measurements of 54,364 stems ≥ 10 cm diameter, and a total sampling effort of 863.8 ha yr. For all plots, family and generic taxonomy has been standardized following the procedures described in Baker *et al.* (2004a). Wood specific gravity data are derived from a variety of publications. These sources and the approach used to match specific gravity data to tree records is also described in Baker *et al.* (2004a).

To make comparisons of rates of AGB change between different landforms, we distinguish two groups of sites, separating 12 plots on old, recent,

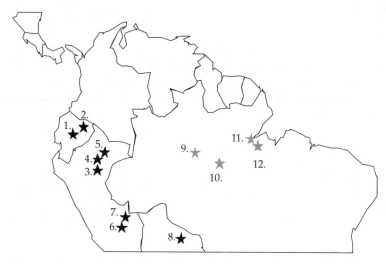

Figure 11.1 Location of forest sites in western (black symbols) and eastern (grey symbols) Amazonia. 1. Jatun Sacha, 2. Bogi, Tiputini, 3. Allpahuayo, 4. Yanamono, 5. Sucusari, 6. Tambopata, 7. Cusco Amazonico, 8. Huanchaca, Las Londras, Chore, Cerro Pelao, Los Fierros, 9. BDFFP, BIONTE, Jacaranda, 10. Tapajos, 11. Jari, 12. Caxiuana.

or contemporary floodplains, from forests growing on older surfaces (Table 11.1). This categorization is somewhat subjective, and the 'floodplain' forests in particular comprise sites growing under a wide range of edaphic conditions. Three of the plots (LSL-01, LSL-02, and TIP-03) are flooded annually and one plot (JAS-05) is likely to have been occasionally flooded in the recent past. However, the other eight plots (all CUS, TAM-01, 02, 04, and 06) have been *terra firme* forests for hundreds or thousands of years, and therefore represent the 'Holocene floodplain'. Fluvial geomorphological features and carbon dating suggest that the youngest of the Holocene floodplain sites, TAM-04, must be at least 900 yr old (Phillips *et al.* 2002a).

The substrates underlying all the other plots are thought to have been deposited prior to the Holocene. Within these forests, we distinguish sites in western and eastern Amazonia to compare regional patterns of AGB change (Table 11.1, Fig. 11.1).

In each plot, all trees greater or equal to 10 cm diameter at 1.3 m (= diameter at breast height, d.b.h.) have been measured during each census, with a consistent effort in all plots for all censuses to measure buttressed trees above the top of the buttress. Increasing steps are being made to standardize all aspects of tree measurements across all sites: the most recent measurements of the 34 western

Amazon plots, from Peru, Bolivia, and Ecuador, have been undertaken by an overlapping group of researchers during 1998–2002, using standard measurement protocols that have been progressively refined (see Phillips *et al.* 2002a). Since 2000, this fieldwork has formed part of the RAINFOR project (Malhi *et al.* 2002a). For buttressed trees, for example, current procedures involve measuring tree diameter 50 cm above the top of the buttress, using ladders if necessary, marking the point of measurement with paint and recording its height.

Even with careful field procedures, some difficulties will always arise in reconciling new plot data with previous measurements and standard procedures were developed to deal with problematic tree records. As a guide, plot data were screened for growth rates that exceed or fall below certain limits (≤ -0.2 cm yr^{-1}, or ≥ 4 cm yr^{-1}, following Sheil 1995b), but final decisions on any alterations to the original data were made on a tree-by-tree basis. Obvious typographical errors, or unusual measurement values in an otherwise steady sequence were corrected by linear interpolation. In some cases, however, the most recent diameter measurement was implausibly less than previous values, occasionally by up to 10 cm. This pattern was probably caused by a lower point of measurement in previous censuses, due to

Table 11.1 Site descriptions, location, and biomass data for 59 forest plots in Amazonia

Name	Code	Country	Region	Lat. (dec)	Long. (dec)	Principal investigator(s)	Institution	Forest type	Plot size (ha)	Census date Initial	Census date Final	Census interval (years)	AGB (Mg ha^{-1}) Start	AGB (Mg ha^{-1}) End	AGB change (Mg ha^{-1} yr^{-1})
Allpahuayo A, clay rich soils[a]	ALP-11	Peru	West	−3.95	−73.43	O. Phillips, R. Vasquez	Leeds, Proy. Flora del Peru	Terra firme	0.44	1990.87	2001.03	10.15	272.35	269.51	−0.28
Allpahuayo A, sandy soils[a]	ALP-12	Peru	West	−3.95	−73.43	O. Phillips, R. Vasquez	Leeds, Proy. Flora del Peru	Terra firme	0.40	1990.87	2001.03	10.15	267.98	266.46	−0.15
Allpahuayo B, sandy soils[a]	ALP-21	Peru	West	−3.95	−73.43	O. Phillips, R. Vasquez	Leeds, Proy. Flora del Peru	Terra firme	0.48	1990.87	2001.04	10.16	285.33	287.61	0.22
Allpahuayo B, clay rich soils[a]	ALP-22	Peru	West	−3.95	−73.43	O. Phillips, R. Vasquez	Leeds, Proy. Flora del Peru	Terra firme	0.44	1990.87	2001.04	10.16	226.71	241.02	1.41
BDFFP, 2303 Faz. Dimona 4,5[b]	BDF-01	Brazil	East	−2.40	−60.00	W. Laurance	Smithsonian	Terra firme	2.00	1985.29	1997.71	12.42	376.48	378.67	0.18
BDFFP, 1101 Gaviao	BDF-03	Brazil	East	−2.40	−59.90	W. Laurance	Smithsonian	Terra firme	1.00	1981.12	1999.29	18.17	330.13	338.90	0.48
BDFFP, 1102 Gaviao	BDF-04	Brazil	East	−2.40	−59.90	W. Laurance	Smithsonian	Terra firme	1.00	1981.12	1999.29	18.17	325.80	250.68	−4.13
BDFFP, 1103 Gaviao	BDF-05	Brazil	East	−2.40	−59.90	W. Laurance	Smithsonian	Terra firme	1.00	1981.21	1999.29	18.08	288.74	304.29	0.86
BDFFP, 1201 Gaviao[b]	BDF-06	Brazil	East	−2.40	−59.90	W. Laurance	Smithsonian	Terra firme	3.00	1981.29	1999.29	18.00	289.83	295.04	0.29
BDFFP, 1109 Gaviao	BDF-08	Brazil	East	−2.40	−59.90	W. Laurance	Smithsonian	Terra firme	1.00	1981.62	1999.46	17.83	329.49	318.90	−0.59
BDFFP, 1301 Florestal	BDF-10	Brazil	East	−2.40	−59.90	W. Laurance	Smithsonian	Terra firme	1.00	1983.46	1997.12	13.67	316.23	326.88	0.78
BDFFP, 1301 Florestal 2[b]	BDF-11	Brazil	East	−2.40	−59.90	W. Laurance	Smithsonian	Terra firme	3.00	1983.46	1997.12	13.67	334.82	354.72	1.46
BDFFP, 1301 Florestal 3[b]	BDF-12	Brazil	East	−2.40	−59.90	W. Laurance	Smithsonian	Terra firme	2.00	1983.46	1997.12	13.67	332.42	348.98	1.21
BDFFP, 3402 Cabo Frio	BDF-13	Brazil	East	−2.40	−60.00	W. Laurance	Smithsonian	Terra firme	9.00	1985.86	1998.87	13.02	321.97	342.19	1.55
BDFFP, 3304 Porto Alegre[b]	BDF-14	Brazil	East	−2.40	−60.00	W. Laurance	Smithsonian	Terra firme	2.00	1984.21	1998.37	14.17	368.40	356.11	−0.87
Bionte 1	BNT-01	Brazil	East	−2.63	−60.17	N. Higuchi	INPA	Terra firme	1.00	1986.50	1999.20	12.70	332.21	370.45	3.01
Bionte 2	BNT-02	Brazil	East	−2.63	−60.17	N. Higuchi	INPA	Terra firme	1.00	1986.50	1999.20	12.70	350.03	389.57	3.11
Bionte 4	BNT-04	Brazil	East	−2.63	−60.17	N. Higuchi	INPA	Terra firme	1.00	1986.50	1999.20	12.70	318.98	331.91	1.02
Bionte T4 B2 SB1	BNT-05	Brazil	East	−2.63	−60.17	N. Higuchi	INPA	Terra firme	1.00	1986.50	1993.50	7.00	306.99	324.03	2.43
Bionte T4 B1 SB3	BNT-06	Brazil	East	−2.63	−60.17	N. Higuchi	INPA	Terra firme	1.00	1986.50	1993.50	7.00	376.03	363.59	−1.78

Plot	Code	Country	E/W	Lat	Long	Author(s)	Institution	Forest type							
Bionte T4 B1 SB4	BNT-07	Brazil	East	−2.63	−60.17	N. Higuchi	INPA	Terra firme	1.00	1986.50	1993.50	7.00	349.44	358.61	1.31
Bogi 1 (PA)	BOG-01	Ecuador	West	−0.70	−76.48	N. Pitman, A. DiFiore	Duke University, NYU	Terra firme	1.00	1996.29	2002.13	5.83	262.71	289.42	4.58
Bogi 2 (PB)	BOG-02	Ecuador	West	−0.70	−76.47	N. Pitman, A. DiFiore	Duke University, NYU	Terra firme	1.00	1996.29	2002.13	5.83	211.43	221.98	1.81
Caxiuana 1	CAX-01	Brazil	East	−1.70	−51.53	S. Almeida	Museu Goeldi	Terra firme	1.00	1994.50	2002.88	8.38	369.60	378.73	1.09
Caxiuana 2	CAX-02	Brazil	East	−1.70	−51.53	S. Almeida	Museu Goeldi	Terra firme	1.00	1995.50	2003.21	7.71	367.48	364.62	−0.37
Chore 1	CHO-01	Bolivia	West	−14.35	−61.16	T. Killeen	Museo Noel Kempff	Terra firme	1.00	1996.50	2001.44	4.94	117.52	124.82	1.49
Cerro Pelao 1	CRP-01	Bolivia	West	−14.54	−61.48	T. Killeen	Museo Noel Kempff	Terra firme	1.00	1994.21	2001.45	7.25	212.54	213.66	0.15
Cerro Pelao 2	CRP-02	Bolivia	West	−14.53	−61.48	T. Killeen	Museo Noel Kempff	Terra firme	1.00	1994.27	2001.46	7.19	220.84	233.83	1.81
Cuzco Amazonico, CUZAM1E	CUZ-01	Peru	West	−12.50	−68.95	O. Phillips, R. Vasquez	Leeds, Proy. Flora del Peru	Terra firme, floodplain	1.00	1989.39	1998.77	9.38	252.19	283.34	3.32
Cuzco Amazonico, CUZAM1U	CUZ-02	Peru	West	−12.50	−68.95	O. Phillips, R. Vasquez	Leeds, Proy. Flora del Peru	Terra firme, floodplain	1.00	1989.42	1998.77	9.35	216.95	248.66	3.39
Cuzco Amazonico, CUZAM2E	CUZ-03	Peru	West	−12.49	−69.11	O. Phillips, R. Vasquez	Leeds, Proy. Flora del Peru	Terra firme, floodplain	1.00	1989.40	1998.77	9.37	217.50	250.26	3.50
Cuzco Amazonico, CUZAM2U	CUZ-04	Peru	West	−12.49	−69.11	O. Phillips, R. Vasquez	Leeds, Proy. Flora del Peru	Terra firme, floodplain	1.00	1989.44	1998.78	9.34	269.52	289.19	2.11
Huanchaca Dos, plot 1	HCC-21	Bolivia	West	−14.56	−60.75	L. Arroyo	Museo Noel Kempff	Terra firme	1.00	1996.52	2001.43	4.91	245.28	249.19	0.80
Huanchaca Dos, plot 2	HCC-22	Bolivia	West	−14.57	−60.74	L. Arroyo	Museo Noel Kempff	Terra firme	1.00	1996.54	2001.43	4.89	263.77	270.88	1.45
Jacaranda, plots 1–5	JAC-01	Brazil	East	−2.63	−60.17	N. Higuchi	INPA	Terra firme	5.00	1996.50	2002.50	6.00	319.46	315.88	−0.60
Jacaranda, plots 6–10	JAC-02	Brazil	East	−2.63	−60.17	N. Higuchi	INPA	Terra firme	5.00	1996.50	2002.50	6.00	315.41	311.52	−0.65
Jatun Sacha 2	JAS-02	Ecuador	West	−1.07	−77.60	D. Neill	Herbario Nacional	Terra firme	1.00	1987.63	2002.04	14.42	248.81	247.96	−0.06
Jatun Sacha 3	JAS-03	Ecuador	West	−1.07	−77.67	D. Neill	Herbario Nacional	Terra firme	1.00	1988.88	2002.04	13.17	231.88	262.78	2.35
Jatun Sacha 4	JAS-04	Ecuador	West	−1.07	−77.67	D. Neill	Herbario Nacional	Terra firme	0.92	1994.50	2002.04	7.54	282.69	318.58	4.79
Jatun Sacha 5	JAS-05	Ecuador	West	−1.07	−77.67	D. Neill	Herbario Nacional	Terra firme, floodplain	1.00	1989.38	2002.04	12.67	268.10	286.83	1.48
Jari 1c	JRI-01	Brazil	East	−1.00	−52.05	N. Silva	CIFOR, EMBRAPA	Terra firme	1.00	1985.50	1996.00	10.50	392.50	387.09	−0.51
Los Fierros Bosque I	LFB-01	Bolivia	West	−14.61	−60.87	T. Killeen	Museo Noel Kempff	Terra firme	1.00	1993.62	2001.40	7.78	221.59	239.95	2.36
Los Fierros Bosque II	LFB-02	Bolivia	West	−14.60	−60.85	T. Killeen	Museo Noel Kempff	Terra firme	1.00	1993.65	2001.40	7.76	271.17	284.99	1.78
Las Londras, plot 1	LSL-01	Bolivia	West	−14.40	−61.13	L. Arroyo	Museo Noel Kempff	Seasonally flooded	1.00	1996.53	2001.48	4.95	164.82	173.32	1.72
Las Londras, plot 2	LSL-02	Bolivia	West	−14.40	−61.13	L. Arroyo	Museo Noel Kempff	Seasonally flooded	1.00	1996.53	2001.48	4.95	176.84	203.55	5.39
Sucusari A	SUC-01	Peru	West	−3.23	−72.90	O. Phillips, R. Vasquez	Leeds, Proy. Flora del Peru	Terra firme	1.00	1992.13	2001.06	8.93	285.61	278.52	−0.79
Sucusari B	SUC-02	Peru	West	−3.23	−72.90	O. Phillips, R. Vasquez	Leeds, Proy. Flora del Peru	Terra firme	1.00	1992.13	2001.07	8.93	298.08	287.49	−1.18
Tambopata plot zero	TAM-01	Peru	West	−12.85	−69.28	O.Phillips, R. Vasquez	Leeds, Proy. Flora del Peru	Terra firme, floodplain	1.00	1983.78	2000.59	16.81	250.49	260.01	0.57

Table 11.1 (Continued)

Name	Code	Country	Region	Lat. (dec)	Long. (dec)	Principal investigator(s)	Institution	Forest type	Plot size (ha)	Census date		Census interval (years)	AGB (Mg ha⁻¹)		AGB change (Mg ha⁻¹ yr⁻¹)
										Initial	Final		Start	End	
Tambopata plot 1	TAM-02	Peru	West	−12.83	−69.28	O.Phillips, R. Vasquez	Leeds, Proy. Flora del Peru	Terra firme, floodplain	1.00	1979.87	2000.58	20.71	241.64	260.07	0.89
Tambopata plot 2 clay	TAM-04	Peru	West	−12.83	−69.28	O.Phillips, R. Vasquez	Leeds, Proy. Flora del Peru	Terra firme, floodplain	0.42	1983.79	1998.75	14.96	268.33	288.62	1.36
Tambopata plot 3	TAM-05	Peru	West	−12.83	−69.28	O.Phillips, R. Vasquez	Leeds, Proy. Flora del Peru	Terra firme	1.00	1983.70	2000.56	16.86	243.37	266.21	1.35
Tambopata plot 4	TAM-06	Peru	West	−12.83	−69.30	O.Phillips, R. Vasquez	Leeds, Proy. Flora del Peru	Terra firme, floodplain	0.96	1983.71	2000.55	16.84	233.51	281.95	2.88
Tambopata plot 6	TAM-07	Peru	West	−12.83	−69.27	O.Phillips, R. Vasquez	Leeds, Proy. Flora del Peru	Terra firme	1.00	1983.76	1998.73	14.97	250.82	257.26	0.43
Tapajos, RP014, 1–4[d]	TAP-01	Brazil	East	−3.31	−54.94	N. Silva	CIFOR, EMBRAPA	Terra firme	1.00	1983.50	1995.50	12.00	262.17	296.14	2.83
Tapajos, RP014, 5–8[d]	TAP-02	Brazil	East	−3.31	−54.94	N. Silva	CIFOR, EMBRAPA	Terra firme	1.00	1983.50	1995.50	12.00	332.04	373.82	3.48
Tapajos, RP014, 9–12[d]	TAP-03	Brazil	East	−3.31	−54.94	N. Silva	CIFOR, EMBRAPA	Terra firme	1.00	1983.5	1995.50	12.00	346.21	377.28	2.59
Tiputini 2	TIP-02	Ecuador	West	−0.63	−76.14	N Pitman	Duke University	Terra firme	0.80	1997.71	2002.13	4.42	257.12	260.84	0.84
Tiputini 3	TIP-03	Ecuador	West	−0.64	−76.15	N. Pitman	Duke University	Seasonally flooded	1.00	1998.13	2002.13	4.00	250.07	255.15	1.27
Yanamono A	YAN-01	Peru	West	−3.43	−72.85	O.Phillips, R. Vasquez	Leeds, Proy. Flora del Peru	Terra firme	1.00	1983.46	2001.05	17.59	290.11	299.20	0.52

[a] Allpahuayo A and B contain two distinctive soil types that are treated separately in these analyses.

[b] These sites comprise non-contiguous 1-ha plots separated by <200 m.

[c] Twenty-five, 10 × 10 m subplots, within each of four, nearby, 1-ha plots.

[d] Twelve, 0.25-ha plots laid out in a randomized fashion over an area of 300 m × 1200 m; treated as 3 × 1 ha units.

Notes: AGB values were calculated using equation 11.2 (Baker et al. 2004a). Plot data are the best available to the lead author at the time of final analyses, but are subject to future revision as a result of additional censuses and continued error-checking.

uncertainty locating the top of the buttress. For these trees, prior growth was estimated using the median growth rate of the appropriate (10–20 cm, 20–40 cm, or >40 cm) size class.

These tree records where the diameter data have been altered following fieldwork clearly introduce uncertainty into estimates of AGB change. Another source of uncertainty is the use of optical methods (digital camera or Relaskop) to measure the diameter of some trees, in some plots. Optical methods tend to underestimate tree diameter, and although we have included a theoretical correction factor (see Phillips and Baker 2002) to account for inevitable parallax effects, these methods are less precise than using a tape measure as they cannot integrate irregularities in bole shape. Therefore, using the 34 western Amazon plots, we evaluate the impact of these trees on conclusions concerning the direction or magnitude of AGB change. This was achieved by comparing AGB change using the whole dataset with values when these records are excluded. Removing records makes the conservative assumption that no excluded stem grew during the census interval.

Stand AGB (kg DW ha^{-1}) for all trees $\geq$10 cm d.b.h., including palms, was calculated using a variety of allometric equations:

$$AGB = \sum_{1}^{n} \exp \left[0.33(\ln D_i) + 0.933(\ln D_i)^2\right.$$
$$\left. - 0.122(\ln D_i)^3 - 0.37\right] \quad (11.1)$$

$$AGB = \sum_{1}^{n} \frac{\rho_i}{0.67} \left\{\exp \left[0.33(\ln D_i) + 0.933(\ln D_i)^2\right.\right.$$
$$\left.\left. - 0.122(\ln D_i)^3 - 0.37\right]\right\} \quad (11.2)$$

$$AGB = \sum_{1}^{n} \exp \left[2.42(\ln D_i) - 2.00\right] \quad (11.3)$$

$$AGB = \sum_{1}^{n} \frac{\rho_i}{0.58} \left\{\exp \left[2.42(\ln D_i) - 2.00\right]\right\} (11.4)$$

$$AGB = 600 \times \left[66.92 + (16.85 \times BA)\right], \quad (11.5)$$

where D_i (cm) and ρ_i (g cm^{-3}) are, respectively, the diameter and wood density of tree i, n is the number of stems per plot, and BA (m^2 ha^{-1}), is plot basal area, calculated as:

$$BA = \sum_{1}^{n} \pi(D_i/200)^2.$$

The different AGB equations reflect different underlying datasets of tree mass data, the inclusion of exclusion of variation in wood specific gravity, and tree-by-tree and stand-level approaches to calculating biomass. Equation (11.1) was obtained from data for 315 trees, harvested in five, 0.04 ha (20 m $\times$ 20 m) plots, as part of the BIONTE project, near Manaus, Brazil (Chambers et al. 2001b). Equation (11.2) is a modified version, incorporating a simple multiplication factor to account for variation in wood specific gravity between species (Baker et al. 2004a). Equation (11.3) was derived from an independent set of tree diameter and mass data of 378 trees (Chave et al. 2001), and equation (11.4) is the same relationship, but including wood specific gravity (Baker et al. 2004a). In contrast, equation (11.5) is based on the same tree harvest data as equations (11.1) and (11.2), but calculates AGB on a stand, rather than tree-by-tree basis, using the relationship between basal area and fresh, AGB of trees >5 cm diameter for the five, 0.04-ha subplots (Phillips et al. 1998b).

We focus on testing whether there have been concerted, within-site changes in AGB since plot establishment, by calculating AGB change between the first and last census for each plot. Errors are expressed as 95% confidence limits of the mean. Issues relating to the distribution of AGB change, the statistical independence of the plots, variation in sampling error, and stem breakage are discussed in Baker et al. (2004b). Here, units of dry mass are used for AGB and AGB change. However, AGB values can be also expressed in terms of carbon by assuming a carbon content of 50%, so carbon change metrics can be calculated simply by dividing the reported values by 2.

Results

The AGB estimates using stand-level and tree-by-tree approaches based on the Chambers et al. (2001b) tree mass data, give very similar estimates across the plot network (318.3 $\pm$ 11.7 and 325.5 $\pm$ 10.2 Mg ha^{-1}, respectively, Fig. 11.3, equations 11.1 and 11.5). However, when variation in wood specific gravity is incorporated into the same tree-by-tree equation, the among-plot AGB estimate drops slightly (Fig. 11.3, equation 11.2). This pattern is due to the lower specific gravity values of western Amazon forests compared to the central Amazon site where the original biomass equation was developed. AGB estimates derived using equations based on the compilation of tree mass data by Chave *et al.* (2001) are substantially lower (225.3 ± 10.3 and 239.0 ± 12.6 Mg ha^{-1}, respectively, Fig. 11.3, equations 11.3 and 11.4). However, despite these significant differences between AGB estimates, estimates of *change* derived using different allometric equations are remarkably similar (Fig. 11.3). For all the subsequent results we use equation 11.2 to estimate AGB, as this equation was developed solely using Amazonian tree mass data and adjusts for the regional scale variation in stand-level wood specific gravity.

Across all plots, AGB change is normally distributed (Kolmogorov–Smirnov test, $D = 0.08$ ns). AGB has increased since plot establishment by 1.22 ± 0.43 Mg ha^{-1} yr^{-1}, or $0.50 \pm 0.17\%$ yr^{-1} (unweighted average, Fig. 11.2(a)) or 0.98 ± 0.38 Mg ha^{-1} yr^{-1} (weighted by hectare years of monitoring). The lower value using the weighted average largely reflects the fact that the three plots with the highest rates of AGB change (BOG-01, JAS-04, and LSL-01) have been monitored for comparatively short periods (4.9–6.9 yr).

The AGB change is significantly positive in both non-floodplain and floodplain sites, and floodplain plots have higher rates of increase than non-floodplain sites (2.32 ± 0.79 Mg ha^{-1} yr^{-1} unweighted, and 2.08 ± 0.74 Mg ha^{-1} yr^{-1} weighted ($n = 12$, Fig. 11.2d), compared to 0.93 ± 0.46 Mg ha^{-1} yr^{-1} unweighted, and 0.80 ± 0.42 Mg ha^{-1} yr^{-1} weighted, $n = 47$). The patterns of AGB change are also broadly spatially consistent. Increases have occurred in non-floodplain forests in both eastern (Fig. 11.2(b)) and western (Fig. 11.2(c)) Amazonia, although the rate of change is only marginally significant when the central and eastern Amazonia plots are considered alone (central and eastern Amazon, $n = 25$, 0.73 ± 0.68 Mg ha^{-1} yr^{-1} unweighted, 0.70 ± 0.58 Mg ha^{-1} yr^{-1} weighted; western Amazonia $n = 22$, 1.17 ± 0.62 Mg ha^{-1} yr^{-1} unweighted, and 1.08 ± 0.59 Mg ha^{-1} yr^{-1} weighted). The tendency

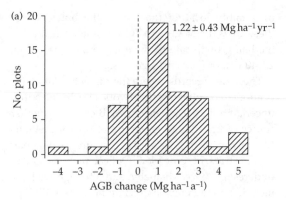

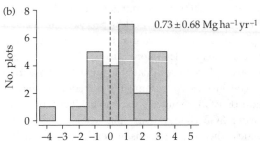

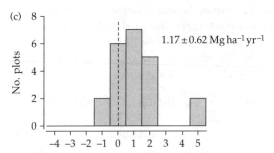

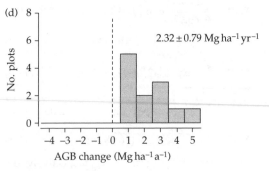

Figure 11.2 Frequency distribution of rates of change in AGB for trees $\geq$10 cm d.b.h for (a) all 59 plots, (b) pre-Holocene central and eastern Amazon forests, (c) pre-Holocene western Amazon forests, (d) Holocene floodplain and contemporary floodplain plots. AGB change calculated using equation 11.2.

for higher absolute rates of AGB change in western Amazon forests is not significant (t-test, $p = 0.36$). Due to the lower overall AGB in western Amazon forests, regional differences in the relative rates of change are greater than the differences in absolute rates (based on unweighted estimates, central and eastern Amazon, $0.23 \pm 0.21\%\,\mathrm{yr}^{-1}$; western Amazon, $0.51 \pm 0.25\%\,\mathrm{yr}^{-1}$), but the regional difference is again not significant (t-test, $p = 0.10$).

Overall, basal area change represents a very good measure of AGB change within Amazonian forest plots (Fig. 11.4; $\Delta \mathrm{AGB} = 9.57(\Delta \mathrm{BA}) + 0.12$, $r^2 = 0.89$, $p < 0.001$). In a multiple regression analysis, change in stand-level wood specific gravity was included as an additional term, but was not individually significant and did not lead to any improvement in predictions of AGB change.

Excluding records of trees measured using optical methods and individuals where growth rates have been estimated following fieldwork does not alter the significance of the direction of AGB change. Of the total western Amazonian dataset of 24,229 trees, 322 trees have been measured with a Relaskop or digital camera and diameter measurements for a partially overlapping set of 492 trees were interpolated or otherwise re-estimated following fieldwork. The total number of trees in at least one of these categories is 609 (2.5% of all stems). If we apply the conservative assumption that all 609 individuals have zero growth over the measurement period, then the AGB change estimate declines by approximately 30% in both floodplain and non-floodplain sites (floodplain: from $2.32 \pm 0.79\,\mathrm{Mg\,ha}^{-1}\,\mathrm{yr}^{-1}$ to $1.70 \pm 0.83\,\mathrm{Mg\,ha}^{-1}\,\mathrm{yr}^{-1}$ (unweighted) and $2.08 \pm 0.74\,\mathrm{Mg\,ha}^{-1}\,\mathrm{yr}^{-1}$ to $1.46 \pm 0.75\,\mathrm{Mg\,ha}^{-1}\,\mathrm{yr}^{-1}$ (weighted); non-floodplain: $1.17 \pm 0.62\,\mathrm{Mg\,ha}^{-1}\,\mathrm{yr}^{-1}$ to $0.79 \pm 0.61\,\mathrm{Mg\,ha}^{-1}\,\mathrm{a}^{-1}$ (unweighted) and $1.08 \pm 0.59\,\mathrm{Mg\,ha}^{-1}\,\mathrm{yr}^{-1}$ to $0.68 \pm 0.59\,\mathrm{Mg\,ha}^{-1}\,\mathrm{yr}^{-1}$ (weighted)). These trees have a disproportionate impact on the stand-level estimates because the most difficult to measure trees tend to be the largest individuals. However, although these factors introduce uncertainty in the magnitude of change, for the western Amazon plots, AGB change remains significantly positive, even when these trees are excluded.

Discussion

The re-analysis of Amazonian forest plot data presented here supports the original findings of Phillips *et al.* (1998b). This study demonstrates that since plot establishment, the AGB of trees ≥ 10 cm diameter has increased by $1.21 \pm 0.43\,\mathrm{Mg\,ha}^{-1}\,\mathrm{yr}^{-1}$ (unweighted), or $0.98 \pm 0.38\,\mathrm{Mg\,ha}^{-1}\,\mathrm{yr}^{-1}$ (weighted by monitoring effort). These values are higher than the comparable unweighted result for stems ≥ 10 cm diameter obtained in the original study ($0.88 \pm 0.52\,\mathrm{Mg\,ha}^{-1}\,\mathrm{yr}^{-1}$). Here we have also shown that this pattern is neither confounded by spatial or temporal variation in wood specific gravity, nor is it dependent on the allometric equation used to estimate AGB. Moreover, the AGB of the western Amazon forests has increased even when the most difficult-to-measure trees are discounted.

It is noteworthy that the stand-level approach used by Phillips *et al.* (1998b) to estimate biomass from inventory data is comparable with tree-by-tree methods, as the stand- and tree-level equations based on the same underlying tree mass data (equations 11.1 and 11.5, Fig. 11.3) give extremely similar results. This similarity is because the basal area of individual trees is roughly linearly related to tree biomass up to relatively large sizes (80–90 cm bole diameter). Even though at the very largest sizes, tree basal area overestimates tree biomass (Chave *et al.* 2004), at the stand-level the linear correlation between basal area and AGB holds (Baker *et al.* 2004a). Therefore, within our plots, AGB can be estimated directly from stand-level basal area regardless of how that basal area is distributed between stems of different sizes.

A limitation of the method of AGB estimation used by Phillips *et al.* (1998b) is that it did not account for wood specific gravity, which varies both between forests and, potentially, over time. For example, reported increases in the rate of forest dynamics (Phillips and Gentry 1994; Lewis *et al.* 2004; Phillips *et al.* 2004) might be expected to favour faster growing species with lower wood specific gravity values. Overall, AGB estimates are slightly lower when specific gravity is included

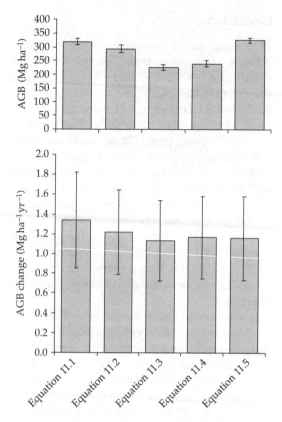

Figure 11.3 Estimates of (a) mean AGB, and (b) mean AGB change based on five different allometric equations for calculating biomass with inventory data. Error bars are 95% confidence limits of the mean values.

(equation 11.2 compared to equation 11.1, Fig. 11.3). This is because the underlying tree mass/diameter relationships were developed in central Amazon forests, where stands have relatively high wood specific gravity values compared to most plots in western Amazonia (Baker *et al.* 2004a). However, AGB change estimates are only weakly affected by the equation used (Fig. 11.3). Spatial variation in wood specific gravity therefore does not confound previously reported increases in AGB. The close correlation between basal area and AGB change (Fig. 11.4), shows that the changes in AGB have been caused by an overall structural change in these plots, and suggests that any compositional shifts between tree species with differing wood specific gravity, have not significantly affected stand-level AGB estimates.

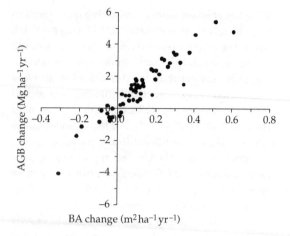

BA change ($m^2 ha^{-1} yr^{-1}$)

Figure 11.4 The relationship between BA change and AGB change.

Uncertainty about some diameter measurements does not influence the significance of the direction of AGB change in western Amazonia, as AGB change remained significantly positive even when problematic tree records were deleted. Removing records makes the assumption that these trees did not grow and therefore introduces a downward bias to stand-level growth estimates. This issue is compounded by the fact that these individuals tend to comprise larger, often buttressed, trees that make a greater relative contribution to overall stand-level productivity as a result of their size. Estimating the growth of these trees is a more satisfactory option for obtaining unbiased estimates of AGB change. We suggest that as diameter increment distributions are strongly skewed, median growth rates within an appropriate size-class will provide the best estimate. Although species-level estimates may be possible for some common species, stand-level values will always be required for rarer species and may be the clearest and most robust method to apply to all stems.

Overall, AGB change ranges from -4.14 to 5.40 $Mg ha^{-1} yr^{-1}$, with a mean value of 1.22 ± 0.43 $Mg ha^{-1} yr^{-1}$ (Fig. 11.2). Some of the variability between plots is doubtless caused by variability in the natural disturbance regime. For example, the greatest decrease in AGB occurs in BDF-04 where 145.4 Mg were lost between censuses in 1987 and

1991, due to mortality caused by flooding. Equally, some of the plots with high rates of AGB increase may be recovering from mortality events prior to plot establishment. None of the plots are, however, obviously strongly successional. The Bolivian plots with low AGB values, for example, are located in forest types that are typically less massive than other Amazon forests. The plot with the lowest AGB, CHO-01, comprises evergreen liana forest, which is a forest type that is found across substantial areas of the southern fringe of Amazonia, possibly as a result of fire or an interaction between poor soils and seasonal drought (Killeen 1998).

Determining why most plots show moderate increases in AGB is difficult when changes in AGB are considered alone, as we have done here, without examining simultaneous changes in growth and mortality rates. In particular, it is not possible to distinguish with certainty whether increases are driven by widespread recovery from a previous disturbance, or by an overall increase in forest productivity. In Amazonia, mega El Niño events (Meggers 1994) provide one mechanism that potentially could drive a broad-scale increase in AGB due to succession, as it is well known that El Niño events cause increased tree mortality (e.g. Condit *et al.* 1995). However the increase in AGB reported here has occurred despite two of the most severe El Niños on record occurring during the monitoring period (Malhi and Wright 2004), suggesting that El Niño events may not necessarily dominate tropical forest dynamics over decadal timescales (Williamson *et al.* 2000). In addition, it is difficult to reconcile the spatial variability of El Niño intensity across Amazonia (Malhi and Wright 2004) with the spatial consistency in the patterns of AGB change.

While successional processes may not explain the overall trend, could they explain the significantly higher rates of increase in AGB of the floodplain plots? Succession obviously dominates patterns of biomass accumulation on young Amazonian floodplains where forest establishes and develops on aggrading river sediments (e.g. Salo *et al.* 1986). Whether it continues to influence patterns of biomass change in the plots studied here depends on the age of the stands, and the time taken by successional forest to reach biomass

values equivalent to old-growth forest. Although both factors are difficult to quantify, current understanding suggests that the age of the forests are far greater than the persistence of successional effects on biomass accumulation. For example, although data are sparse, studies of forest recovery following complete human clearance for agriculture suggest that biomass approaches old-growth values after 100 yr (Guariguata and Ostertag 2001). In contrast, geomorphological features and carbon dating suggest that the Holocene floodplain sites in southern Peru are at least 900 yr old (Phillips *et al.* 2002a). The AGB and stand-level wood specific gravity of the floodplain plots also suggest that they are structurally no different from plots on older land surfaces (floodplain versus non-floodplain sites for western Amazon plots: AGB 256.7 ± 20.4 versus. 257.8 ± 16.3 Mg ha^{-1}, wood specific gravity (stems basis) 0.61 ± 0.03 versus. 0.62 ± 0.02 g cm^{-3}). Given these patterns, it is difficult to attribute a significant role for primary succession in the dynamics of these forests.

An alternative explanation for the observed increase in AGB is that stand-level growth rates have increased. Compelling evidence for an increase in Amazonian forest productivity has emerged from combined analyses of stem and basal area dynamics of an overlapping set of plots. These indicate (Lewis *et al.* 2004b, Chapter 12) that increases in stem recruitment, stem mortality, and total stem density, and stand-level growth rates, basal area mortality, and total basal area, have on average all occurred, with stem recruitment gains generally leading stem mortality gains (Phillips *et al.* 2004, Chapter 10). It is argued that these patterns are incompatible with forest succession, but are most plausibly driven by an enhancement of stand-level growth rates (Lewis *et al.* 2004b, Chapter 4). In this context, the higher rates of AGB change in floodplain forests may be associated with the potential for greater increases in growth on the more fertile soils that are typically found in these sites. However, such an explanation, among such a heterogeneous group of post-Pleistocene substrates, remains tentative.

An important question for the overall carbon balance of these plots is whether the increase in the biomass of the trees ≥ 10 cm dbh, might be offset by changes in the biomass of other compartments (e.g. small trees, lianas, coarse woody debris, fine litter, or soil carbon). Trees <10 cm d.b.h. and lianas only comprise approximately 5% and 2%, respectively of total above-ground biomass (Nascimento and Laurance 2002), but small trees can have a significant influence on calculations of stand-level patterns of biomass change (Chave *et al.* 2003). However, the increasing recruitment rates in Amazonian forest plots (Phillips *et al.* 2004) and the increases in the abundance of large lianas (Phillips *et al.* 2002b), suggest that the biomass of small trees and lianas are also increasing, and any changes in these compartments are therefore unlikely to counteract the increase in the biomass of trees ≥ 10 cm d.b.h.

Changes in the stocks of coarse woody debris (CWD), fine litter, and soil carbon are broadly controlled by inputs from living above-ground biomass. Therefore, the question of whether changes in their biomass can alter the overall trend in forest carbon balance determined from trees ≥ 10 cm d.b.h., depends on their rate of turnover and the timescale of the study. In short term studies (e.g. 2 yr, Rice *et al.* 2004), pools and fluxes of coarse woody debris may be partially independent of simultaneous changes in the biomass of larger trees, and be substantially controlled by mortality events prior to the measurement period. However, over longer timescales, stocks of CWD must be closer to equilibrium with inputs from mortality. The turnover rate of CWD is approximately 7–10 yr (Chambers *et al.* 2000), similar to the median length of plot monitoring (10.2 yr) in this study. Although a fraction of the CWD will derive from mortality prior to monitoring, this component will have much less impact on the ecosystem carbon balance than in short-term studies. In fact, increasing rates of mortality (Lewis *et al.* 2004b; Phillips *et al.* 2004, Chapter 4, Chapter 10), suggest that stocks of CWD would have increased in our plots.

Fine litter has a short turnover time (Clark *et al.* 2002), so fine litter carbon should follow decadal trends in AGB and productivity. In contrast, soil carbon is very heterogeneous, and deep-soil

carbon turns over at timescales substantially longer than the scale of this study (Trumbore 2000). Changes in this pool may still therefore be responding to events that occurred prior to the establishment of these plots. However, any changes in soil C stocks over the timescale of this study are likely to be quite small. Telles *et al.* (2003) reported no measurable change in organic C stocks over 20 yr in oxisols near Manaus, and, using a model of soil C dynamics, found a limited potential for these soils to act as carbon sinks following any increase in ecosystem productivity.

Overall, longer-term monitoring of these plots and specific studies of other components of the total biomass of these forests are required to examine changes in total biomass. However, since trees ≥ 10 cm diameter represent such a large fraction of total AGB, the plots in this study have been monitored for a relatively long period, and there has been a concurrent acceleration in forest dynamics, we suggest that changes in carbon of other compartments are unlikely to counteract the increase in trees ≥ 10 cm diameter. The likelihood is that the increase in total carbon storage has been greater than the increase in carbon stored in trees ≥ 10 cm diameter.

If the carbon pool stored in these Amazonian forest plots has increased, can its rate of increase be extrapolated to a regional scale? A key issue is whether biomass loss from relatively rare, but high intensity, disturbance events, that occur beyond the scale of current sampling, may offset any increase in the biomass of other regions. Major disturbance events occur in tropical forest as a result of, for example, fire, windstorms and landslides (Whitmore and Burslem 1998), but obtaining data on their frequency, distribution, and magnitude to quantify their importance for regional-scale patterns of carbon cycling is extremely difficult. In the context of the Amazon, analysis of satellite images has, however, provided some quantitative data on the frequency of destructive blowdown events due to storms (Nelson *et al.* 1994). Where such events are most concentrated, return times are estimated at 5000 yr (Nelson *et al.* 1994). This type of disturbance would have to be much more frequent to substantially alter the observed mean increase in carbon storage. For instance, if the current mean rate of AGB change persists for a total monitoring effort of 5000 ha yr, this would equate to a total accumulation of 6100 Mg. If all the biomass were then destroyed by a severe storm in one, 1 ha plot, with a mean AGB of 300 $Mg\,ha^{-1}$, then the total biomass accumulation would decline by 5% and the estimated mean rate of increase in AGB would fall from 1.22 to 1.16 $Mg\,ha^{-1}\,yr^{-1}$. This analysis is clearly a crude simplification, and quantifying the frequency and intensity of the full range of disturbance events that occur across Amazonia will be necessary to accurately extrapolate these plot-based trends to a regional scale. However, it does show that the rarity of blow-down events in Amazonian forests means that their effects on regional-scale carbon cycling will be small (Nelson et al. 1994).

This study demonstrates a significant increase in the carbon content of forest plots across Amazonia, and an important challenge is to integrate these trends in old-growth forest into regional scale models of carbon flux. Equally, future trends in the carbon storage of these plots remains uncertain, and careful monitoring of Amazonian forest plots remains a high priority, particularly in the face of predicted regional drying which may enhance tree mortality and reduce growth rates. In addition, biotic feedbacks may ultimately limit biomass accumulation (Phillips et al. 2002b), and given the recent acceleration in forest dynamics (Lewis et al. 2004b; Phillips et al. 2004), potential changes in tree composition may also have important implications for carbon cycling and biodiversity within these forests.

Acknowledgements

Development of the RAINFOR network, 2000–2, was funded by the European Union Fifth Framework Programme, as part of CARBONSINK, part of the European contribution to the Large Scale Biosphere–Atmosphere Experiment in Amazonia (LBA). TRB acknowledges financial support from the Max Planck Institut für Biogeochemie. RAINFOR field campaigns were funded by the National Geographic Society, CARBONSINK-LBA and the Max Planck Institut für Biogeochemie.

CHAPTER 12

Late twentieth-century trends in the structure and dynamics of South American forests

Simon L. Lewis, Oliver L. Phillips, Timothy R. Baker, Jon Lloyd,
Yadvinder Malhi, Samuel Almeida, Niro Higuchi,
William F. Laurance, David A. Neill, J. Natalino M. Silva,
John Terborgh, Armando Torres Lezama,
Rodolfo Vásquez Martínez, Sandra Brown, Jerome Chave,
Caroline Kuebler, Percy Núñez Vargas, and Barbara Vinceti

Several widespread changes in the ecology of old-growth tropical forests have recently been documented for the late twentieth century, in particular an increase in stem turnover (pantropical), and an increase in above-ground biomass (neotropical). Whether these changes are synchronous and whether changes in growth are also occurring is not known. We analysed stand-level changes *within* 50 long-term monitoring plots from across South America spanning 1971–2002. We show that: (1) basal area (BA: sum of the cross-sectional areas of all trees in a plot) increased significantly over time (by 0.10 ± 0.04 m^2 ha^{-1} yr^{-1}, mean $\pm 95\%$ CI); as did both (2) stand-level BA growth rates (sum of the increments of BA of surviving trees and BA of new trees that recruited into a plot); and (iii) stand-level BA mortality rates (sum of the cross-sectional areas of all trees that died in a plot). Similar patterns were observed on a per-stem basis: (1) stem density (number of stems per hectare; 1 ha is 10^4 m^2) increased significantly over time (0.94 ± 0.63 stems ha^{-1} yr^{-1}); as did both (2) stem recruitment rates; and (3) stem mortality rates. The gain terms (BA growth, stem recruitment) consistently exceeded the loss terms (BA loss, stem mortality) throughout the period, suggesting that whatever process is driving these changes was already acting before the plot network was established. Large long-term increases in stand-level BA growth and simultaneous increases in stand-level BA and stem density imply a continent-wide increase in resource availability which is increasing net primary productivity and altering forest dynamics. Continent-wide changes in incoming solar radiation, and increases in atmospheric concentrations of CO_2 and air temperatures may have increased resource supply over recent decades, thus causing accelerated growth and increased dynamism across the world's largest tract of tropical forest.

Introduction

Significant areas of tropical forest have been conventionally considered to represent a 'natural', 'pristine', or 'undisturbed' state. However, these areas may be impacted by human-induced changes to major biogeochemical cycles such as the global carbon, water, and nitrogen cycles, or other 'unseen' impacts such as the impacts of habitat fragmentation or increased hunting pressure (Lewis *et al.* 2004a). Tropical forests store and process large quantities of carbon and harbour more than 50% of the world's species (Malhi and Grace 2000;

Groombridge and Jenkins 2003), so if consistent biome-wide changes are occurring, these could have profound impacts on the global carbon cycle, the rate of climate change, and biodiversity.

Several studies have reported results from long-term monitoring plots in tropical forests that suggest large-scale ecological changes:

(1) A pantropical increase in stem turnover rates since the 1950s (Phillips and Gentry 1994; Phillips 1996).
(2) A neotropical increase in above-ground woody biomass since the 1980s (Phillips *et al.* 1998b, 2002a; Baker *et al.* 2004b).
(3) An Amazon-wide increase in stem recruitment since the 1970s (Phillips *et al.* 2004).
(4) An Amazon-wide increase in stem mortality since the 1970s (Phillips *et al.* 2004).
(5) A western Amazon increase in large liana biomass and density since the 1980s (Phillips *et al.* 2002b).

These results have generated an evolving debate about whether tropical forests are showing widespread directional changes caused by one or more widespread changes to the environment, or whether the observed patterns can be explained by methodological problems, mathematical artefacts or statistical errors regarding the data and analyses (Phillips 1995; Sheil 1995a; Condit 1997; Phillips and Sheil 1997; Clark 2002a, 2004; Phillips *et al.* 2002a,b, 2004; Baker *et al.* 2004b; Chambers and Silver 2004; Lewis *et al.* 2004b). Several basic questions remain. Have there been widespread increases in forest growth? Are these structural and dynamic changes synchronous within the same plots? Are the changes all widespread? If so, do they share a common cause?

In this chapter we document changes in stand-level processes of growth, recruitment, and mortality within individual plots from across a network of plots in South America. This provides new information in three respects. First, previous studies on stem turnover and above-ground biomass (Phillips and Gentry 1994; Phillips *et al.* 1998b) did not analyse the same plot dataset, so we do not know if these changes were occurring simultaneously within the same plots. Second, by choosing only plots with at least two census intervals, it is possible

to specifically look at rates of change without potentially confounding the results by including different plots in the dataset over time (Condit 1997). Third, we have standardized our dataset to use only three inventories with two approximately equal census interval lengths, so that any changes detected cannot be attributable to census interval effects on rate parameter estimation (Sheil 1995a; Sheil and May 1996; Lewis *et al.* 2004c). Thus in this study, we have eliminated three factors that potentially complicate the interpretation of previous studies.

When considering potential changes in forest dynamics it can be useful to view the forest, at the stand level, as a simple system consisting of a basal area (BA) pool with the size of the pool changing as BA is added to the pool by stem increment growth and new recruitment fluxes, and subtracted from mortality losses. Thus, it can be seen that the documented increase in the BA pool (above-ground biomass) in South American forests

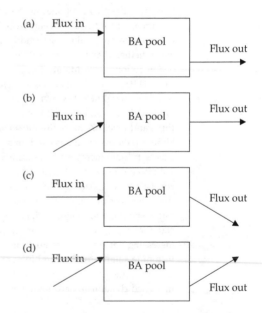

Figure 12.1 Schematic representations of four different scenarios (a–d) where the BA of a forest (BA pool) increases. The box represents the BA pool, the line represents a flux in (additions from tree growth and recruitment) or a flux out (from mortality). The heights of the lines represent the size of the flux, and the slopes of the lines represent the rate of change of the flux over time. Knowledge of the residence time of BA in the pool is also needed to estimate the change in pool size.

(Phillips *et al.* 1998b, 2002a; Baker *et al.* 2004b) must be caused by BA growth rates exceeding BA mortality rates. However, both the size of these growth and mortality fluxes and whether these flux rates are themselves changing is not known. For example, the increase in the BA pool may be caused by an increase in stand-level BA growth with no change in stand-level BA mortality, or by no change in stand-level BA growth and a decrease in stand-level BA mortality rates. However, many different combinations of sizes and rates of change in growth and mortality fluxes could also lead to the same BA increment response (Fig. 12.1). Thus, a key objective is to document how growth and mortality fluxes have changed in South American tropical forests over recent decades.

Methods

Forest monitoring sites

We compiled data (Table 12.1) from the RAINFOR network of long-term forest monitoring plots across South America where all trees had been measured at least three times (Malhi *et al.* 2002a). If more than three censuses were available we selected the mid-census inventory that provided two approximately equal census intervals, with a bias towards choosing slightly longer second census intervals where possible. This ensures that any results showing increases in the fluxes cannot be attributed to census interval-length effects (Sheil and May 1996). Full details of our measurement criteria are detailed in Lewis *et al.* (2004b). Figure 12.2 shows the locations of the plot clusters across South America.

Approach and definitions

We consider the forest as a simple system of a pool of BA with a flux into (from growth) and flux out of (from mortality) the BA pool (Fig. 12.1). We define the pool of BA as the sum of the cross-sectional areas of all trees with a d.b.h. ('diameter at breast height', 1.3 m or above deformities) of 10 cm or more in a plot (in square metres per hectare). We define growth, the flux in, as the sum

of the increments of BA of all surviving trees *and* the sum of BA of all newly recruited trees into a plot over the census interval (in $m^2 ha^{-1} yr^{-1}$). We call this stand-level BA growth. We define stand-level BA mortality, the flux out, as the sum of BA of all trees of d.b.h. of 10 cm or more that died in a plot over the census interval (in $m^2 ha^{-1} yr^{-1}$). Thus, for the forests studied we seek to discover:

(1) the size of the BA pool;
(2) the direction and rate of change in the size of the pool;
(3) the mean flux into the BA pool, that is, stand-level BA growth rate;
(4) the mean flux out of the BA pool, that is, stand-level BA mortality rate;
(5) the direction and rate of change of stand-level BA growth rates;
(6) the direction and rate of change of stand-level BA mortality rates.

Overall, we are interested in determining the rate at which the fluxes determining the size of the BA pool are themselves changing. Most importantly, we are interested in knowing if the observed changes in BA pool are due to an acceleration or deceleration of the rates of growth and/or mortality, and whether changes in growth rates precede changes in mortality rates or vice versa (schematically shown in Fig. 12.1).

We treat stems in the same way, that is, as a pool of stems with fluxes into and out of the pool. The pool is the number of stems of greater than $10 cm d.b.h. ha^{-1}$. The flux into the stem pool is the stem recruitment rate, the number of stems attaining 10 cm or more d.b.h. over the census interval. The flux out of the stem pool is the stem mortality rate, the number of stems dying, over the census interval. We report the same information for stem pool and fluxes as for BA pools and fluxes.

Statistical analysis

For each of our analyses we checked that our data were normally distributed. Overall, the data analysed were all approximately normal but tended to have a weak right skew. No category of parameter

Table 12.1 Plots used in the analyses, initial values, and flux rates

Plot name	Code	Country	Lat (dec)	Long (dec)	Size (ha)	Census			BA Start, (m²ha⁻¹)	Stems Start (ha⁻¹)	Stand-level BA growth (% yr⁻¹)		Stand-level BA mortality (% yr⁻¹)		Stem recruits (% yr⁻¹)		Stem mortality (% yr⁻¹)	
						First	Mid	Final			interval 1	interval 2	interval 1	interval 2	interval 1	interval 2	interval 1	interval 2
BDFFP, 2303 Dimona 4–5[a]	BDF-01	Brazil	−2.40	−60.00	2.00	1985.29	1990.62	1997.71	30.15	688	1.17	1.58	1.73	1.05	0.84	1.60	1.37	1.11
BDFFP, 1101 Gaviao	BDF-03	Brazil	−2.40	−59.90	1.00	1981.13	1991.13	1999.29	28.39	593	1.21	1.56	1.08	1.24	0.83	1.40	0.90	1.55
BDFFP, 1102 Gaviao	BDF-04	Brazil	−2.40	−59.90	1.00	1981.13	1991.13	1999.29	28.08	590	1.13	2.44	3.77	1.81	0.69	4.64	3.25	2.19
BDFFP, 1103 Gaviao	BDF-05	Brazil	−2.40	−59.90	1.00	1981.21	1991.62	1999.29	25.28	650	1.30	1.64	1.27	1.11	0.54	1.41	1.47	1.32
BDFFP, 1201 Gaviao[a]	BDF-06	Brazil	−2.40	−59.90	3.00	1981.29	1991.37	1999.29	25.48	632	1.50	1.68	1.10	1.95	0.84	1.52	1.18	1.78
BDFFP, 1109 Gaviao	BDF-08	Brazil	−2.40	−59.90	1.00	1981.63	1991.55	1999.46	28.47	590	1.13	1.10	1.46	1.06	0.87	2.25	1.67	1.53
BDFFP, 1301.1	BDF-10	Brazil	−2.40	−59.90	1.00	1983.46	1987.21	1997.13	27.47	621	1.00	1.57	0.60	1.45	1.00	1.67	0.83	1.61
BDFFP, 1301.4, 5, 6 Florestal[a]	BDF-11	Brazil	−2.40	−59.90	3.00	1983.46	1987.21	1997.13	28.85	629	0.90	1.16	0.33	0.84	0.53	0.65	0.49	0.78
BDFFP, 1301.7, 8 Florestal[a]	BDF-12	Brazil	−2.40	−59.90	2.00	1983.46	1987.21	1997.13	28.45	617	0.76	1.15	0.93	0.75	0.35	0.70	0.37	0.70
BDFFP, 3402 Cabo Frio	BDF-13	Brazil	−2.40	−59.90	9.00	1985.86	1991.16	1998.88	26.52	568	1.34	1.40	0.84	0.84	1.24	1.34	0.87	1.04
BDFFP, 3304 Porto Alegre[a]	BDF-14	Brazil	−2.40	−60.00	2.00	1984.21	1992.29	1998.38	32.05	651	1.00	1.67	1.62	1.55	0.81	1.77	1.22	1.29
Bionte 1	BNT-01	Brazil	−2.63	−60.17	1.00	1986.50	1991.50	1999.50	28.04	561	1.61	1.43	0.62	0.69	1.54	0.82	0.99	0.89
Bionte 2	BNT-02	Brazil	−2.63	−60.17	1.00	1986.50	1991.50	1999.50	30.14	692	1.43	1.38	0.56	0.75	1.06	0.41	0.61	0.68
Bionte 4	BNT-04	Brazil	−2.63	−60.17	1.00	1986.50	1991.50	1999.50	27.76	608	1.69	1.69	1.84	1.05	1.60	0.81	1.53	1.08
Caxiuana 1	CAX-01	Brazil	−1.70	−51.53	1.00	1994.50	1999.50	2002.84	30.07	524	0.98	1.15	0.61	1.04	0.51	1.16	0.86	0.93
Jacaranda 1–5	JAC-01	Brazil	−2.63	−60.17	5.00	1996.50	2000.50	2002.50	27.51	593	1.13	1.90	1.24	2.14	1.01	2.80	0.92	1.24
Jacaranda 6–10	JAC-02	Brazil	−2.63	−60.17	5.00	1996.50	2000.50	2002.50	26.60	573	1.01	1.98	1.14	2.18	0.97	2.09	0.87	1.57
Jari 1[b]	JRI-01	Brazil	−1.00	−52.05	1.00	1985.50	1990.50	1996.00	32.99	572	1.18	1.38	0.79	1.66	1.52	1.59	0.97	1.28
Tapajos, RP014, 1–4[c]	TAP-01	Brazil	−3.31	−54.94	1.00	1983.50	1989.50	1995.50	23.61	527	1.99	1.80	0.49	1.13	1.63	1.46	0.68	0.67
Tapajos, RP014, 5–8[c]	TAP-02	Brazil	−3.31	−54.94	1.00	1983.50	1989.50	1995.50	27.82	479	2.00	1.27	0.66	0.63	1.86	1.45	0.78	0.50
Tapajos, RP014, 9–12[c]	TAP-03	Brazil	−3.31	−54.94	1.00	1983.50	1989.50	1995.50	31.25	491	1.63	1.37	0.64	0.77	1.64	1.33	0.91	0.70
Jatun Sacha 2	JAS-02	Ecuador	−1.07	−77.60	1.00	1987.63	1994.54	2002.04	30.18	724	1.83	3.28	1.54	3.73	1.14	2.61	1.08	2.74
Jatun Sacha 3	JAS-03	Ecuador	−1.07	−77.67	1.00	1988.88	1994.29	2002.04	27.96	648	3.30	2.89	2.45	2.33	1.79	2.39	2.40	1.69
Jatun Sacha 5	JAS-05	Ecuador	−1.07	−77.67	1.00	1989.38	1994.46	2002.04	30.90	534	2.65	3.54	1.61	2.50	1.87	2.97	1.76	2.32
Allpahuayo A dayey[d]	ALP-11	Peru	−3.95	−73.43	0.44	1990.87	1996.13	2001.03	27.36	580	1.80	2.49	1.54	2.58	2.05	2.30	2.13	3.12
Allpahuayo A, sandy[d]	ALP-12	Peru	−3.95	−73.43	0.40	1990.87	1996.13	2001.03	25.19	570	2.14	2.57	0.93	4.53	1.36	2.05	1.03	3.97
Allpahuayo B, sandy[d]	ALP-21	Peru	−3.95	−73.43	0.48	1990.87	1996.13	2001.04	26.88	575	2.14	3.07	2.72	2.12	1.80	3.13	1.73	2.34

Site	Plot code	Country	Lat (dec)	Long (dec)													
Allpahuayo B, clayey[d]	ALP-22	Peru	−3.95	−73.43	0.44	1990.87	1996.13	2001.04	25.49	614	2.59	2.06	1.64	2.06	2.09	2.71	1.46	2.57
Cuzco Amazonico, CUZAM1E	CUZ-01	Peru	−12.50	−69.95	1.00	1989.39	1994.63	1998.77	25.41	489	2.31	2.95	1.40	1.55	2.79	2.24	1.80	1.55
Cuzco Amazonico, CUZAM1U	CUZ-02	Peru	−12.50	−69.95	1.00	1989.42	1994.63	1998.77	25.27	509	2.53	3.55	1.01	2.86	1.92	2.36	1.37	1.77
Cuzco Amazonico, CUZAM2E	CUZ-03	Peru	−12.49	−69.11	1.00	1989.40	1994.62	1998.77	21.69	470	3.17	3.93	1.76	1.40	2.76	2.70	2.20	1.97
Cuzco Amazonico, CUZAM2U	CUZ-04	Peru	−12.49	−69.11	1.00	1989.44	1994.62	1998.78	27.26	571	2.44	3.51	2.34	1.91	1.47	3.90	1.92	2.35
Manu, Trail 3	MNU-01	Peru	−11.88	−71.35	0.97	1975.00	1990.75	2000.75	28.56	549	2.09	2.04	1.72	2.28	2.15	2.50	2.19	2.45
Manu, terra firme terrace	MNU-03	Peru	−11.88	−71.35	2.00	1991.75	1996.75	2001.63	25.90	578	2.71	3.27	2.60	2.99	3.24	3.55	3.00	3.34
Manu, terra firme ravine	MNU-04	Peru	−11.88	−71.35	2.00	1991.75	1996.75	2001.75	27.12	587	2.43	2.36	1.73	2.08	2.84	1.78	2.02	2.34
Manu, Trail 12	MNU-05	Peru	−11.88	−71.35	2.00	1989.99	1994.99	1999.99	33.59	599	1.77	1.27	1.11	1.32	2.03	1.53	1.24	1.92
Manu, Trail 2 and 31	MNU-06	Peru	−11.88	−71.35	2.25	1989.80	1994.80	1999.80	32.21	511	1.77	1.90	0.98	1.71	1.79	2.06	1.67	1.92
Cocha Salvador Manu	MNU-08	Peru	−11.88	−71.35	2.00	1991.80	1996.83	2001.87	36.81	563	1.47	1.61	1.40	1.06	1.52	1.55	1.20	1.48
Sucusari A	SUC-01	Peru	−3.23	−72.90	1.00	1992.13	1996.08	2001.06	28.25	612	2.44	2.00	2.75	2.01	1.86	1.53	2.11	1.80
Sucusari B	SUC-02	Peru	−3.23	−72.90	1.00	1992.13	1996.08	2001.07	29.46	606	2.45	2.09	1.92	3.70	2.33	2.07	2.09	2.77
Tambopata plot zero	TAM-01	Peru	−12.85	−69.28	1.00	1979.87	1991.53	2000.59	26.91	555	2.40	2.89	3.01	1.60	2.15	2.87	2.31	2.15
Tambopata plot one	TAM-02	Peru	−12.83	−69.28	1.00	1983.78	1991.58	2000.58	27.44	576	1.82	2.05	1.42	1.59	1.68	2.55	1.35	1.57
Tambopata plot two	TAM-04	Peru	−12.83	−69.28	0.42	1983.79	1990.76	1998.75	28.56	705	2.48	2.91	3.10	1.76	2.28	2.59	2.77	2.46
Tambopata plot three	TAM-05	Peru	−12.83	−69.28	1.00	1983.70	1991.54	2000.56	24.27	548	2.21	2.76	1.87	2.06	2.13	3.19	2.32	2.27
Tambopata plot four	TAM-06	Peru	−12.83	−69.30	0.96	1983.71	1991.54	2000.55	30.54	520	1.98	2.39	1.58	0.90	2.73	3.37	1.69	1.47
Tambopata plot six	TAM-07	Peru	−12.83	−69.27	1.00	1983.76	1991.54	1998.73	27.36	548	2.50	2.71	2.25	2.19	2.18	2.96	2.22	2.91
Yanamono A	YAN-01	Peru	−3.43	−72.85	0.50	1983.46	1991.29	2001.05	30.95	570	2.82	2.48	2.45	2.31	2.00	3.06	2.54	2.69
Dorado, 91-km plot	ELD-12	Venezuela	6.50	−61.50	0.50	1971.55	1981.18	1994.44	27.69	492	1.76	1.62	1.01	0.51	1.09	0.76	0.93	0.47
El Dorado, 98-km plot EDL-03, 04[e]	ELD-34	Venezuela	6.50	−61.50	0.50	1971.56	1976.21	1981.19	25.29	622	2.58	2.95	1.01	4.29	1.47	1.74	1.06	2.00
Rio Grande RIO-01, 02[e]	RIO-12	Venezuela	8.00	−61.75	0.50	1981.20	1994.46	1994.46	29.45	570	1.81	1.73	0.82	1.94	1.27	1.15	0.91	1.46

Data are best available as of 1 March 2003, but are subject to future revision as a result of additional censuses and continued error-checking. dec, decimal.

[a] These sites comprise non-contiguous 1-ha plots separated by <200 m; treated as one plot.

[b] Twenty-five 10 m × 10 m subplots, within each of four, nearby, 1 ha plots.

[c] Twelve 0.25 ha plots laid out in a randomized fashion over an area of 300 m × 1200 m; treated as 3 × 1 ha units.

[d] Allpahuayo A and B contains two distinct soil types that are treated separately in these analyses.

[e] These sites comprise two nearby non-contiguous 0.25-ha plots.

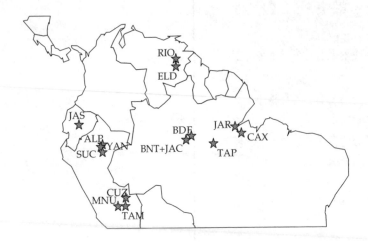

Figure 12.2 The location of the study plots. Codes RIO, JAR, CAX, and YAN have 1 plot each, EDL, JAC, and SUC have 2 plots, TAP, BNT, and JAS, have 3 plots, ALP and CUZ have 4 plots, MNU and TAM have 6 plots, and BDF has 11 plots.

was grossly non-normal. Explorations of a variety of transformations did not consistently move the data to become more normal. Thus, we used untransformed data and mostly employed parametric t-tests on paired observations in our statistical analyses. When analysing the full set of 50 plots we used two-tailed tests of significance, as either increases or decreases in parameters were expected.

We compare the flux rates over the first and second census intervals, first as BA, and second on a stem basis. We report BA, stand-level BA growth and mortality rates in square metres per hectare per year and stem density, stem recruitment, and stem mortality rates in stems per hectare per year. In addition, for direct comparisons between fluxes we calculate each flux employing the commonly used 'exponential population decline' in mortality studies (λ), corresponding recruitment equations (μ), and corresponding equations on a per BA basis for BA growth and BA mortality fluxes, as per cent per year (Sheil *et al.* 1995; Phillips *et al.* 2004; Lewis *et al.* 2004b). Thus we allow for direct comparisons of fluxes both on a per stem and a per BA basis.

To compare *changes* in the fluxes we use two methods: (1) absolute changes, the rate over the first census interval subtracted from the rate over the second census interval; and (2) relative changes, by applying the method used to calculate relative growth rates (Evans 1972). To compare *changes* in the pools (BA or stems) we calculate

absolute changes in the same way as for the fluxes. To calculate relative changes, we use the difference in the pool scaled by the initial size of the pool and the census interval, as has commonly been used elsewhere, again expressed in per cent per year (Malhi and Grace 2000; Hamilton *et al.* 2002).

Results

Basal area

The average date of the first, mid- and final censuses was late 1985, early 1992, and early 1999, respectively. The earliest start date was 1971; 5 plots started in the 1970s, 33 in the 1980s, and 12 in the 1990s. The final census was between 1994 and 2002 for all plots except ELD-34, which was concluded in 1981. The first census interval was, on average, 6.4 ± 0.7 yr ($\pm 95\%$ CI; range of 3.8–15.8), and the second 6.8 ± 0.7 years (range of 2.0–13.3). The average plot size was 1.50 ± 0.41 ha (range of 0.4–9): 8 plots were 0.4–0.5 ha, 29 plots were 0.96–1 ha, and 13 plots were 2 ha or more. The mean BA of the 50 plots was 28.2 ± 0.75 m^2 ha^{-1} (range of 21.7–36.8) at the first census interval and was significantly higher at 29.5 ± 0.88 m^2 ha^{-1} by the final census interval ($t = 4.93$, $p = 0.0001$, d.f. 49; 41 out of 50 plots increased). The mean size of the flux into the BA pool, stand-level BA growth, was 0.51 ± 0.04 m^2 ha^{-1} yr^{-1}. The mean flux out, stand-level BA mortality, was 0.41 ± 0.04 m^2 ha^{-1} yr^{-1}. Hence, for the study period, BA increased by

0.10 ± 0.04 m^{2}ha^{-1}yr^{-1}, or a relative increase of $0.38 \pm 0.15\%$ yr^{-1} (change in the BA pool scaled by the initial BA pool and the census interval).

Across the 50 plots the stand-level BA growth rate increased significantly between the first and second census intervals, shown by subtracting the rate over interval one from that over interval two for each plot and noting that the 95% CI for this change parameter does not cross zero (Fig. 12.3; Table 12.2; $t = 3.89$, $p = 0.0003$; 34 out of 50 plots increased). The wide confidence intervals associated with comparing the mean stand-level BA

growth rates over intervals one and two are caused by the wide range of tropical forests sampled, for example, CAX-01 increased stand-level BA growth from 0.29 to 0.35 m^{2}ha^{-1}yr^{-1}, while CUZ-03 increased from 0.68 to 0.94 m^{2}ha^{-1}yr^{-1} (Fig. 12.3). Mean stand-level BA growth rate increased from 0.50 ± 0.04 m^{2}ha^{-1}yr^{-1} in the first census interval to 0.58 ± 0.05 m^{2}ha^{-1}yr^{-1} in the second. In relative terms, the annual rate of increase in stand-level BA growth rate was $2.55 \pm 1.45\%$ yr^{-1} (log model, as used to calculate relative growth rates).

Stand-level BA mortality rates increased significantly between the first and second census intervals (Fig. 12.3; Table 12.2; $t = 2.01$, $p = 0.05$; 29 out of 50 plots increased). Mean stand-level BA mortality rates increased from 0.40 ± 0.05 m^{2}ha^{-1}yr^{-1} in the first interval to 0.48 ± 0.06 m^{2}ha^{-1}yr^{-1} in the second, a relative increase of $4.03 \pm 2.71\%$ yr^{-1} (Table 12.2). The rates of increase in stand-level BA growth rates and stand level BA mortality rates were very similar. However, stand-level BA growth rates were significantly higher than stand-level BA mortality rates in both intervals (by 0.11 ± 0.05 and 0.10 ± 0.06 m^{2}ha^{-1}yr^{-1} for census intervals one and two, respectively).

In summary, over the past two to three decades the study plots have experienced a net increase in BA of 0.10 ± 0.04 m^{2}ha^{-1}yr^{-1}. This is not solely attributable to the documented increase in stand-level BA growth rates (which increased, on average, by 0.08 ± 0.04 m^{2}ha^{-1}yr^{-1} between the two censuses), as stand-level BA mortality rates, while much more variable, also increased by a similar amount (by 0.08 ± 0.07 m^{2}ha^{-1}yr^{-1} between the two censuses). The BA pool increased

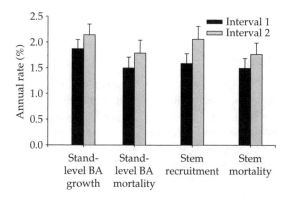

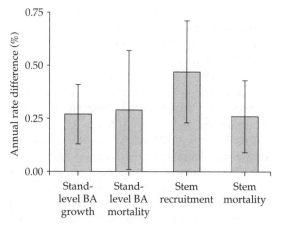

Figure 12.3 (a) Annualized rates of stand-level BA growth, stand-level BA mortality, stem recruitment, and stem mortality from two consecutive census intervals (black bars, interval 1; grey bars, interval 2), (b) stand-level BA growth, stand-level BA mortality, stem recruitment, and stem mortality over census interval one subtracted from that over interval 2 (rate difference), each from 50 plots with 95% CIs. The average mid-year of the first and second censuses was 1989 and 1996, respectively.

Table 12.2 Mean and 95% CIs for flux rates of BA and stems from 50 South American forest plots (in per cent per year)

Flux	Interval 1	Interval 2	Relative change
Stand-level BA growth	1.87 ± 0.18	2.14 ± 0.21	2.55 ± 1.45
Stand-level BA mortality	1.50 ± 0.21	1.79 ± 0.25	4.03 ± 2.71
Stem recruitment	1.59 ± 0.19	2.06 ± 0.25	4.23 ± 2.53
Stem mortality	1.50 ± 0.19	1.77 ± 0.22	3.08 ± 1.81

in this group of forests as growth rates exceeded mortality rates by 0.10 ± 0.05 m^2 ha^{-1} yr^{-1} in the first census period, and this difference between growth and mortality rates continued over the second census period. These results correspond to the situation represented schematically as Fig. 12.1(d).

Stems

The mean number of stems per hectare across the 50 plots was 581 ± 16 (range of 470–724) at the first census, and was significantly higher at 592 ± 14 ha^{-1} at the final census ($t = 2.46$, $p = 0.017$, d.f. 49; 32 out of 50 plots increased). The mean size of the flux into the stem pool, stem recruitment, was 9.4 ± 0.88 stems ha^{-1} yr^{-1} across the 50 plots over the entire monitoring period. The mean size of the flux out of the stem pool, stem mortality, was 8.4 ± 0.89 stems ha^{-1} yr^{-1}. Hence, for the study period, stem number increased by 0.94 ± 0.63 stems ha^{-1} yr^{-1}, or relatively speaking by $0.18 \pm 0.12\%$ yr^{-1}.

Across the 50 plots stem recruitment rates increased significantly between the first and second census intervals (Fig. 12.3; Table 12.2; $t = 3.86$, $p = 0.0003$, d.f. = 49; 37 out of 50 plots increased). Mean stem recruitment rates increased from 8.7 ± 0.98 stems ha^{-1} yr^{-1} in the first census interval to 11.3 ± 1.3 stems ha^{-1} yr^{-1} in the second, a relative increase of $4.23 \pm 2.53\%$ yr^{-1} (Table 12.2).

Stem mortality rates increased significantly between the first and second census intervals (Fig. 12.3; Table 12.2; $t = 2.97$, $p = 0.005$; d.f. = 49; 32 out of 50 plots increased). Mean stem mortality increased from 8.2 ± 1.0 stems ha^{-1} yr^{-1} in the first census interval to 9.6 ± 1.2 stems ha^{-1} yr^{-1} in the second interval, a relative increase of $3.08 \pm 1.81\%$ yr^{-1} (Table 12.2). The rate of increase in stem recruitment, although greater than the rate of increase in stem mortality, was not significantly so (an increase of 2.6 ± 1.3 stems ha^{-1} yr^{-1} between census intervals one and two for stem recruitment compared with an increase of 1.4 ± 0.98 stems ha^{-1} yr^{-1} between census intervals one and two for stem mortality, with a mean difference of 1.2 ± 1.6 stems ha^{-1} yr^{-1}).

In summary, over the past two to three decades the study plots have, on average, been characterized by an increase in stem density of 0.94 ± 0.63 stems ha^{-1} yr^{-1}. This is partly attributable to the increase in recruitment rates, which increased, on average, by 2.6 ± 1.3 stems ha^{-1} yr^{-1} between the two censuses, but stem mortality also increased over the monitoring period, on average, by 1.4 ± 0.98 stems ha^{-1} yr^{-1} between the two censuses. Stem density also increased because stem recruitment exceeded mortality by a statistically insignificant amount over the first census interval (0.5 ± 0.9 stems ha^{-1} yr^{-1}), then by a much greater and significant amount over the second census interval (1.7 ± 1.1 stems ha^{-1} yr^{-1}).

Discussion

We found a concerted, widespread, and consistent directional change in the structure and dynamics of the 50 forest plots spanning South America (Fig. 12.3). We have shown for the first time that growth is increasing and that simultaneous increases in growth, recruitment, and mortality rates have occurred within the same plots. Overall, the structure and dynamics of these forests have altered substantially over the 1971–2001 period of monitoring.

The flux into the BA pool, stand-level BA growth, exceeded the flux out of the pool, stand-level BA mortality, and therefore the BA pool increased. This result is consistent both in direction and magnitude with other recent estimates of increases in above-ground biomass in South American tropical forests (Phillips et al. 1998b, 2002a; Baker et al. 2004b). In addition, both the growth and mortality fluxes increased significantly and similarly to each other. Therefore, the size of the BA pool increased because stand-level BA growth rates were higher than stand-level BA mortality rates at the outset of the study and this difference was maintained through the study period (cf. the schema in Fig. 12.1(d). A similar, albeit slightly more complicated pattern, was shown for the stems pool and fluxes. Stem recruitment rates exceeded stem mortality rates, thus stem density increased. Again, both of these fluxes increased significantly over time, and the rates of change of these fluxes were not significantly different from one another. This increase in stem recruitment rates and stem mortality rates is consistent both in direction and magnitude, with previous estimates across the tropics showing that these fluxes approximately doubled from the 1950s to the 1990s (Phillips and Gentry 1994; Phillips 1996; Phillips *et al.* 2004). However, the increase in stem density was partly attributable to stem recruitment rates increasing faster than the increase in stem mortality rates (but not significantly so), and partly attributable to stem recruitment rates being higher than stem mortality rates at the beginning of the study (again not significantly so). The changes in both the stem and BA fluxes indicate that the current imbalance of additions and losses were occurring before the onset of monitoring the plots.

It has previously been suggested that *individual* patterns of change documented from long-term plot data may have been caused by: (1) statistical problems; (2) biases and artefacts in the data; (3) widespread recovery from past disturbances; or (4) a widespread environmental change or changes. We briefly discuss each of these four options.

We compiled the dataset to remove statistical problems associated with analysing forest plot data with irregular census intervals. First, we deliberately chose census intervals so that the second census interval was, on average, slightly longer than the first. Therefore, census interval effects (Sheil 1995a; Sheil and May 1996; Lewis *et al.* 2004c) cannot explain the flux results. Second, we monitored change within the same group of 50 plots over time, so potential biases associated with a lack of continuity of monitoring ('site-switching') cannot be driving the results (Condit 1997). Previously identified statistical problems cannot, therefore, explain the suite of results reported here.

A variety of other methodological, analytical, and artefactual biases may also potentially affect long-term monitoring data (Sheil 1995b). Although it is possible that an individual pattern of change may be caused by one of these artefacts, it is difficult to conceive artefacts that are causing the suite of changes we document. Furthermore, the most commonly discussed issues that we have not accounted for in this study have been addressed in other recent publications, and the changes in stem turnover (Phillips *et al.* 2004) and above-ground biomass (Baker *et al.* 2004b) both hold. Although biases must affect the confidence we have in parameter estimates, we do not know of a bias or artefact, or set of biases and artefacts, that could plausibly cause the suite of changes within the same plots shown in this study.

Several authors have suggested that the impacts of disturbance, and recovery from disturbance, may account for either the increase in stem turnover rates (Sheil 1995a), or the increase in above-ground biomass (Körner 2003b; Chambers and Silver 2004). However, it is difficult to explain the simultaneous increases in stand-level BA growth and stem recruitment rates, *while* mortality rates are simultaneously increasing, through internal disturbance- and-succession processes. Furthermore, we know of no continent-wide disturbance event that may cause these patterns. The most obvious candidate, El Niño Southern Oscillation (ENSO) events, show a strong spatial pattern across South America, running approximately northeast (strongly affected) to southwest (little affected) across the continent (Malhi and Wright 2004). However, elsewhere we show that simultaneous changes in forest structure and dynamics have occurred proportionately similarly across the continent (Lewis *et al.* 2004b). Disturbance and recovery cycles certainly contribute to short/medium-term changes in a forest plots' growth, recruitment and mortality fluxes but these cycles appear as an implausible mechanism to explain the suite of changes within forest plots we document across such a geographically widespread area.

We suggest a parsimonious explanation of our results. The data appear to show a coherent fingerprint of increasing growth, that is, increasing net primary productivity (NPP), across tropical South America, probably caused by a long-term increase in resource availability, in turn caused by one or more widespread environmental changes (see additional analyses in Lewis *et al.* 2004b, not reported here). The argument runs: increasing resource availability increases NPP, which increases stem growth rates. This accounts for the increase in stand-level BA growth rates and stem recruitment rates, and the fact that these show the 'clearest' signal in our dataset. Over time some of these faster-growing, larger trees die, as do some of the 'extra' recruits. This accounts for the increase in the fluxes out of the system, stand-level BA mortality rates, and stem mortality rates. Thus, the system has increasing additions of BA and stems, while the losses lag behind, causing an increase in the BA and stem pools. Overall, the suite of results may be qualitatively explained by a long-term increase in a limiting resource.

Is a long-term increase in resource availability that increases NPP and growth, and thus accelerate forest dynamics a plausible scenario? First, stand-level BA growth is 10–30% of total NPP for mature tropical forests (Clark et al. 2001b). Hence, it is reasonable to assume that the large increase in stand-level BA growth we document may reflect an increase in NPP. However, the increase in NPP may be smaller than the increase in stand-level BA growth if allocation patterns also change. Second, this scenario implicitly assumes that tropical forests are resource-limited systems. This may or may not be the case, although there is a body of evidence suggesting they are (see Phillips et al. 2004, for a discussion of this). Third, there are 'smoking guns': incoming solar radiation may have increased across tropical South America over the past two decades (Wielicki et al. 2002; Nemani et al. 2003), and air temperatures (Malhi and Wright 2004) and atmospheric CO_2 concentrations have increased (Prentice et al. 2001), each of which may increase NPP (Lewis et al. 2004a).

However, the reader should note that for none of the three candidates for widespread environmental changes increasing resource availability—solar radiation, temperature, CO_2—do we have good evidence to say both that the driver has actually changed and evidence that such a change will cause an increase in flux rates and pools (Lewis et al. 2004a). The increase in incoming solar radiation comes from a single satellite dataset (Wielicki et al. 2002), which may contradict land-based sensors that show a decrease in incoming solar radiation (Stanhill and Cohen 2001; see Lewis et al. 2004a, for a discussion). If the satellite data are correct, a modelling study suggests that the

increase in NPP would increase, and the increase is similar to the rate of increase of the BA pool which we find in this study (Nemani et al. 2003). Temperature increases are undisputed, but evidence as to whether the ca. 0.5°C increase in temperature over the monitoring period would be expected to increase or decrease NPP is unclear: photorespiration and respiration costs may increase as temperatures rise, which may reduce NPP, or higher temperatures may increase soil nutrient availability which may increase NPP (Lewis et al. 2004a). Finally, the ca. 10% increase in CO_2 concentrations between 1980 and 2000 is undisputed. Carbon dioxide is a key substrate for photosynthesis, and higher CO_2 concentrations increase $CO_2:O_2$ ratios thereby reducing photorespiration and also increase the optimum temperature for photosynthesis, so CO_2 appears as a promising candidate. Experiments show that elevated CO_2 concentrations increase plant growth under many conditions (Curtis and Wang 1998), including in situ tropical seedlings (Würth et al. 1998a) and whole stands of temperate trees (Hamilton et al. 2002; Norby et al. 2002), but experiments on whole stands of tropical forest trees have not been conducted, thus this interpretation is currently a topic of active debate (Chambers and Silver 2004; Clark 2004; Lewis et al. 2004a).

In conclusion, we believe that we have shown a consistent 'fingerprint' of increasing growth across a large sample of geographically widespread South American tropical forests over the 1980s and 1990s. These forests, on average, simultaneously increased growth, recruitment and mortality rates, and accumulated both stems and BA and hence have increased in biomass, and have been a carbon sink (see Baker et al. 2004b). The simplest explanation of this concerted increase in forest dynamics across South America is that increasing resource availability has increased NPP, which, in turn, has accelerated BA and stem dynamics and increased above-ground woody biomass. Whether these changes will persist, stabilize, or reverse, and over what timescales, is unknown. What the consequences of these changes have been for biodiversity is also not known. Whatever the mechanism, over recent decades, profound changes seem to have occurred across the world's largest tract of tropical forests.

Acknowledgements

The authors acknowledge the essential contributions of numerous field assistants over the past 25 yr, as detailed in Lewis *et al.* (2004b). These analyses were supported by the UK Natural Environment Research Council (NER/A/S/2000/00532) and the EU Fifth Framework programme (CARBONSINK-LBA). Simon Lewis is supported by a Royal Society University Research Fellowship.

CHAPTER 13

Error propagation and scaling for tropical forest biomass estimates

Jerome Chave, Guillem Chust, Richard Condit,
Salomon Aguilar, Andres Hernandez, Suzanne Lao, and
Rolando Perez

The dry above-ground biomass (AGB) of tropical forests is a crucial variable for ecologists, biogeochemists, foresters, and policy-makers. Permanent tree inventories are an efficient way of assessing this variable. In order to make correct inferences about long-term changes in biomass stocks, it is essential to know the uncertainty associated with AGB estimates, yet this uncertainty is seldom evaluated carefully. Here we quantify four types of uncertainties that could lead to statistical error in AGB estimates: (1) error due to tree measurement; (2) error due to the choice of allometric model relating AGB to other tree dimensions; (3) sampling uncertainty, related to the size of the study plot; (4) representativeness of a network of small plots across a forest landscape. In previous studies, some of these sources error were reported, but they are not integrated in a consistent framework. We estimate all four terms in a 50-ha plot on Barro Colorado Island, and in a network of 1-ha plots scattered across the Panama Canal Watershed, central Panama. We find that the most important source of error is currently related to the choice of the allometric model. More work should be devoted to improving the predictive power of allometric models for biomass.

Introduction

Permanent sampling plots have long been used in ecological studies for assessing how much biomass is held in ecosystems (Olson *et al.* 1983; Fearnside 1996; Brown 2002). Tree above-ground biomass (AGB) is strongly correlated with trunk diameter (Brown 1997; Parresol 1999; Clark *et al.* 2001b), and it is therefore possible to use forest inventory data to estimate the stocks and changes in AGB in those inventories. Recently, plot data have been influential in carving new hypotheses on the dynamic coupling between tropical forests and the atmosphere (Phillips and Gentry 1994; Phillips *et al.* 1998b; 2002a). This interest in tropical forest inventories has also motivated a new literature on tree census methods (Sheil 1995b; Condit 1998b; Clark *et al.* 2001b; Brown 2002; Clark 2002a; Phillips *et al.* 2002a).

Nevertheless, difficulties in assessing data quality in forest inventories lead to continuing debate on the functional response of tropical forests to global change (Clark 2002a; Chave *et al.* 2003; Baker *et al.*, Chapter 11, this volume). The biomass stocks of tropical forests remain poorly resolved at the regional scale (Fearnside 1996; Houghton *et al.* 2001). Integrating site-specific and heterogeneously collected data to draw regional-scale conclusions about tree densities, turnover rates, or biomass stocks of tropical forests is a challenging task, and it thus seems precarious to extrapolate local data to larger scales without assessing how representative these data are.

Figure 13.1 depicts one strategy for converting forest plot data into regional-scale AGB estimates (Brown *et al.* 1989; Brown 1997; Houghton *et al.*

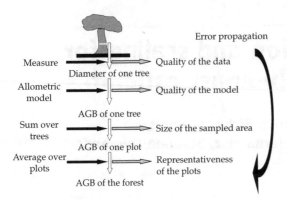

Figure 13.1 Error propagation for estimating the AGB of a tropical forest from permanent sampling plots.

2001). Each tree in a plot is measured, tagged, and identified (Clark 2002a; Phillips *et al.* 2002a); an allometric equation is used to relate its diameter to an AGB estimate (Brown 1997). The plot-level estimate is then summed over all the trees to get a stand-level AGB estimate. For carbon sequestration issues, the quality of this estimate depends on the plot size. In addition, the landscape-scale environmental variability should be integrated by replicating the measurement in other plots of the same forest (Clark and Clark 2000; Keller *et al.* 2001; Nascimento and Laurance 2002). These steps integrate a variety of techniques that all contain some uncertainty, yet there is no consistent methodology for propagating uncertainty across scales for tropical forests (but see Parresol 1999 and reference therein, for a related discussion). Since errors due to these problems add up, each needs to be quantified carefully and independently. In the present contribution, we assess the different sources of error associated with AGB estimates from forest inventories, and present worked-out examples for a moist tropical forest of central Panama.

Methods

Uncertainty on tree-level AGB estimate

The first potential source of error is the tree measurement process. Stems ≥ 10 cm in diameter at 130 cm above-ground or above any trunk deformity, are tagged, located, and their diameter

D is measured. As any ecological variable, D is likely to be measured with some imprecision, and special conventions should apply (Sheil 1995b; Condit 1998b; Phillips *et al.* 2002). Let the standard error associated with the diameter measurement be denoted as σ_D, an increasing function of D. When a height measurement H is also taken, the related error is denoted σ_H. These two error terms covary as D and H are positively correlated. Finally, a wood specific gravity value ρ (oven-dry weight over green volume) can be associated to each tree, either by direct measurement (e.g. from tree cores) or using databases that provide the mean ρ for the species to which the tree belongs. The corresponding error σ_ρ could be due to a misidentification of the tree, or to a variation in ρ within a tree or among conspecific trees.

Errors in trunk diameter, height, or density measurement all contribute to an increased error in the AGB estimate. The conversion into AGB, is usually performed using an allometric model of the form $AGB = f(D, H, \rho)$. This error is propagated to the AGB estimate through the allometric model by expanding the model function f in Taylor series. Assuming an allometric model of the form $f(D, H, \rho) = aD^{\alpha}H^{\beta}\rho^{\gamma}$, the measurement standard deviation for the AGB estimate, σ_M, is estimated from the errors on the different measured terms (Chave *et al.* 2004)

$$\sigma_M = \langle AGB \rangle \left(\alpha^2 \frac{\sigma_D^2}{D^2} + \beta^2 \frac{\sigma_H^2}{H^2} + \delta^2 \frac{\sigma_\rho^2}{\rho^2} + 2\alpha\beta \frac{\sigma_{DH}^2}{DH} \right)^{1/2},$$

(13.1)

where $\sigma_{DH} = \langle \delta D \delta H \rangle$ represents the covariance between D and H.

A second source of error is due to the construction of the allometry. The regression model's parameters are usually estimated from directly felled and weighed trees, by a regression on the log-transformed variables. The residuals should be normally distributed, and the log of AGB is estimated with no bias. Back-transforming into an AGB estimate is not straightforward, however, as $\langle AGB \rangle$ differs from $\exp(\langle \ln(AGB) \rangle)$. In fact, the following formula holds (Baskerville 1972):

$$\langle AGB \rangle = \exp(\sigma^2/2) \times \exp(\ln\langle AGB \rangle)$$
$$= CF \times \exp(\ln\langle AGB \rangle).$$

(13.2)

The term $\exp(\sigma^2/2)$ is often called the *correction factor*, CF (Brown *et al.* 1989). The uncertainty on the estimate of AGB associated with the allometric model is measured by the variance $\sigma_A = \sqrt{CF^2 - 1} \times \langle AGB \rangle$. Thus, for a given allometric model, one can estimate both the expected AGB held in a tree, and the standard deviation of this estimate.

We assume that the measurement and the allometric uncertainties are independent sources of variability. The overall uncertainty on the AGB estimation of a single tree therefore is $\sqrt{\sigma_A^2 + \sigma_M^2}$.

Allometric model selection error

Ideally, as is the case for temperate trees, each species should have its own biomass equation, based on a large sample size. This is unrealistic for tropical forests, however. Tropical forest allometric models used for AGB estimation suffer from three important shortcomings: they are constructed from limited samples, they are sometimes applied beyond their valid diameter range, they rarely take into account available information on wood specific gravity.

Many of the published models are based on harvest experiments carried out in a single forest, and based on few harvested trees. Here we suggest that the number of trees used to calibrate allometric models is a major source of variation of the AGB estimate when different models are selected. We selected six published allometric models reconstructed directly from the original datasets (trees ≥ 10 cm). In order to test for the variation among allometric models' predictions, we used a pantropical equation that relates the AGB (in kg) to the trunk diameter (in cm) and the wood specific gravity (in g cm^{-3}) deduced from a compilation of 634 trees ≥ 10 cm d.b.h. (J. Chave *et al.* unpublished results):

$$f(D, \rho) = \frac{\rho}{0.6} \exp(-3.742 + 3.450 \ln(D) - 0.148 \ln(D)^2). \qquad (13.3)$$

Many published allometric models lack reliable data for the largest diameter classes, so they cannot be used to estimate the AGB held in large trees. We assessed the potential error caused by extrapolating the models beyond their range of applicability. This uncertainty does not average out for large sample sizes, since it reflects our limited knowledge of the model itself for the largest diameter classes. We used the pantropical equation to correct for the AGB of extrapolated trees for the other published models.

Including wood specific gravity as a predictive variable greatly improves the quality of the AGB estimate. We corrected the published allometric models by including a dependence of the form

$$\langle AGB \rangle = CF \times \frac{\rho}{\rho_{av}} f(D), \qquad (13.4)$$

where the AGB estimate is a linear function of tree-level wood specific gravity divided by the average specific gravity for the plot where the allometric equation was constructed. The parameter ρ_{av} is the ratio of the total oven-dry weight of the trees used to construct the equation, over their fresh volume.

Finally, it may be dangerous to convert oven-dry AGB into carbon content simply by assuming that oven-dry biomass contains 50% of carbon. Elias and Potvin (2003) showed that this fraction varies across species.

Minimal single plot size

Potential sources of sampling error include the incorrect estimation of the plot area, trees missed or measured twice, or dead trees counted as alive. Tree-level errors average out in large plots, and for this reason too it is advisable to set up large permanent sampling plots. Since rare large trees contribute an important fraction of the overall AGB, the distribution of AGB across 10 m × 10 m subplots is non-normal (Chave *et al.* 2003).

Lianas, multistemmed trees, resprouting trees, are often not recorded during the censuses. This leads to an underestimation of the stand-level AGB. Correction factors have been computed in studies where this information is available, and the uncertainty on these correction factors contributes to the stand-level error. Similarly, below-ground biomass (BGB) is usually estimated from other studies' averages (Houghton *et al.* 2001), more rarely from diameter-BGB allometries (Cairns *et al.* 1997). In this study we have no data on liana or on BGB, and we cannot consider errors in those areas further. In the Panama data, trees <10 cm (and ≥1 cm) were measured in the whole 50-ha Barro Colorado Island (BCI) plot and in subplots of the Marena plot network of plots. For trees <10 cm, we used a single equation modified from the model devised by Hughes *et al.* (1999) for a moist tropical forest of south Mexico (Los Tuxtlas). The model was

$$f(D, \rho) = \frac{\rho}{\rho_{av}} \exp(-1.970 + 2.117 \ln(D)). \quad (13.5)$$

Landscape-scale representativeness

A single plot is unlikely to represent the whole landscape-scale environmental variability. Among the possible biases, there is a tendency for researchers to select nice-looking forests. The landscape-level AGB estimate should be assessed by setting up a network of plots randomly distributed over the landscape, to assess the variability of forest types (Clark and Clark 2000; Keller *et al.* 2001; Chave *et al.* 2003). However, this estimate is invalid if the landscape is a mosaic of forest types.

We used a generalized linear model (GLM) to estimate the AGB stock of the landscape in relation to the following variables: geographical location (X and Y universal transverse mercator (UTM) coordinates, as well as second-and third-order moments thereof), and spectral values for the bands b1–b5 and b7 of a Landsat Thematic Mapper image acquired on this zone on 27 March 2000. Spectral data were based on the average value of a 3×3 pixels square window (90 m × 90 m) centred on each tree plot. The best GLM was selected by a stepwise selection method that minimized Akaike's Information Criterion, and tested using a jackknife validation procedure (Lobo and Martín-Piera 2002).

Material and study sites

We quantified the uncertainty associated with the estimation of AGB of a single tree, assuming that the allometric method is unbiased, for the forest of

the Panamá Canal Zone, central Panamá. We have already provided estimates of the AGB held in the 50-ha permanent sampling plot on BCI. We used diameter measurements for over 200,000 trees ≥1 cm, combined with tree heights modelled from diameter-height regressions that had been developed for 80 common tree species. We used literature data on wood specific gravity for 123 species occurring in the BCI plot (Chave *et al.* 2003). Here, we reassess various sources of error in the previously published AGB estimate on the most recent census of the BCI plot, conducted during the year 2000.

We also investigated the landscape-scale sampling problem, by using a network of 53 plots distributed across the watershed of the Panamá Canal, henceforth called the Marena plots. These plots were originally set up to study the variation of floristic composition in forests across the north–south climatic gradient of this region (Pyke *et al.* 2001), spatial turnover in diversity (Condit *et al.* 2002), and differential forest response to drought (Condit *et al.* 2004). Each plot is 0.32 or 1 ha in size and has all trees ≥10 cm tagged, mapped, and identified to species or morphospecies, except for 154 trees out of 22,955 (0.7%) that remain unidentified. A total of 775 species or morphospecies were identified. This represented a total sampling effort of ca. 49 ha. These plots spanned a variety of environmental types and of successional ages (Chave *et al.* 2004 appendix 2). Two plots located to the south-west of the Canal (plots 38 and 39) were excluded from the present analysis.

Results

Uncertainty on tree-level AGB estimate

The uncertainty associated with the diameter measurement in the BCI forest was discussed in Condit *et al.* (1993). To estimate rates of error, we did a double-blind remeasurement of 1715 trees in 1995 and 2000 (Condit 1998b) and fit the discrepancies with a sum of two normal distributions. The first describes small errors and has a standard deviation (SD_1) proportional to the trunk diameter; the second has a fixed larger standard deviation

(SD$_2$). The 1715 errors were best fit with SD$_1 = 0.0062 \times D + 0.0904$, SD$_2 = 4.64$, (all units in cm), with 5% of the trees subject to the larger error. For example, the diameter of a 30-cm tree has a typical error of 0.27 cm (95% probability) or of 4.63 cm (5% probability). The uncertainty associated with the height estimate is due to the inherent measurement problem of tree height. Tree heights were measured with a laser rangefinder for over 1000 trees ≥ 10 cm in 80 different species. Based on this dataset, we assume that the error on height is about 10% of the estimated value.

We assume a standard deviation of 10% of the mean wood specific gravity for all species. This figure is based on 50 neotropical tree species for which >6 different estimates were available from a total of 43 literature sources. A detailed report on this dataset is beyond the scope of the present work and will be the topic of a forthcoming publication. For species missing wood specific gravity estimates, we used a mean of 0.58 g cm^{-3}, and the same error of 10%.

The measurement error on the AGB can be deduced from the equations provided in Chave *et al.* (2004, appendix 1) for the pantropical model used for trees ≥ 10 cm. We find $\sigma_M = 0.165 \langle AGB \rangle$, and $\sigma_A = 0.313 \langle AGB \rangle$. Hence, the uncertainty on the AGB estimation of a single tree ≥ 10 cm is 48% of the estimated AGB, partitioned into 31% due the allometric model and 16% due to the measurement

uncertainty. However, this error averages out at the stand level. For the model used for trees <10 cm (equation 13.5), the uncertainties are $\sigma_M = 0.234 \langle AGB \rangle$ and $\sigma_A = 0.547 \langle AGB \rangle$, and this model predicts that 7.66 Mg ha^{-1} are in trees <10 cm (1 Mg $= 10^6$ g).

Allometric model selection error

The different allometric models estimated AGB from 215 and 461 Mg ha^{-1} with a mean of 347 Mg ha^{-1} and a standard deviation of 77 Mg ha^{-1}, before correcting for variation in wood specific gravity (Table 13.1). Equations that included wood specific gravity predicted 218–334 Mg ha^{-1} in trees ≥ 10 cm (mean: 284 Mg ha^{-1}, SD: 37 Mg ha^{-1}, or 13% of the mean).

Next we assessed the uncertainty due to the AGB estimation in large trees for the same eight allometric models. The extrapolated AGB represented 7–30% of the total AGB, depending on the model. We used the pantropical equation to correct for the AGB of extrapolated trees. BCI has an unusually high abundance of very large trees, and this problem is unlikely to be as important in other forests. When corrected for very large trees, the predicted AGB estimate for the BCI forest ranged between 220 and 315 Mg ha^{-1} (mean: 277 Mg ha^{-1}, SD: 30 Mg ha^{-1}, or 11% of the mean).

Table 13.1 AGB estimates (in Mg ha^{-1}) for the BCI 50-ha forest based on eight different equations involving only diameter, developed for different forests

Reference	Location	Nb trees	Max d.b.h.	ρ_{av}	Uncorrected AGB	Correction ρ/ρ_{av}	Correction large trees + ρ/ρ_{av}
Araujo *et al.* (1999)	Para, Brazil	127	138	0.68	375	307	315
Chambers *et al.* (2001b)	Manaus, Brazil	161	120	0.69	330	266	278
Overman *et al.* (1994)	Colombia	51	98.2	0.62	351	292	274
Yamakura *et al.* (1986)	Kalimantan, Indonesia	38	130.5	0.7	461	334	310
A. Joyce, in Brown 1997	Costa Rica	92	116	0.52	215	218	220
Lescure *et al.* (1983)	French Guiana	187	118	0.66	428	322	288
Chave *et al.* (in press)	Pantropical	634	148	0.6	324	278	268
Chave *et al.* (in press)	Pantropical	634	148	0.6	293	260	260

The AGB estimate for trees ≥ 10 cm varied significantly among equations, even when the problem of using an equation beyond its acceptable range was corrected. We provide the reference for the original data, the region of this study, the number of trees ≥ 10 cm used to construct the allometry, the maximal diameter in this sample, and the mean wood density of the trees. See Chave *et al.* (2004) for further details.

The 'best estimate' equation predicted 260 Mg ha^{-1} for trees $\geq$10 cm, very close to the value reported in Chave *et al.* (2003).

Minimal single plot size

We used the BCI 50-ha plot to evaluate the stand-level sampling uncertainty. The tree-level uncertainties average out at the stand scale. In a typical plot of ¼ ha in size, the error on the AGB estimate is 10% of the mean. The AGB held in the subplots of 50-ha plot are not autocorrelated, even for very small subplots (Chave *et al.* 2003). Hence, two neighbouring subplots of size 10 m × 10 m to 100 m × 100 m are not significantly more similar in their AGB stock than two randomly chosen plots. Although this figure of a minimal plot size of ¼ ha might slightly vary with the stem density in the plot, it can be taken as a reasonable rule of thumb.

Landscape-scale representativeness

The landscape-scale variation of the AGB estimate for the BCI forest was assessed using the Marena dataset. First, we estimated the mean and standard deviation of basal area and AGB by averaging across plots. Landscape-scale AGB was 245 ± 57 Mg ha^{-1}. This high variance reflects the variability of environmental conditions, and of variations in forest disturbance history. In particular, climatic gradients are large across this landscape: annual rainfall ranges from 1900 to 4000 mm yr^{-1} from south to north.

We then constructed a GLM using all available spatial information (geographical location and Landsat imagery). The best model explained 76% of the deviance, and the coefficient of determination using the jackknife validation procedure was $r^2 = 0.664$, $p < 0.001$ (mean error = 9.46%,

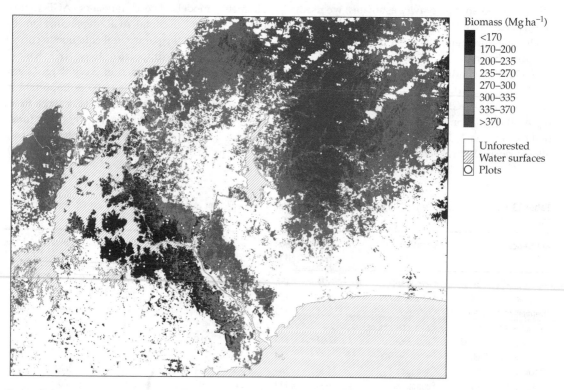

Biomass (Mg ha^{-1})

- <170
- 170–200
- 200–235
- 235–270
- 270–300
- 300–335
- 335–370
- >370

- Unforested
- Water surfaces
- Plots

Figure 13.2 Predicted forest biomass of Panama Canal Watershed based on a generalized linear model. The model is extrapolated to the rainforest (young secondary forest, old secondary forest, and old-growth forest) which was previously classified with a maximum likelihood classification algorithm (classification error: 1%). In white, non-rain forested land: agricultural and urban land, mangroves, shrubland, and clouds. (See also Plate 4.)

percentage of agreement in model selection = 91%). Significant factors were geographical position, and bands 2 (green) and 4 (near infrared) of the Landsat image ($p < 0.05$). We then used this model to construct an AGB map for the Panama Canal Watershed (Fig. 13.2). North of Panama City, the AGB is overestimated due to a haze on the Landsat image. Cumulating the estimate for each pixel, we found a total of 0.045 Pg for 1715 km^2 of forested area (dry AGB, 1 Pg = 10^{15} g). This corresponds to a mean of 262.6 Mg ha^{-1}, larger than the plot-based average. We also estimated that the potential forested area in the Panama Canal Watershed was 3299 km^2, and that 48% had been deforested in the recent past.

Discussion

Relative importance of the sources of error

In the present study, we did not examine the importance of biases such as the measurement of trees at breast height when the stem is buttressed, although those can be present in some datasets. However, we did consider the error terms that are unavoidable in ecological studies: imprecision on the measurements, and on the estimate of wood specific gravity. These are obvious sources of error, yet we contend that they are not the largest ones. Analysing the structure of the existing biomass regression models, we found an intrinsic source of error not due to the size of the census plot, but to the sample available to construct the model itself. For the 50-ha plot, an error of over 20% on the AGB estimate was due to the choice of the allometric equation. Insofar as possible, we correct these models by including wood specific gravity. This reduced the error to about 13% of the mean. Since none of these equations were designed to estimate the AGB of trees beyond a limited range, we used a pantropical regression model to estimate the AGB of the largest trees. This also led to a significant reduction of the error, to about 10% of the mean AGB.

In general, useful biomass regression models should reflect inter-site variation as well as within-site variation. Hence, allometric biomass models based on regional or pantropical compilations

should be preferred to site-specific models based on small sample sizes. Sampling issues were also considered. We reinforce previous results advocating the use of plots at least 0.25 ha in size (Laurance *et al.* 1999; Clark and Clark 2000; Keller *et al.* 2001). Our plot-based averages of AGB were lower than the landscape-scale averages. This should be an important source of discrepancy across plot-level AGB averages (Houghton *et al.* 2001).

A summary of our results are reported in Table 13.2 For a total sampled area of 5 ha, the cumulated uncertainty on the estimate is about 20% of the mean, with only a small fraction due to

Table 13.2 Summary of the sources of error in the AGB estimation of a tropical forest

Error type	SE (in % of the mean)	Type of data
1—Tree level error		
Trees >10 cm	48	BCI plot—pantropical allometric model
Trees <10 cm	78	
2—Allometric model		
Before ρ correction	22	BCI plot—eight allometric models
After ρ correction	13	
After large tree correction	11	
3—Within plot uncertainty		
0.1-ha plot	16	BCI plot—pantropical allometric model
0.25-ha plot	10	
1-ha plot	5	
4—Among plot uncertainty	11	Marena plots— pantropical allometric model
Total		
Fifty 1-ha plots, after ρ and large tree corrections	24	—

Type 1 error refers to the error made in the estimation of the AGB held in a single tree; this error averages out in plots. Type 2 error is that caused by the choice of the allometric model. Types 3 and 4 are two types of sampling errors, which can be minimized by large-sized, multi-plot, censuses. The reported values hold for the forests of the Panama Canal Watershed.

measurement error (assuming unbiased measurement), 10% due to the allometric error, and 10% due to the sampling error. With larger plots, one can reduce the sampling uncertainty, but not the allometric uncertainty. We stress that such conclusions may vary, depending on the forest under study.

Comparison with other landscape-scale studies

Several studies report results on AGB estimates for forests at the landscape scale. We compare our results to three similar studies. The first study, carried out at Los Tuxtlas Biological Station, southern Mexico, used a nested sampling design to estimate the AGB in four plots ca. 0.79 ha in size (Hughes *et al.* 2000, table 5). A mean AGB of 403 ± 50 Mg ha^{-1} was found for this forest. They report a very high density of large trees (as much as 23 trees >70 cmha^{-1}), lmost twice higher than the values commonly found in neotropical rainforests. However, tree diameters were measured correctly, (R. F. Hughes, personal communication). The second study took place at La Selva Biological Station, Costa Rica. Clark and Clark (2000) used three sampling designs to assess the landscape-scale variability. They report a low AGB estimate (160.5–186.1 Mg ha^{-1}) and an among-site sampling error of 4.2–8.4 Mg ha^{-1}, based on trees ≥ 10 cm. Their sampling error is consistent with that found in the present study, and measurement error was minimized as much as possible. Though they did not account for the allometric error, and their allometric equation may significantly underestimate tree AGB, it is possible that the La Selva forest holds less AGB than the forests of central Panama.

The third study took place in the forests of central Amazon, in the biological dynamics of forest fragments (BDFF) project north of Manaus. Nascimento and Laurance (2002) reported an AGB estimate of 325 ± 31 Mg ha^{-1} ($n = 20$ 1-ha plots). The AGB held in small trees (<10 cm) represented 21 Mg ha^{-1}. This study used an allometry comparable to ours, and suggests that the central Amazonian forests hold on average about 20% more AGB than the forests of central Panama, although the abundance of large trees is much

lower. This is a clear illustration of the importance of including wood specific gravity in pantropical allometric models.

The lack of standardization in AGB procedures largely hampers comparative exercises. We hope that collaborative efforts will help resolve this problem (e.g. Malhi *et al.* 2002a). Baker *et al.* (2004a) report results for 59 forest plots across Amazonia, for a total sampled area of about 80 ha. They consider three regions: north-western Amazonia, south-western Amazonia, and central and eastern Amazonia, and find significant difference among these regions: using the equation of Chambers *et al.* (2001b) they predicted 288, 258, and 347 Mg ha^{-1} in the regions, respectively. Using another equation, however, they consistently find AGB figures 20% lower, confirming the utmost importance of the choice of the allometric model.

Recommendations

Using allometric regression models to convert tree diameter data into stand-level AGB estimates often leads to errors. Plots where very many large trees are recorded (e.g. over 15 trees ≥ 70 cm ha^{-1}) should be double-checked. Only stands ≥ 0.25 ha should be included in the analysis. These points have fortunately been taken into account in the most recent AGB estimation procedures (Baker *et al.* 2004a; Chapter 11, this volume). Allometric models constructed from small sample sizes and from trees spanning a small diameter range should be avoided, and only equations based on at least 100 weighed trees should be used. Pantropical allometric models are for the moment the best available ones. The AGB of large trees should be carefully estimated, especially if their diameter exceeds the range for which the use of the allometric equation is valid. In this case, only a 'best guess' estimate can be produced. Wood specific gravity should be included in the allometric equation wherever possible. AGB held in life forms other than trees ≥ 10 cm should be estimated as well (in particular trees <10 cm, lianas, and bamboos, when present). These life forms may represent as much as 10% of the total AGB stock. The landscape-scale variability and issues of spatial autocorrelation of the data should be carefully investigated.

Acknowledgements

The US Department of Defense Legacy Program supported the census of the Marena plots at Fort Sherman and Cocoli, both on former US military land. The BCI plot has been supported by the National Science Foundation, the Smithsonian Scholarly Studies Program, the John D. and Catherine T. MacArthur Foundation, the World Wildlife Fund, the Earthwatch Center for Field Studies, the Geraldine R. Dodge Foundation, and the Alton Jones Foundation. We thank the many field workers who tagged and measured over a third of a million trees, and the Smithsonian Tropical Research Institute for logistical and financial support. JC was supported by a grant from the Institut Français de la Biodiversité; GC was supported by a postdoctoral grant from the Spanish government. This is a scientific contribution from the Center for Tropical Forest Science.

PART IV

The past and future of tropical forests

CHAPTER 14

The longevity and resilience of the Amazon rainforest

Mark Maslin

The aim of this chapter is to highlight and explain the impressive story of the persistence of the Amazonian rainforest throughout the Cenozoic. Palaeoclimate and palaeoecological records suggest that the Amazon rainforest originated in the late Cretaceous and has been a permanent feature of South America for at least the last 55 million years. During the late Palaeocene the 'rainforest' or 'megathermal moist forest' (MTMF) may have stretched as far south as 45°S in South America. The main climatic feature of the last 55 million yr has been global cooling and the general constriction of the megathermal moist forests to the tropics. The Amazon rainforest has, however, survived the high temperatures of the early Eocene climate optimum and the aridity and low carbon dioxide levels of the Quaternary glacial periods. The Amazon rainforest should, therefore, not be viewed as a geologically ephemeral feature of South America, but rather as a constant feature of the global Cenozoic biosphere. The forest is now, however, entering a set of climatic conditions with no past analogue. A future combination of a frost-free zone still restricted to the tropics with a hotter and more arid tropical climate would have a disastrous affect on the Amazon rainforest; particularly as there will be no mitigating presence of MTMF at higher latitudes as occurred in the geological past.

Introduction

Tropical rainforests display a biodiversity unmatched by any other vegetation type (e.g. Morley 2000; Willis and McElwain 2002). As such they have stimulated debates concerning both their origin as well as the cause of this unparalleled biodiversity. This chapter provides an overview of the long-term history of the Amazon rainforest in the context of both plate tectonics and global climate change. The primary thesis of this study is that the Amazon rainforest has existed since the Cretaceous, and despite huge changes in the global climate, has prevailed relatively intact. This resilience also extends to the Quaternary and its characteristic glacial periods, as increasing evidence indicates the Pleistocene refuge hypothesis to be incorrect (e.g. Mayle and Bush, Chapter 15, this volume). Hence, the Amazon rainforest is not an ephemeral feature of South America; rather it developed during the

Cretaceous and has been a permanent feature of this continent for at least the last 55 million yr.

There are a number of definitions of tropical rainforests but, evergreen to partially evergreen forests currently occur in tropical regions receiving more than 2000 mm of rain annually with less than four consecutive months with less than 100 mm of rain, a mean monthly temperature minimum of 18°C and a small annual variation in temperature. In more broad terms tropical rainforests exist in areas which are 'frost- free', have an abundance of rainfall, and lack a significant dry season. In the present climate regime these conditions only occur within the tropics. However, in earlier geological periods such conditions occurred at much higher latitudes, making 'tropical rainforest' an inappropriate term. Morley (2000) suggests the term 'megathermal moist forests' (MTMFs) when discussing the longer-term evolution and distribution of tropical

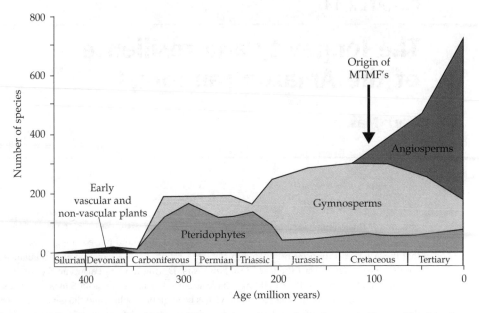

Figure 14.1 Evidence for the appearance and major expansion of the angiosperms in the Cretaceous. The possible origin of megathermal moist forests is indicated (adapted from Willis and McElwain 2002).

rainforest and this together with the simpler 'rainforest' are the terms used in this chapter.

The history of megathermal rainforests is inexorably linked to the history of angiosperms (flowering plants). Angiosperms are the dominant plants in the world today accounting for between 300 and 400 families and between 250,000 and 300,000 species (Fig. 14.1). It is believed that angiosperms first evolved in the early Cretaceous, ∼140 Ma (e.g. Crane *et al.* 1995; Willis and McElwain 2002; Schneider *et al.* 2004), though an earlier origin in the late Permian–early Triassic could be possible depending on how one interprets the recent genetic data (Schneider *et al.* 2004). One interesting question is 'did they evolve in the tropics'? It has been suggested that angiosperms could have evolved in the middle latitude uplands (e.g. Axelrod 1970). This might be consistent with the generally semi-arid and arid conditions of the tropical Cretaceous at a time when the middle latitudes were moist and warm (Figs 14.2 and 14.3). The currently favoured hypothesis, however, based primarily on pollen records, is that angiosperms instead evolved in the tropics (0–30°) and radiated out to colonize high latitude

environments over a 30 million yr period (e.g. Hickey and Doyle 1977; Barrett and Willis 2001). Based on studies integrating both pollen and macrofossil data (e.g. Lidgard and Crane 1990) five main phases of angiosperm diversification and migration have been suggested (Muller 1981; Morley 2000), with the first four all occur during the Cretaceous. These key stages also compare well with the most recent phylogenetic reconstructions of angiosperms (Schneider *et al.* 2004).

1. *Hauterivian to Cenomanian (136–99 Ma)*. Initial radiation occurred in which weeds and woody shrubs evolved in aseasonal sub-humid low latitudes while early successional vegetation spread through the low and middle latitudes. Localized occurrence of closed canopy forests may have occurred in the middle latitude convergence zones (see Figs 14.2 and 14.3). Within this period genetic evidence suggests Eudiocts, Asterids, and Rosids all evolved (Schneider *et al.* 2004).

2. *Late Cenomanian (∼94 Ma)*. Ascendancy of angiosperms over all other groups, migration to the polar regions and development of clear latitudinal zonation, and increased species diversity.

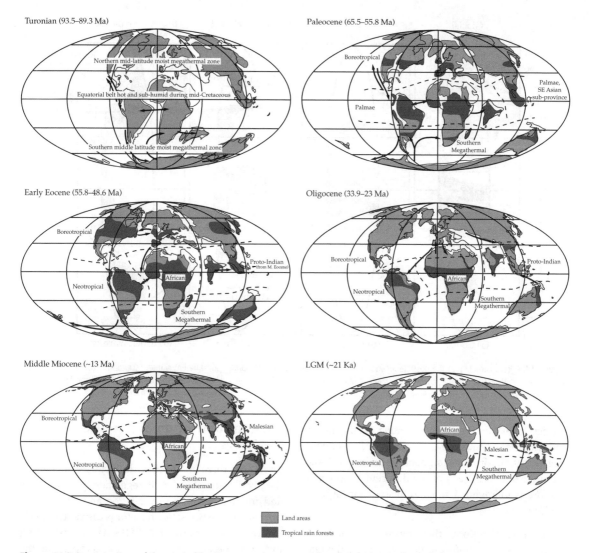

Figure 14.2 Reconstructions of the exposed land masses and occurrence of megathermal moist forests for six key periods. Arrows indicate possible dispersal of rainforest taxa (adapted from Morley 2000; Willis and McElwain 2002; and Cowling *et al.* 2001, 2004).

3. *Turonian to Santonian (~94–84 Ma).* Widespread dispersal due to low sea level and position of tectonic plates. Development of more zonal climate and increased seasonality may have aided the evolution of new families including a number of key rainforest representatives (e.g. Aquifoliaceae, Palmae, Myrtaceae, Sapindaceae, and Zingiberaceae).

4. *Campanian to Maastrichtian (~83–65 Ma).* More zonal climate becomes established, culminating in a wet equatorial zone. Closed canopy rainforests

become widespread both in the equatorial convection zone as well as at higher latitudes. Important rainforest families which evolved during this period include Caesalpinioidae, Celastraceae, Euphorbiaceae, Malvaceae, Meliaceae, and Olacaceae. By 65 Ma at least 95% of all angiosperm lineages have been established (Schneider *et al.* 2004).

5. *Paleocene (65–56 Ma).* The removal of giant herbivores briefly after the K/T (Cretaceous/Tertiary) boundary event may have allowed MTMF

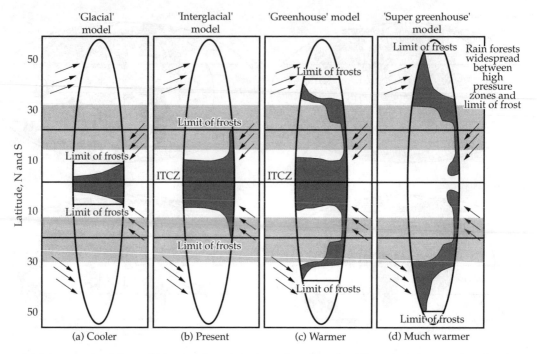

Figure 14.3 Reconstruction of the possible location of megathermal moist forest during four different climatic scenarios. Note the importance of the low-latitude ITCZ 'easterly' convection driven rainfall and the high-latitude 'westerly' convergence driven rainfall and the limit of frost-free conditions (adapted from Morley 2000).

species to expand throughout much of the lower and middle latitudes (Morley 2000). Subsequent fragmentation of the continents through tectonic activity resulted in each of the different continent's rainforests following a separate history throughout the Cenozoic.

In essence, therefore, the angiosperm lineages required to form rainforest had evolved by the late Cretaceous and definitely by the beginning of the Cenozoic. Evidence for the existence South American MTMF during the Palaeocene is limited, but both pollen and leaf physiognomy suggest extensive and diverse MTMF in South America from the early Eocene (Burnham and Johnson 2004). The evidence discussed below suggests that the Amazon rainforest has remained essentially geographically coherent for at least the last 55 million years, forming a permanent feature of the landscape of South America, even though evolution (and therefore changes in species composition) proceeded apace throughout.

Climate controls on occurrence of MTMFs

There are three main climatic factors that control the distribution of different vegetation types in space and time: temperature, precipitation, and atmospheric carbon dioxide (CO_2). Each factor is discussed briefly below in the context of MTMFs as understanding these controls is essential if we are to comprehend the distribution of rainforests in the past.

Temperature

A simplified approach to the possible distribution of MTMFs in the past with respect to temperature is to assume that it can occur within the frost-free zone. Today, the frost-free zone extends approximately 20°S–20°N restricting the present-day rainforest to the tropics. The position of the frost-free zone is primarily related to the current position of the continents and the relatively low atmospheric CO_2 which result in relatively low average global temperatures and a steep pole–equator temperature gradient ($-30°C$ to $+30°C$). However, the average temperature and latitudinal temperature gradient have varied greatly in the past (Hay 1996). In the mid-Cretaceous for example, the pole–equator temperature difference was half what it is today ($0-+30°C$) extending the frost-free zone to the high latitudes (Fig. 14.3), allowing the potential expansion of MTMFs over most continents (Morley 2000). The magnitude of the pole–equator gradient is primarily controlled by tectonics (Hay 1996). First the position of land masses over or around each of the poles can allow continental ice and sea ice to build-up, as has occurred at both poles during the Neogene. This ice build-up can not occur if there is an open ocean over the poles as ocean currents continually remove cold surface water replacing it with warmer water from lower latitudes (Hay 1996). Second, ocean gateways play an important part in redistributing heat around the planet. For example, the opening up of Tasmania–Antarctic Passage (~34 Ma) and the Drake Passage (~30 Ma) initiated the circum-Polar current that thermally isolated Antarctica and helped initiate its glaciation (Fig. 14.4). The formation of ice on the continent enhances the polar albedo, causing further cooling, leading to a positive feedback on cooling via further ice formation and enhanced albedo. Not only does this effect cause regional cooling, but it also results in global cooling as the planet's net energy balance is shifted. Average global temperatures have been decreasing since the early Eocene climate optimum (~52 Ma), moving the frost-free limits equatorward, and eventually into the tropics during much of the Pleistocene.

There is also an upper limit to temperature. During the Cenomanian large parts of the tropics were extremely hot and experienced long dry seasons. This was due to high global temperatures and second the limited marine influence due to the restricted opening of the Atlantic Ocean (Gradstein et al. 2004) which may have prevented the development of multistoreyed forests (Fig. 14.2). It is not until the Paleocene/early Eocene that tropical temperature reduced slightly and the South Atlantic Ocean opened sufficiently to provide a warm but moist climate to the Amazon Basin. This

coincides with the pollen and leaf physiognomy evidence which suggests there was extensive MTMF in South America from the early Eocene (Burnham and Johnson 2004). Temperature during the Quaternary also plays another important role in the distribution of rainforest. The lower temperatures during glacial periods have been shown to mitigate the worst effects of aridity and low carbon dioxide. Hence, Cowling et al. (2001) suggests the lower glacial temperatures would have allowed the rainforests to still be competitive with savanna, possibly explaining the lack of palaeoclimatic evidence for huge excursions of savanna into the Amazon Basin during the last ice age (Mayle and Bush, Chapter 15, this volume). This is discussed in more detail later in this chapter.

Precipitation

The global climate system contains two main global regions where rainfall is high enough to support MTMFs. The first is the tropical convection zone (Barry and Chorley 1992). Intense solar radiation in the tropics heats the land and oceans which in turns warms the lower atmosphere, causing air pressure to drop and air to rise and cool, releasing its stored water as heavy tropical rainfall. Over oceanic regions this convection is usually focused on a tight belt known as the Inter-Tropical Convergence Zone (ITCZ), which approximately girdles the equator but shifts north and south with the seasons in response to the shifting patterns of solar heating. Over continental regions the ITCZ spreads into a more diffuse convective zone, but still shifts north and south with the seasons. The north–south shift in the convection zone results in highly seasonal rainfall in much of Amazonia and thus the usual presence of at least one dry season. During the Cenomanian this dry season may have been so long, due to both high average temperatures and the restricted marine influence, that rainforests were unable to survive in the equatorial belt.

The other major global high precipitation regions are the middle latitude convergence zones, where subtropical warm, moist air meets cold, dry polar air, producing a large amount of generally aseasonal rainfall (except at the margins). Under current

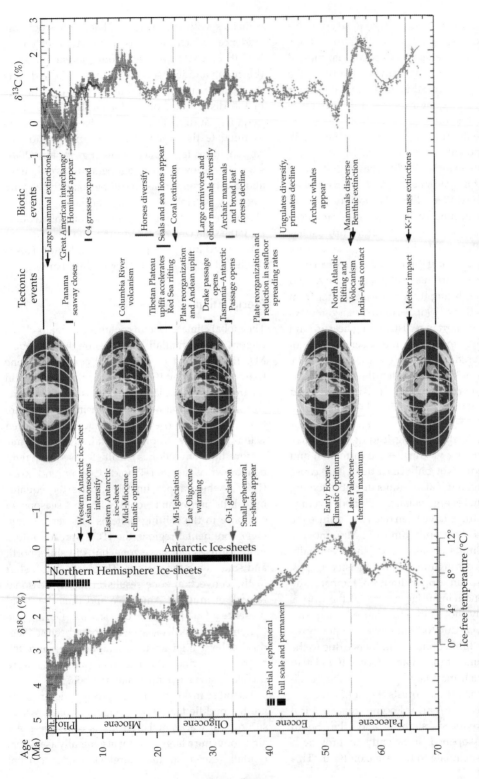

Figure 14.4 Summary of the major climatic, tectonic, and biological events over the last 70 Ma (Zachos *et al.* 2001). The benthic foraminifera oxygen isotope record on the left-hand side provides an indication of global temperature trends and also indicates the expansion of both the Antarctic and Northern Hemisphere ice sheets.

global climate these convergence zones lie between 40° and 70°N and S. Megathermal moist forests have therefore been found at very high latitudes within these convergence belts as long as temperatures remain high and the region frost-free.

Another important element is the role of topography. Orographic and convergence rainfall falls on the windward side of mountain ranges, whereas rain shadows lie on the lee side. In the equatorial belt air masses generally move westward and orographic rainfall falls on the eastern side of mountains, whereas winds in the middle latitudes generally blow eastward and orographic rainfall occurs on the westward side of a mountain range. A prime example of this is the Andes, on the western edge of the South American continent. Convergence of equatorial air masses forced by the physical barrier of the Andes enhances rainfall in western Amazonia and the eastern flank of the Andes, orographic rainfall in the mid-latitudes (Chile) falls on the western mountain flanks and is lost rapidly to the ocean, leaving much of Patagonia fairly dry.

The position of the continents can also be very important. For example, the uplift of the Tibetan plateau approximately 20 Ma which strengthened the South east Asian monsoons, by producing a much stronger convection site during the Northern Hemisphere summer so pulling even more moist oceanic air across the continent.

Carbon dioxide

The third major control on vegetation is the atmospheric concentration of CO_2 (e.g. Cowling 1999a; Cowling and Sykes 1999; Levis et al. 1999). This is an essential raw material for photosynthesis, and experimental studies of plants grown under double the current atmospheric levels of carbon dioxide show that many plants grow faster at higher concentration (e.g. Chambers and Silver, Chapter 5, this volume). CO_2 is extremely well-mixed in the atmosphere, so has very limited spatial variation at any one point in time, but the total quantity of CO_2 in the atmosphere has varied substantially in the Mesozoic and Cenozoic. Reconstruction of past atmospheric carbon dioxide

content beyond the Pleistocene ice core records is difficult but has been attempted using carbon budget models (e.g. Berner 1991, 1997), fossil stomatal densities (e.g. McElwain 1998), stable carbon isotopes of organic molecules (e.g. Pagani et al. 1999), and boron isotopes (e.g. Pearson and Palmer 2000). The general view is that from the late Triassic to the early Cretaceous levels of atmospheric CO_2 were four to five times pre-industrial Holocene levels (~280 ppmv). There is some limited evidence that CO_2 levels increased through the early Cretaceous coeval with the origin and dispersal of angiosperms suggesting a possible link; but this is currently just a supposition. Since about 100 Ma it has been inferred that levels of atmospheric CO_2 have dropped, though there is considerable debate about when and how this occurred. The general consensus is that levels of carbon dioxide did not approach modern values until the Miocene. Up to this time levels of CO_2 were sufficient to have had little effect on the distribution of MTMFs. It is only when they drop close to or below modern levels that CO_2 is likely to become a physiological constraint on C3 plants (Cowling and Sykes 1999). The drop in atmospheric CO_2 in the Miocene may have also resulted in the evolution of C4 plants (Cerling et al. 1997) which are adapted to deal with much lower levels of CO_2 (Ehleringer et al. 1997; Retallack 2001b). CO_2 becomes an important issue when dealing with the extent of rainforests during glacial periods as levels of atmospheric CO_2 dropped to 200 ppmv, which resulted in considerable stress to plants (Cowling and Sykes 1999; Cowling et al. 2001).

Tectonic setting of Amazonia

South America was part of the super continent Pangaea during the Permian, the Triassic, and the Jurassic. During the late Jurassic and early Cretaceous Pangaea started to break up and Gondwana (South America, Africa, India, Antarctica, Australia) progressively became separated from the other continents (Gradstein et al. 2004). Since the early Cretaceous the South American continent has drifted only slightly westward and northward. The northward drift resulted in the South America Plate separating from the Antarctic Plate in the mid-Cretaceous. However, some southward drift did subsequently occur, causing reconnections between Antarctica and South America throughout the Paleocene and Eocene. South America finally became separated from Antarctica during the early Oligocene with the opening of the Drake passage and the initiation of the circum-Antarctic current (Fig. 14.4). The heat loss from Antarctica to the tropics due to the circum-Antarctic current and a significant drop in global atmospheric CO_2 (DeConto and Pollard 2003) resulted in the onset of permanent glaciation on Antarctica, as demonstrated by a significant enrichment of ocean oxygen isotopes (see Fig. 14.4).

The westward movement of the South American plate was caused by the opening of the South Atlantic during the Cretaceous, causing its western South America to override the subducting Nazca (east Pacific) Plate. This collision resulted in the uplift and thus the formation of the Andean Cordillera along the whole western margin. The most intense period of uplift, especially in the northern part of the Andes, occurred during the late Oligocene (~25 Ma). Not only has this influenced the climate of the continent but it also resulted in changing the drainage pattern of the Amazon Basin. For example, both the Orinoco and Amazon have shifted their positions in response to tectonic changes (Hoorn *et al.* 1995).

The north of the South American plate was also strongly affected by tectonic changes in the Caribbean. During the early Cretaceous the proto-Caribbean Ocean separated North and South America. It has been postulated that there were land bridges which existed across this water way between the Campanian (84–74 Ma) and Paleocene (66–54 Ma) due to the formation of the Antillean Island Arc. This land bridge was severed after the Paleocene as the Antilles Arc moved north-east. In the late Cretaceous the Panama–Costa Rica Arc was formed which, by the end of the Eocene, was a continuous feature but had not yet emerged and lay to the west. Through the Cenozoic this Arc has moved eastward, colliding with North and South America. This caused the closure of the Panama Isthmus, which started at about 4.8 Ma and finished about 1.8 Ma (Haug and

Tiedemann 1998), producing a land bridge and the 'Great American Interchange' of fauna and flora.

In summary the South American plate has been in approximately the same position for the last 100 million yr, with the Amazon Basin always in the tropics. It has remained a single tectonic plate with most of its surface above sea level. For the majority of the Cenozoic South America has been separated from both North America and Antarctica. The two dominant effects on the climate of South America during this period have been the uplift of the Andes and changes in global climate (as summarized in Fig. 14.4). The effects of these changes on the extent and diversity of the Amazon rainforest are discussed below.

History of the Amazon rainforest

Based on the discussion of the origins of angiosperms, the history of the Amazon rainforest can be said to have started near the end of the Cretaceous. This is because for much of the middle Cretaceous the tropics were semi-arid and it was only after the opening of the South Atlantic that more humid conditions prevailed. However, the only evidence for rainforest during the late Cretaceous and the Palaeocene is the appearance of moderately high pollen diversity (Burnham and Johnson 2004). However, Morley (2000) argues that the true start of the Amazon rainforest occurred after the K/T impact/extinction event. The meteorite impact which brought to the end the Cretaceous and the dominance of the dinosaurs may have also caused a substantial drop in diversity of fauna and flora in South America (e.g., Wolfe 1990; Morley 2000). However the K/T event did not, according to pollen and macrofossil evidence, produce a comparable mass extinction in plants (Willis and McElwain 2002). It did though have a significant effect on the vegetation of both southern North America and northern South America. There is some limited evidence for a mass die-off of vegetation in these areas, initial recolonisation by ferns, and then secondary succession of megathermal moist vegetation (e.g., Upchurch and Wolfe 1987). So may be combination of wetter climates and the removal of very large generalist herbivores (i.e., dinosaurs) allowed MTMF to spread out and develop throughout the continent (Barrett and Willis 2001). In fact it has been suggested that the small mammals that replaced the herbivore dinosaurs may have enhanced this proliferation due to their greater focus on frugivory and thus seed dispersal (Morley 2000). During the Paleocene in the low latitudes, or 'Amazonia', the wet tropical climate may have enabled the development of a diverse MTMF vegetation while in southern South America the vegetation consisted of a mixture of general megathermal and endemic southern hemisphere taxa. Especially noticeable was the considerable diversity of palms throughout both of these South American rainforests. The extent of rainforest in South America during the Paleocene is still contentious, but there is strong agreement that by the beginning of the Eocene there were extensive rainforest throughout this whole region (Burnham and Johnson 2004).

There is some pollen and genetic evidence of rapid low-latitude vegetation species turnover during the Paleocene–Eocene Thermal Maximum (PETM, ~55 Ma), when there is a documented rise in ocean temperatures of at least 5°C (Zachos et al. 2001). For example, there is a clear reduction of Palmae in South America and replacement by dicotyledonous angiosperms several of which have bizarre pollen types that disappear by the middle Eocene (Morley 2000) and there are several phylogenetic lineages (Schneider et al. 2004) both in derived ferns (Pteridoids and Eupolypods) and angiosperms (Asterids and Rosids) which occur coeval with PETM. However, in general MTMFs continued to flourish throughout this period and tropical and paratropical rainforest extended as far south as 45°S (Fig. 14.2). Following the PETM there is a gradual cooling throughout the Eocene. South America pollen records show continuation of palms such as *Mauritia* and primitive Palmae through the Oligocene, suggesting the terminal Eocene cooling event had a limited effect on tropical South America compared with other tropical areas (Morley 2000). In contrast, the Oligocene cooling event did have a significant affect on the southern South America MTMF, with its disappearance and the expansion of sub-humid

wooded savannas. Tropical South American rain forest was unaffected by this savanna expansion until the Pliocene, when there is the first evidence of grass-dominated vegetation and open woodland expanding into parts of the tropics, particularly with the radiation of C4 grasses after ~8 Ma (Ehleringer et al. 1997; Retallack 2001b). The significant expansion of grass-dominated savannas into tropical South America occurred about 1 Ma, but even after this savanna expansion the majority of the Amazon Basin remained covered with tropical rainforest. There is one other remnant of the once ubiquitous MTMF of South America, which is the orographic rainfall maintained Atlantic rainforest of the Brazilian coast.

In summary from ~65 to 34 Ma rainforest dominated much of the South American continent as far as 45°S. From 34 to 1 Ma tropical rainforest dominated the Amazon region and much of the rest of northern South America. The climatic fluctuations of the last million years are dominated by the long glacial–interglacial climate cycle of approximately 100,000 years which allowed savanna to expand and to compete with rainforest in tropical South America.

Amazon rainforest refuge hypothesis refuted?

The 'Pleistocene tropical rainforest refuge hypothesis' has been advanced as one possible explanation for both the immense diversity and species endemism of the Amazon Basin (e.g. Haffer 1969). During each glacial period, it is suggested, that lower temperatures and precipitation in the tropics allowed savanna to replace the majority of the tropical rainforest (e.g. Colinvaux et al. 1996; Colinvaux and de Oliveira 2000). However some of the tropical rainforests would have survived in small refugia. These isolated islands of rainforest would have become hotbeds of evolution by allopatric speciation, producing many new species. At the end of each glacial period the patchwork of rainforests merges back together with higher levels of species diversity and endemism than previously. However, very little palaeoecological evidence exists for

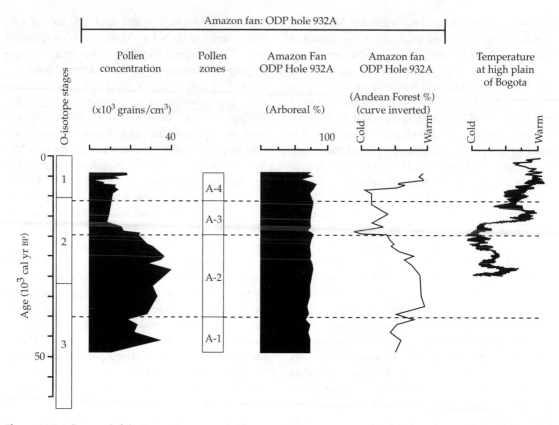

Figure 14.5 Pollen record of the Amazon Fan compared with reconstructed temperatures at the High Plain of Bogotá (adapted from Haberle and Maslin 1999).

this inferred massive incursion of savanna into the Amazon Basin during the last glacial period (Mayle and Bush, Chapter 15, this volume). In fact many lake pollen records show no reduction of rainforest pollen at all (Colinvaux *et al.* 1996; Bush 2002). Other evidence for this lack of savanna incursion include a marine pollen record from the Amazon Fan that shows large changes in the concentrations of pollen over the last 45 Ka but almost no variation in the amount of arboreal pollen (Haberle and Maslin 1999; Fig. 14.5). These records suggest that a significant Amazon rainforest still existed during the last glacial period. This interpretation is supported by biomarker work on Amazon Fan sediments which showed a consistency of the carbon isotopic composition of organic matter and ratios of different phenols received by the Amazon Fan throughout the

last glacial–interglacial cycle (Kastner and Goñi 2003).

These results have also been supported by detailed modelling work using two different types of vegetation models (Cowling *et al.* 2001, 2004; Chapter 16, this volume). Figure 14.6 shows the results of applying the glacial conditions on the Amazon Basin to the BIOME3 vegetation model (Cowling *et al.* 2001), showing a maximum of 20% reduction in the area of the Amazon rainforest during the last glacial maximum. Cowling *et al.* (2001) were able to demonstrate that the survival of at least 80% of the Amazon rainforest during the last glacial period was due to the *interactions* of the three key parameters: temperature, precipitation, and atmospheric CO_2. Individually, both low precipitation and atmospheric CO_2 are detrimental to tropical 'C3' rainforest and

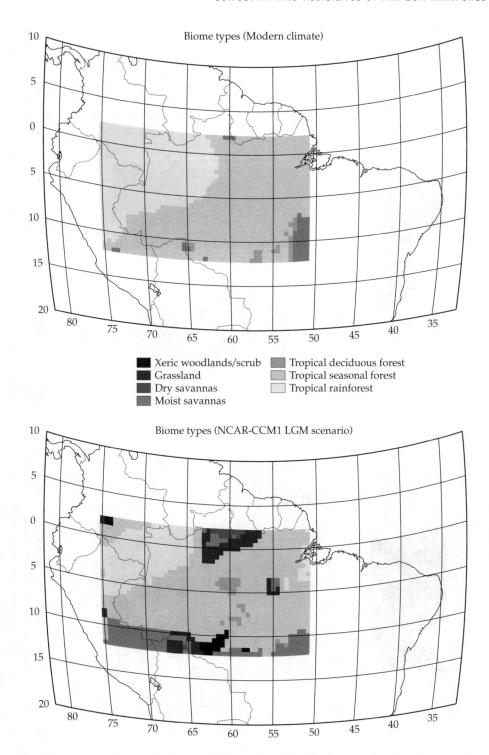

Figure 14.6 BIOME3 vegetation type reconstructions for both the modern and last glacial maximum (LGM). The LGM scenario uses the NCAR-CCM1 LGM climate output (adapted from Cowling *et al.* 2001).

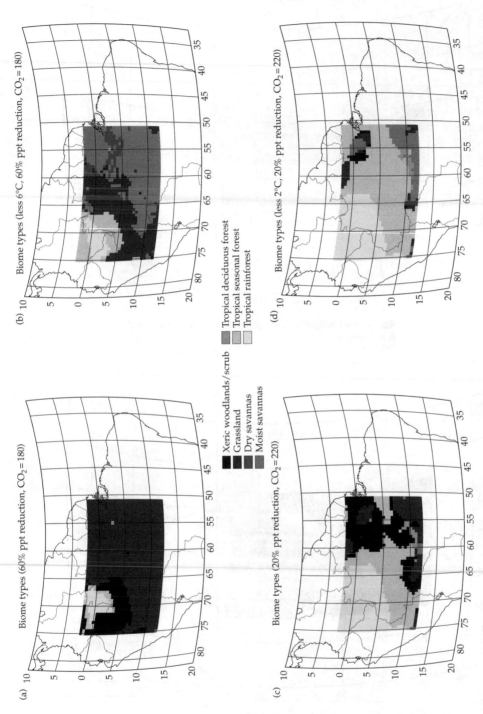

(a) Biome types (60% ppt reduction, $CO_2 = 180$)

(b) Biome types (less 6°C, 60% ppt reduction, $CO_2 = 180$)

(c) Biome types (20% ppt reduction, $CO_2 = 220$)

(d) Biome types (less 2°C, 20% ppt reduction, $CO_2 = 220$)

Xeric woodlands/scrub
Grassland
Dry savannas
Moist savannas
Tropical deciduous forest
Tropical seasonal forest
Tropical rainforest

Figure 14.7 BIOME3 vegetation type sensitivity analysis (Cowling *et al.* 2001; Cowling personal and communication 2003). (a) 20% reduction in precipitation and atmospheric CO_2 level at 220 ppmv, (b) 2°C drop in temperature, 20% reduction in precipitation and atmospheric CO_2 level at 220 ppmv, (c) 60% reduction in precipitation and atmospheric CO_2 level at 180 ppmv, (d) 6°C drop in temperature, 60% reduction in precipitation and atmospheric CO_2 level at 180 ppmv. Note in each case that the reduced temperature enables rainforest to cover a larger area than if precipitation and carbon dioxide changes are considered alone.

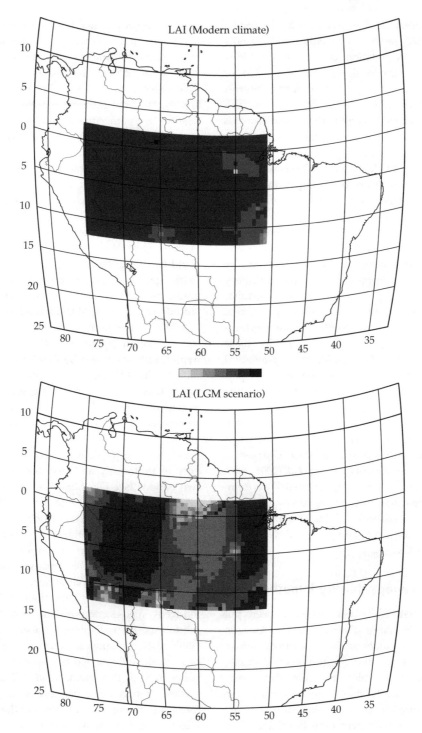

Figure 14.8 BIOME3 LAI reconstruction for modern and the LGM, adapted from Cowling *et al.* (2001). Note the modern homogeneous LAI compared to the LGM heterogeneous LAI.

favour C4 grasslands. This is shown by the results of the BIOME3 model when either 20% or 60% reduction of precipitation is applied combined with either a carbon dioxide level of 220 or 180 ppmv there is a large reduction in rainforest (see Fig. 14.7(a) and (c)). In the first case the rainforest is reduced to 50% and in the second case it is reduced by 90%. However, when the temperature effects are included many of the detrimental effects are negated (Fig. 14.7(b) and (d)). This is because of two processes which are operating in the model which may mimic how plants would have responded in reality: (1) the decrease in temperature reduces evapotranspirative demand and therefore increases moisture availability for carbon uptake and leaf development and (2) temperature decreases result in lower rates of photorespiratory carbon loss in C3-plants (Brooks and Farquhar 1985), so enhancing carbon use efficiency of trees relative to C4 grasses. Despite the assumptions inherent in any vegetation/climate model, these results show that the combined physiological effects of temperature, carbon dioxide, and precipitation must be considered when evaluating the distribution of past vegetation.

In summary the current consensus is that Amazon rainforest survived and even flourished during glacial periods (Mayle and Bush, Chapter 15, this volume). The work of Cowling et al. (2001, 2004; Chapter 16, this volume) suggests that lower temperatures aided trees to compete successfully with grasses in the majority of the Amazon Basin by mitigating the worst effects of aridity and low atmospheric CO_2.

Rainforest canopy density hypothesis

Lots of alternative explanations have been put forward for the large diversity of the Amazon rainforest. In addition to the Pleistocene refuge hypothesis there are other suggestions that should be considered. The tropical 'museum' theory suggests that a relatively stable climate through the Cenozoic resulted in low extinction rates allowing species to accumulate over time (Stebbins 1974). This fell out of favour when the variability of tropical Pleistocene climates became clear. Recent

support, however, has been provided by genetic divergence data showing that the speciation of some rainforest animals predates the Pleistocene (Glor et al. 2001). The results of plant genetic divergence data of seasonally dry tropical forest are less clear-cut showing pre-Pleistocene diversification in South America and Pleistocene speciation in Central America (Pennington et al. Chapter 17, this volume). In contrast, for rainforest plant species, there is strong evidence of Pleistocene diversification of Inga, a species rich neotropical tree genus (Richardson et al. 2001), and for rainforest species of Ruprecthia (Pennington et al. Chapter 17, this volume). A second group of theories suggest that Amazon rainforest diversity could be due to both recent and/or ancient parapatric speciation caused by steep environmental gradients due to variations in soil, topography, or latitude (Buzas et al. 2002; Hillebrand 2004).

Added to this body of thought is a new allopatric speciation hypothesis (the 'canopy density hypothesis') suggested by Cowling et al. (2001) to replace the Pleistocene rainforest refuge hypothesis. Cowling et al. (2001) suggest that both the vegetation type and structure varied during the last glacial period. Reconstruction of modern and glacial period Leaf Area Index (LAI), a common proxy for canopy density, shows a major difference in uniformity and distribution between the two periods (see Fig. 14.8). The modern rainforest LAI is homogeneous with very little variation, whereas the glacial reconstruction is heterogeneous, with large variation across very small distances. This may in part be a structural response of the rainforest due to the sensitivity of the vegetation to small-scale environmental changes while climatically stressed. That there were climate-induced species-compositional changes within rainforests is undisputed. Haberle and Maslin (1999) showed significant increases in the number of Andean cold-adapted tree species in the glacial pollen record. Pennington et al. (Chapter 17, this volume) report an expansion of dry deciduous forest species during key periods of the Quaternary. Hence, modelling pollen and genetic data suggests a much more varied glacial rainforest than at present. Cowling et al. (2001) argued that variations

Canopy-density hypothesis of neotropical allopatry

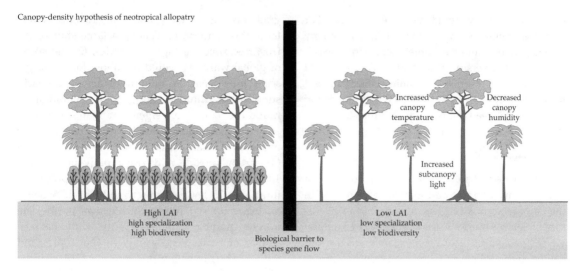

Figure 14.9 Cartoon explaining the biological barrier that would generate refuge in the canopy density hypothesis suggested by Cowling *et al.* (2001).

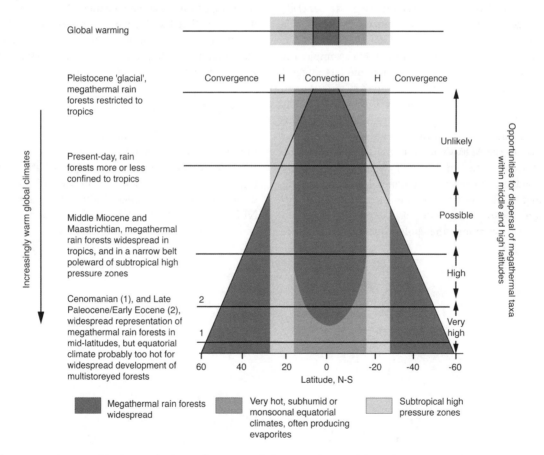

Figure 14.10 Summary of the changing distribution of MTMFs over the last 100 million years (adapted from Morley 2000). Top bar provides the suggested distribution of the non-analogue global warming future based on the work of Cox *et al.* (2000) and Cowling *et al.* (2004).

in canopy density could produce 'refugia'. For example, rainforest with high LAI or canopy density would imply a higher degree of specialization and diversity than forest with low LAI (see Fig. 14.9). Both plant and animal species might have difficulty crossing between habitats, restricting gene flow between the two types of forest.

Conclusions

Palaeoclimate and palaeoecological records suggest that the Amazon rainforest may have originated in the late Cretaceous and has been a permanent feature of South America for at least the last 55 million yr. The geological record is a testament to the longevity and resilience of the Amazon rainforest. However, there is extreme concern about the future of the Amazon rainforest both from the threat of deforestation and climate change (e.g. Cramer *et al.*, Chapter 2, this volume; Laurance, Chapter 3, this volume; Lewis, Chapter 4, this volume; Phillips and Malhi, Chapter 18, this volume). For example, the most extreme climate/vegetation models suggest the possible loss of half the Amazon rainforest in the next 50 yr (Betts *et al.* 2000; Cox *et al.* 2000; Cowling *et al.* 2004). We are entering a non-analogue future. Figure 14.10 shows the compression of the MTMFs in response to global cooling. However, the future will clearly not simply involve a transition to warmer Miocene-type climates. This is because there are significant differences to past climates: (1) despite the predicted global warming the pole–equator temperature

gradients are still large, and will remain relatively large this century, preventing a large shift of the frost-free zones to higher latitudes, (2) the speed of global warming would not allow for the large-scale movement of rainforests across the arid subtropical latitudes, (3) even if migration of rainforest to the convergence rainfall zone were possible the middle latitudes are already dominated by human activities such as farming, and (4) carbon dioxide may rise rapidly in the next 100 yr to levels (>700 ppmv) without precedent in at least 25 and possibly 65 million yr. A better analogue for our future climate may be the late Eocene (number 2 on Fig. 14.10), with the possibility of widespread aridity in the tropics, but without the mitigating presence of MTMFs in the high latitudes. Understanding these earlier periods of tropical aridity is essential to our understanding of future climate change. At the moment it appears that the Amazon rainforest which has survived for 55 million yr in the face of enormous global climate changes, may now risk destruction within as short a space as a century.

Acknowledgements

The author thanks Sharon Cowling, Oliver Phillips, Yadvinder Malhi, Peter Cox, Richard Betts, Virginia Ettwein, and George Swann for stimulating discussions and ideas. The Department of Geography drawing office drafted the diagrams. NERC grant NER/A/2001/01160 provided financial support.

CHAPTER 15

Amazonian ecosystems and atmospheric change since the last glacial maximum

Francis E. Mayle and Mark B. Bush

The aims of this chapter are to review previously published palaeovegetation and independent palaeoclimatic datasets to: (1) determine the responses of Amazonian ecosystems to changes in temperature, precipitation, and atmospheric CO_2 concentrations that occurred since the last glacial maximum (LGM), ca. 21,000 yr ago, and (2) use this long-term perspective to predict the likely vegetation responses to future climate change. Amazonia remained predominantly forested at the LGM, although savannas expanded at the margins of the basin. The combination of reduced temperatures, precipitation, and atmospheric CO_2 concentrations resulted in forests structurally and floristically quite different from those of today. Evergreen rainforest distribution increased during the glacial–Holocene transition due to ameliorating climatic and CO_2 conditions. However, reduced precipitation in the early–mid Holocene (ca. 8000–3600 years ago) caused widespread, frequent fires in seasonal southern Amazonia, with increased abundance of drought-tolerant dry forest taxa and savanna in ecotonal areas. Rainforests expanded once again in the late Holocene due to increased precipitation. The plant communities that existed during the early–mid Holocene may constitute the closest analogues to the kinds of vegetation responses expected from similar increases in temperature and aridity posited for the twenty-first century.

Introduction

The Amazon Basin is predicted to experience an increase in temperature by 3–5°C (Chapter 2) coupled with a possible reduction in precipitation by ca. 20% over the twenty-first century (Houghton *et al.* 2001). These climatic changes would reduce plant-water availability and thereby increase drought stress for many Amazonian species (Chapter 7). Insights into the nature of such anticipated vegetation responses can be obtained from understanding how Amazonian ecosystems have responded to environmental changes in the past. The aim of this chapter is therefore to determine the responses of Amazonian ecosystems to the significant changes in temperature, precipitation, and atmospheric CO_2 concentrations that occurred since the last glacial maximum (LGM), ca. 21,000 calendar years before present (cal yr BP), and synthesize this information to facilitate predictions about likely vegetation responses to future climate change. Towards this aim, we review previously published palaeovegetation datasets together with independent multi-proxy palaeoclimatic data.

Amazonia at the LGM

Climate and CO_2 concentrations

Although it is clear from polar ice-core studies (e.g. Monnin *et al.* 2001) that LGM atmospheric CO_2 concentrations were ca. 30–35% below pre-industrial values (180–200 ppm versus 280 ppm, respectively), the patterns of precipitation and

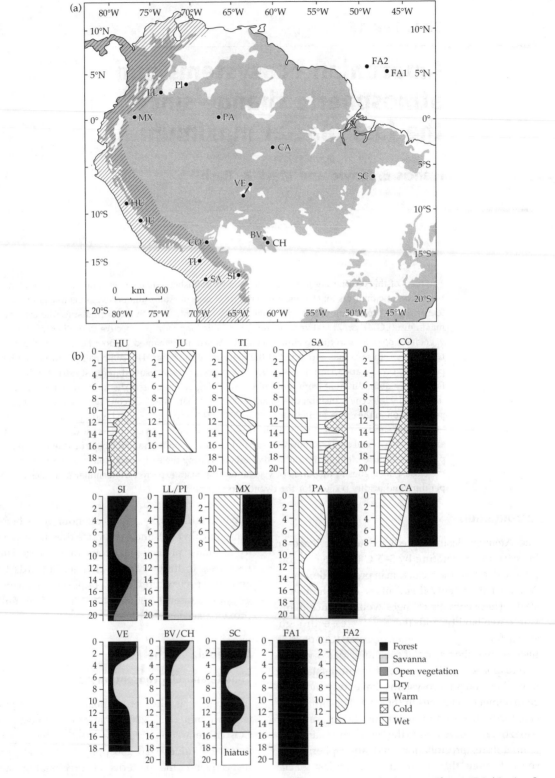

Figure 15.1 (*Continued*).

temperature across Amazonia at this time remain poorly understood. Independent multi-proxy data point to an Amazon basin ca. 5°C cooler than today (Mayle *et al.* 2004). However, Bush and Silman (2004) argue that the nature of this cooling was complex, with the relative importance of differing climatic parameters varying across Amazonia. They suggest the lowered concentrations of greenhouse gases (CO_2 and CH_4) would have caused a basin-wide general cooling of ca. 2–3°C due to increased radiative heat loss on clear nights. 'Event-based' cooling phenomena, such as incursions of polar fronts, would have been superimposed on this 'background cooling' and varied in severity and frequency across Amazonia. Bush and Silman (2004) hypothesize that the more southerly position of the north polar jet, caused by the Laurentide ice-sheet, would have increased both the frequency and severity of cold airmass incursions from North America into northern Amazonia. Southern Hemisphere polar incursions (e.g. *surazos, friajes, friagens*) were unlikely to have been any more severe than today, but were probably more frequent.

Although both empirical data (e.g. lake level) (reviewed by Mayle *et al.* 2004) and model simulations (e.g. CLIMAP 1981; Ganopolski *et al.* 1998; Kutzbach *et al.* 1998; Hostetler and Mix 1999) point to reduced LGM precipitation compared with the present, the seasonal and spatial distribution of precipitation across the basin is likely to have been complex, as it is today. Significantly, with regard to predicted impact on vegetation, all the model simulations suggest that December–January–February (DJF) (the wet season over most of Amazonia) was drier than today, while June–July–August (JJA) (the dry season over most of Amazonia) was as wet as, or wetter than, present. Given that there is a significant water surplus in most of the Amazon Basin during the rainy season, one would predict that the simulated reduction in wet season precipitation would be insufficient to cause any noticeable compositional or structural change to the forest, except perhaps in ecotonal areas. Although there is empirical evidence for lowered lake levels across much of lowland Amazonia at the LGM (Mayle *et al.* 2004), these climate model results indicate that a drop in lake level (which can only provide information about total annual precipitation, rather than its seasonal distribution) need not necessarily imply any change in vegetation.

Vegetation responses

Accumulating palaeoecological evidence suggests that most of the Amazon Basin remained forested at the LGM (Fig. 15.1), contrary to the rainforest refugia hypothesis (Haffer 1969; Haffer and Prance 2001). Colinvaux *et al.* (1996) and Bush *et al.* (2002) provide pollen data to show that the Lake Pata catchment in north-western Amazonia was forested at this time, and that, furthermore, it has been continuously forested over the last 170,000 years. Soil stable carbon isotope analyses by De Freitas *et al.* (2001) show that savanna islands within southern Amazonian rain forests (between Porto Velho, Rondonia State, and Humaita, Amazonas State, Brazil) were no more extensive at the LGM

Figure 15.1 (a) Map showing the location of sites discussed in the text. The shaded area shows the current distribution of humid evergreen broad-leaf forest (rainforest). The hatched area shows the Andes mountains. Lowland unshaded areas represent seasonally dry forests and savannas. (b) Schematic broadscale summary trends of climatic change and/or vegetation response for each site since 21,000 cal yr BP. NB: These profiles are intended as simplified cartoons to summarize the information discussed in the text and should not be viewed as absolute quantitative depictions. Key for sites: HU, Huascaran ice-core (Thompson *et al.* 1995, 9°07'S, 77°37'W); JU, Lake Junin (Seltzer *et al.* 2000, 11°S, 76°W); TI, Lake Titicaca (Baker *et al.* 2001, 17°S, 69°W); SA, Sajama ice-core (Thompson *et al.* 1998, 18°06'S, 68°53'W); CO, Lago Consuelo (Bush *et al.* 2004, 13°57'S, 68°59'W); SI, Siberia (Mourguiart and Ledru 2003, 17°50'00''S, 64°43'08''W); PI, Laguna El Pinal (Behling and Hooghiemstra 1999, 4°08'N, 70°23'W); LL, Laguna Loma Linda (Behling and Hooghiemstra 2000, 3°18'N, 73°23'W); MX, Maxus 4 (Weng *et al.* 2002, 0°27'S, 76°37'W); PA: Lake Pata (Colinvaux *et al.* 1996, Bush *et al.* 2002, 0°16'N, 66°41'W); CA, Lago Calado (Behling *et al.* 2001, 3°16'S, 60°35'W); VE, Porto Velho/Humaita (De Freitas *et al.* 2001, 8°43'S, 63°58'W to 7°38'S, 63°04'W); BV, Laguna Bella Vista (Mayle *et al.* 2000, Burbridge *et al.* 2004, 13°37'S, 61°33'W); CH, Laguna Chaplin (Mayle *et al.* 2000; Burbridge *et al.* 2004, 14°28'S, 61°04'W); SC, Serra dos Carajas (Absy *et al.* 1991; Sifeddine *et al.* 2001, 6°35'S, 49°30'W); FA1, Amazon Fan (Haberle and Maslin 1999, 5°12.7'N, 47°1.8'W); FA2, Amazon Fan (Maslin and Burns 2000, 5°45'N, 49°06'W). Modified from Mayle *et al.* (2004).

than today. Precipitation was sufficiently high for cloud forest to persist in the southern Peruvian Andes (Bush *et al.* 2004), although these Pleistocene cloud forest communities cannot be considered as refugia (as argued by Haffer 1969; Haffer and Prance 2001) because their floristic composition was considerably different from that of present-day cloud forests.

However, in contrast to these sites in northwestern and southern Amazonia and the tropical Andes, records from sites at the ecotonal margins of the Basin do reveal changes in forest/savanna distribution between the LGM and today. The 50,000-yr pollen record from Laguna Chaplin (Mayle *et al.* 2000; Burbridge *et al.* 2004) shows that Amazonian rainforest communities at the LGM were located at least 30 km north of their current southern limit in eastern Bolivia (Fig. 15.1), and that this ecotonal area was then dominated by a mix of open savannas and semi-deciduous dry forest. Pollen evidence from Laguna El Pinal, in the Colombian savannas of the Llanos Orientales at the opposite end of the Basin (Behling and Hooghiemstra 1999) (Fig. 15.1), shows that this site, which is presently bordered by localized gallery forests, was then surrounded by virtually tree-less savanna.

While it would be unwarranted to extrapolate these isolated pollen records to the entire Amazon Basin, pollen data from Amazon Fan sediments can be considered a more reliable indicator of Basin-wide changes in vegetation, since these pollen assemblages have been deposited from the entire Amazon River catchment. Pollen spectra from these Amazon Fan cores show no significant changes in the relative proportions of forest versus savanna pollen taxa between the LGM and the Holocene (Haberle 1997; Hoorn 1997; Haberle and Maslin 1999). These findings corroborate the isolated terrestrial pollen records, showing that while there is evidence for more widespread savannas at the ecotonal northern and southern Amazonian margins relative to today, most of the Basin remained forested at the LGM. Some have argued that the Amazon Fan signal merely reflects localized *varzea* rainforest lining the rivers, and therefore cannot be used to test the rainforest refugia hypothesis. However, Haberle (1997) and Haberle

and Maslin (1999) convincingly rebut this criticism with detailed studies of modern fluvial pollen assemblages throughout the Amazon drainage system, which are clearly shown to be derived, not only from local *varzea* vegetation, but also regional *terra firme* vegetation distant from the rivers.

Evidence from a variety of sources indicates that these environmental changes during the LGM had significant impacts upon the floristic composition and structure of the forest communities. The fossil pollen data (albeit from very few sites) indicate that the species composition of these glacial-age forests has no modern analogue. The climate cooling caused taxa that are presently largely restricted to the Andes (e.g. *Podocarpus* and *Alnus*) to spread throughout much of the Amazon Basin and form mixed communities with lowland Amazonian forest taxa (Colinvaux *et al.* 2000). Vegetation modelling by Cowling *et al.* (2001) suggests that carbon limitation and physiological water stress due to lowered atmospheric CO_2 concentrations may have caused a 34% reduction in leaf area index, that is, canopy density, of Amazon forest trees (Chapter 16).

Notwithstanding these floristic and structural changes, it has generally been assumed (e.g. Colinvaux *et al.* 1996) that these forest communities remained dominated by evergreen rainforest taxa. However, Prado and Gibbs (1993) and Pennington *et al.* (2000) raise the possibility that, under a more arid climate, suitably fertile, base-rich soils could have supported predominantly semi-deciduous and/or deciduous dry forests instead, arguing that the Pleistocene fossil pollen spectra are consistent with pollen signatures of both types of forest.

Amazonia during the late-glacial and Holocene

Climate and CO_2 concentrations

Changes in climate and greenhouse gases since the LGM are discussed in detail by Mayle *et al.* (2004). Following the LGM, CO_2 concentrations rose to near pre-industrial levels by the onset of the Holocene, after which they rose only gradually until the Industrial Period. Multi-proxy data from Lake Titicaca (Cross *et al.* 2000, 2001; Baker *et al.* 2001; Paduano *et al.* 2003), Lake Junin (Seltzer *et al.* 2002), and Sajama Mountain (Thompson *et al.* 1998) are indicative of high-frequency fluctuations in temperature and precipitation on the Peruvian–Bolivian Altiplano during the glacial–Holocene transition (Fig. 15.1). However, fossil pollen evidence (Bush *et al.* 2004) from the Andean slopes immediately below Lake Titicaca suggests that temperature changes, though not necessarily precipitation oscillations, were relatively gradual on the Andean slopes and in the Amazon lowlands below. Bush *et al.* (2004) applied pollen-derived climate transfer functions to a 48,000-year fossil pollen record from the cloud forest site Lago Consuelo (1360 m elevation) (Fig. 15.1) on the eastern flank of the Peruvian Andes, which showed that temperatures rose only gradually over 8000 years between the LGM and the onset of the Holocene, possibly averaging <1°C per millennium, with no evidence for high-frequency climatic oscillations superimposed on this climatic trend.

In contrast, climatic aridity during the early–mid Holocene was widespread, affecting both the Altiplano and much of southern Amazonia. Evidence for this comes from the −90 m low-stand of Lake Titicaca between ca. 8500 and 3600 cal yr BP (Fig. 15.1), when water levels were at their lowest over the last 25,000 yr (Cross *et al.* 2000; Baker *et al.* 2001; D'Agostino *et al.* 2002), coincident with charcoal peaks in the Bolivian Andes (Mourguiart and Ledru 2003; Paduano *et al.* 2003), lowland north-eastern Bolivia (Burbridge *et al.* 2004), and Para State, Brazil (Turcq *et al.* 1998).

Vegetation responses

Over the last glacial–Holocene transition (ca. 21,000–10,000 cal yr BP) forests expanded in southern and eastern Amazonia. Cloud forests exhibited directional successional changes over this period in response to the gradual 6°C rise in temperature: progressive elimination from the forests around Lago Consuelo of higher altitude montane elements such as *Alnus*, *Bocconia*, *Hedyosmum*, and *Vallea*, concomitant with increasing abundance of

lowland elements such as *Acalypha*, *Alchornea*, *Celtis*, *Trema*, and Moraceae/Urticaceae (Bush *et al.* 2004). Pollen data from the Bolivian Andes further show that cloud forest moved further upslope between 12,500 and 10,000 ^{14}C yr BP (14,900 and 11,500 cal yr BP, Stuiver *et al.* 1998) (Baker *et al.* 2003; Mourguiart and Ledru 2003) in response to these rising temperatures and CO_2 concentrations. There was simultaneous forest expansion at forest/savanna ecotones in the lowlands; that is, Rondonia/Amazonas states (based on stable carbon isotope data) (De Freitas *et al.* 2001) and Para state (Absy *et al.* 1991). Expanded populations of *Podocarpus* are documented in many late-glacial pollen records (e.g. Mayle *et al.* 2000; Behling 2001; Ledru *et al.* 2001; Sifeddine *et al.* 2003; Burbridge *et al.* 2004), suggesting that the combination of increasing CO_2 levels and precipitation, together with sufficiently cool temperatures, had reduced water stress sufficiently to allow expansion of this predominantly Andean genus in the lowlands.

The early–mid Holocene episode of aridity appears to have had a significant impact on plant communities, especially in ecotonal areas of Amazonia. Stable carbon isotope data from soil organic matter show expansion of savanna islands at the border of Amazonas and Rondonia states, Brazil (Pessenda *et al.* 1998; De Freitas *et al.* 2001). The increased fire frequencies between 7000 and 3000 ^{14}C yr BP (7800 and 3200 cal yr BP) in southern Amazonia would be expected to have caused significant structural and compositional changes to the forests, especially increases in drought-tolerant lianas and semi-deciduous tree species. For example, the peaks in the dry forest taxa *Anadenanthera colubrina*, *Astronium urundueva*, and *Astronium fraxinifolium*, together with the savanna indicator *Curatella americana*, and Poaceae pollen, in combination with high charcoal concentrations, from Laguna Chaplin and Laguna Bella Vista (Burbridge *et al.* 2004) show that the ecotonal area of Noel Kempff Mercado National Park (NKMNP) in south-western Amazonia, which is presently dominated by evergreen rainforest, was instead covered by a combination of seasonally flooded savannas and semi-deciduous dry forests throughout most of the Holocene, until ca. 3000–1000 cal yr BP. Similarly, at the northern Amazon ecotone of Colombia (e.g. Laguna Loma Linda) there was greater extent of savannas during the early–mid Holocene than in the late Holocene (Behling and Hooghiemstra 2000, 2001). Forests in the Ecuadorian Amazonian lowlands, which today experience the highest precipitation in Amazonia, remained as early successional communities, dominated by the pioneer tree *Cecropia*, between ca. 8700 and 5800 cal yr BP. Weng *et al.* (2002) argue that repeated droughts were the most likely reason for tree mortality, and hence increased gap formation. This early–mid Holocene climatic aridity also caused reduction of Andean cloud forest species, both near their lower (Lago Consuelo, Peru, Bush *et al.* 2004) and upper elevational limits (Siberia, Bolivia, Mourguiart and Ledru 2003).

Although increases in savanna and/or dry forest taxa clearly occurred both at the LGM and mid-Holocene at lowland ecotonal areas in response to increased water stress, the structural and compositional changes to the plant communities associated with these vegetation responses would be expected to have been significantly different due to the markedly different climatic and CO_2 regimes driving these changes. Increased water stress at the LGM appears to have been caused primarily by 'ecophysiological drought' caused by the low glacial CO_2 concentrations (Cowling *et al.* 2001), whereas the increased water stress in the early–mid Holocene was caused by 'climatic drought' due to a reduction in precipitation and/or increased length of dry season.

The precipitation increase in the late Holocene caused renewed expansion of moist evergreen forest at the expense of savannas and seasonally dry forest in lowland ecotonal areas of southern and eastern Amazonia (e.g. Absy *et al.* 1991; De Freitas *et al.* 2001; Burbridge *et al.* 2004), as well as cloud forest on the Amazon flank of the Andes (Mourguiart and Ledru 2003; Bush *et al.* 2004).

Anthropogenic influences

There is archaeological evidence from the heart of the Amazon Basin (quartz spear points at Monte Alegre) that indigenous peoples have lived in Amazonia since at least 11,200 ^{14}C yr BP (13,150 cal yr BP) (Roosevelt *et al.* 1996), while Bush *et al.* (1989) show that maize has been cultivated in the Ecuadorian Amazon since ∼6000 cal yr BP. These findings raise the possibility that ecosystem changes through the Holocene could, at least in part, have been caused by anthropogenic disturbance rather than climatic change. However, the sizes of these Pre-Conquest populations, and hence their ability to significantly alter the landscape, has long been a contentious issue. There is incontrovertible evidence from ceramics, artificial earthmounds, and earthen causeways (e.g. Erickson 1995, 2000; Denevan 1966; Langstroth 1999) throughout the seasonally flooded Moxos savannas of the Bolivian Amazon for major landscape modification by Pre-Conquest native peoples. Although many forest islands in these savannas have been demonstrated to be natural remnants or fragments of gallery forests on ancient river levees (Langstroth 1999), other forest islands are clearly the result of human activity, whereby forests have invaded artificial earthmounds following abandonment by palaeoindians (e.g. Erickson 1995; Denevan 1966). Although Burbridge *et al.* (2004) argue against an anthropogenic explanation for the recent rainforest expansion in NKMNP, north-eastern Bolivia, the possibility that extensive areas of Amazonian forest ecosystems contain a legacy of centennial–millennial scale human disturbance (e.g. small-scale deforestation, burning) should not be discounted. It is certainly possible, for example, that some of the charcoal records reflect burning due to human activity (e.g. Behling 1996) rather than climate change, although disentangling the relative importance of these disturbance agents is far from being straightforward.

Predicted vegetation responses to future climatic and CO_2 changes

The combination of continued large-scale deforestation in Amazonia and global warming due to the enhanced greenhouse gas effect is predicted to have a major detrimental effect upon Amazonian ecosystems, due in large part to positive feedback between vegetation loss, climate warming, and increased aridity. Fifty per cent of precipitation in western Amazonia is recycled from eastern Amazonia via repeated cycles of convective activity due to evapotranspiration from the rainforest, which ultimately receives its precipitation from the Atlantic Ocean. Interruption of this flow of moisture from the Atlantic, by deforestation in eastern Amazonia, would markedly reduce precipitation in central and western parts of the Basin (e.g. Shukla *et al.* 1990; Nobre *et al.* 1991). Several studies modelling the outcome of complete deforestation of Amazonia, with various GCMs, simulate a decrease of precipitation in western Amazonia by ~20–30% and a temperature rise of up to 3°C (e.g. Shukla *et al.* 1990; Zhang *et al.* 2001; Avissar *et al.* 2002; Werth and Avissar 2002). Such a scenario is not beyond the realm of possibility. Laurance *et al.* (2001a) simulate that, under their 'nonoptimistic' scenario of current deforestation trends, southern and eastern Amazonia will be largely deforested by 2020 AD, with extensive forest fragmentation in central and northern parts of the Basin, and pristine forests confined largely to the western quarter of Amazonia. Furthermore, by 2050 AD, when CO_2 concentrations may have doubled, Cox *et al.* (2000) simulate, using a fully coupled carbon-climate model, that the Amazon Basin will have changed from a net carbon sink to a net carbon source, thus accelerating climate warming and forest degradation.

Interestingly, the deforestation and rising CO_2 concentrations over the past 40 yr have yielded no empirical evidence for overall drying of Amazonia (see Chapter 1), suggesting that the feedbacks between these different variables are more complex than previously thought (Curtis and Hastenrath 1999), and may relate to opposite precipitation anomalies between northern and southern Amazonia, which lie in opposite hemispheres (Marengo *et al.* 1993, 1998; Marengo 1995). Perhaps a critical threshold of deforestation (i.e. reduction in precipitation recycling via evapotranspiration) needs to be achieved before noticeable precipitation reduction ensues (e.g. Avissar *et al.* 2002).

What can palaeovegetation records tell us about the likely responses of Amazonian ecosystems to these predicted future climate changes? The data for the early–mid Holocene (centred around 6000 cal yr BP), when aridity was higher than present, suggest that future precipitation decreases coupled with warming would result in frequent, widespread fires throughout southern and eastern Amazonia, which would lead to competitive replacement of lowland evergreen rainforest taxa by drought/fire-tolerant semi-deciduous dry forest taxa and cerrado savannas. However, the early–mid Holocene should not be considered a perfect analogue for the future for two key reasons. First, atmospheric CO_2 concentrations are projected to be at least twice mid-Holocene levels by 2050 AD, suggesting that mixed liana forest and/or semi-deciduous forest (dominated by C_3 plants) may be competitively favoured over savannas, which are dominated by C4 grasses. Second, the amplitude and frequency of ENSO (El Niño/Southern Oscillation), which causes increased aridity in most of Amazonia during strong El Niño years, is significantly greater today than it was in the early–mid Holocene (Tudhope *et al.* 2001; Moy *et al.* 2002).

The pollen data from the eastern flank of the tropical Andes suggest that the base of the cloud forest would be forced upslope (Bush 2002). Such a displacement is already underway in Costa Rica where downslope deforestation is influencing cloud formation at Monteverde (Lawton *et al.* 2001). The maximum rate of climate warming over the last glacial–Holocene transition in the eastern Andes was an order of magnitude slower than that predicted by the most conservative estimate of temperature increase (1°C) over the twenty-first century (Bush *et al.* 2004). These authors argue that, while Andean plants with broad elevational distributions should be able to remain in equilibrium with this projected climate warming, taxa with narrow elevation ranges would be expected to lie completely outside their climatic envelopes within only one or two generations and thus become extinct. If temperatures were to rise by the higher IPCC (International Panel on Climate Change) estimate of 3–5°C (Chapter 2; Houghton *et al.* 2001) over the 21st century, this warming would cause the cloud base and frost limit to rise up the Andes by 600 m (almost half the vertical range currently occupied by cloud forest) (Bush 2002). Disappearance of cloud cover would produce climatic conditions favourable for agriculture and settlement, and hence one would expect that in this 'land-hungry' part of westernmost Amazonia, any land below the cloud base would be rapidly deforested. As with the rest of Amazonia, rates of ongoing deforestation far exceed the natural ability of species to respond to climate change by migration. Consequently, ecotonal areas (e.g. between the puna, cloud forest, and upland rainforest on the Andes, and between rainforest, cerrado savannas, and semi-deciduous dry forests in the lowlands) should be the target of conservation strategies to allow 'natural' species movements and plant community reassortments to occur.

Conclusions

Fossil pollen and stable carbon isotope data show that most of the Amazon Basin remained forested at the LGM, although there was some replacement of forest by savannas at the ecotonal areas towards the northern and southern margins of the Basin. The combination of a 5°C cooling, 32% reduction of atmospheric CO_2 concentrations, and a possible 20% reduction in precipitation, relative to Holocene levels, caused marked compositional and structural changes to these glacial-age forests.

Rainforests expanded in response to ameliorating climatic conditions during the last glacial–Holocene transition, although not as far southward as today. During the early–mid Holocene (ca. 8500–3600 cal yr BP) a marked decrease in precipitation affected much of the Amazon Basin, causing increased fires and consequently greater ecosystem disturbance, which probably caused expansion of drought-tolerant lianas and semi-deciduous taxa in the more seasonal forests of southern Amazonia, as well as expansion of savannas at forest-savanna ecotones.

Increased precipitation in the late Holocene (ca. 4000–3000 cal yr BP) caused reduction in fire frequencies and consequently renewed expansion of rainforests. The critical factor for these vegetation responses would not have been the increase in MAP per se, but instead the decreasing length/severity of the dry season. The magnitude of impact of palaeoindians upon Amazonian vegetation through the Holocene is poorly understood, with the exception of the Bolivian Beni in south-western Amazonia, where there is well-documented archaeological evidence that much of the forest/savanna mosaic in these seasonally flooded plains is an artificial landscape created by palaeoindians.

A 3°C rise in temperature and 20% reduction in precipitation in Amazonia over the twenty-first century, would, under natural conditions, be expected to cause similar vegetation responses to those of the early–mid Holocene; that is, renewed expansion of drought-adapted plants such as semi-deciduous/deciduous dry forest trees, lianas, and savannas in response to increased fires and water stress brought about by an increase in aridity and/or length of the dry season. The lower limit of tropical cloud forests would be expected to increase in elevation in response to the rising cloud base. Continued deforestation would no doubt accelerate these vegetation changes since a decrease in forests would itself lead to reduced evapotranspiration and hence increased temperatures and reduced precipitation.

Acknowledgement

We thank Yadvinder Malhi and Oliver Phillips for inviting us to submit this chapter. FEM is grateful for a Leverhulme Trust Research Fellowship.

CHAPTER 16

Modelling the past and the future fate of the Amazonian forest

Sharon A. Cowling, Richard A. Betts, Peter M. Cox,
Virginia J. Ettwein, Chris D. Jones, Mark A. Maslin, and
Steven A. Spall

Modelling simulations of palaeoclimate and past vegetation form and function can contribute to global change research by constraining predictions of potential earth system responses to future warming, and by providing useful insights into the ecophysiological tolerances and threshold responses of plants to varying degrees of atmospheric change. We contrasted HadCM3LC simulations of Amazonian forest at the Last Glacial Maximum (21 kya) and a Younger Dryas-like period (13–12 kya) with predicted responses of future warming to provide estimates of the climatic limits under which the Amazon forest remains relatively stable. Our simulations indicate that despite lower atmospheric CO_2 concentrations and increased aridity during the LGM, Amazonia remains mostly forested, and that the cooler climate of the Younger Dryas-like period in fact causes a trend toward *increased* above-ground carbon balance relative today. The vegetation feedbacks responsible for maintaining forest integrity in past climates (i.e. decreased evapotranspiration and reduced photorespiration) cannot be maintained into the future. Although elevated atmospheric CO_2 contributes to a positive enhancement of plant carbon and water balance, decreased stomatal conductance and increased plant and soil respiration cause a positive feedback that amplifies localised drying and climate warming. We speculate that the Amazonian forest is presently near its critical resiliency threshold, and that even minor climate warming may be sufficient to promote deleterious feedbacks on forest integrity.

Introduction

The maximum extent to which the Amazonian forest can be developed for human activity without compromise of rainforest form and function is an issue of much importance and controversy (Nobre *et al.* 1991; Fearnside 2000; Laurance *et al.* 2001a; Malhi 2002). Not only does the Amazon forest harbour a great proportion of the world's biological diversity, but perturbations to forest cover can have far-reaching regional and global climate consequences (Cox *et al.* 1999; Douville *et al.* 2000; Pielke *et al.* 2002). The role of vegetation in mediating the exchange of water and carbon between the biosphere and atmosphere is accentuated in the Amazon Basin, primarily owing to its size and high productivity (Richards 1996). Being able to predict the fate of the Amazonian rainforest with continued deforestation and potential future warming requires a deepened understanding of the ecophysiological and biophysical responses of tropical rainforests to both natural and anthropogenically induced environmental change.

Predictions of the range of possible responses of tropical rainforest to future scenarios of rising atmospheric CO_2 and global warming can be made using

General Circulation Models (GCMs) (Costa and Foley 2000; Cox *et al.* 2000; Levis *et al.* 2000, Zhang *et al.* 2001). GCM-simulations of past climates and palaeoplant distributions can be useful for investigating tropical ecology in general (Prentice *et al.* 1998; de Noblet-Ducoudri *et al.* 2000; Prentice *et al.* 2000; Harrison and Prentice 2003). Palaeomodelling research provides opportunities for model validations, for constraining our predictions of potential earth system responses to future warming, and for offering unique insights into the ecophysiological tolerances and threshold responses of tropical forests to varying degrees of atmospheric change.

Vegetation–climate interactions occur over a broad range of spatio-temporal scales, with first principle mechanisms centring on plant carbon and water balance. How vegetation is affected by environmental stimuli can be viewed as a continuum from smaller scale responses in plant structure like leaf area production, to larger scale modifications of vegetation composition like species replacements. Being able to identify when a particular climatic catalyst (or combination of catalysts) is likely to cause a shift in vegetation response, from simple effects on plant carbon balance to complex vegetation–climate interactions, is of utmost concern to ecologists and the like.

Our objectives in conducting vegetation–climate simulations using the fully coupled earth system model, HadCM3LC, were threefold. (1) To simulate the response of Amazonian forest to past changes in climate such as long-term cooling during the last glacial maximum (LGM; 21 Kya) and short-term (abrupt) cooling during the Younger Dryas (YD; 12.7–11.6 Kya). (2) To provide an estimate of the climatic limits under which the Amazon Basin has remained predominantly forested. (3) To investigate under what climatic and biophysical conditions the Amazonian forest becomes de-stabilized.

Since the HadCM3LC model does not distinguish between evergreen and deciduous (broad-leaf) trees, 'forest' will be used loosely to refer to both tropical seasonal (dry) and rainforest types. Simulations of the mid-Holocene climatic optimum are currently being conducted and interpreted by another Hadley Centre collaboration (Paul Valdes, personal communication, 2002), therefore we chose to simulate a YD-like climate in order to contrast the

effects of *long-term* versus *abrupt* climate change on forest processes in the Amazon Basin. Because the climatic processes responsible for producing the YD-climate anomaly are still under investigation, we recognize that our simulations are intended to investigate responses of a 'YD-like' climate anomaly. This terminology will be used throughout the chapter to distinguish sustained LGM climate effects from abrupt climatic changes.

Methods

HadCM3LC model

HadCM3LC couples the latest version of the Hadley ocean-atmosphere GCM (HadCM3; Gordon *et al.* 2000) to an ocean (HadOCC; Palmer and Totterdell 2001) and a terrestrial carbon cycle model (TRIFFID; Cox 2001). It uses a horizontal resolution of 2.5° latitude by 3.75° longitude, and requires the use of flux adjustments (Johns *et al.* 1997). Increasing the sophistication of representation of biological processes within GCMs is the focal point of much recent climate modelling research (Cox *et al.* 2000; Friedlingstein *et al.* 2001; Joos *et al.* 2001). The latest version of the Hadley ocean carbon cycle model, for example, now contains a four component (nutrient–phytoplankton–zooplankton–detritus) ocean ecosystem model that simulates the effects of light penetration, alkalinity, and nutrient availability on biological carbon uptake (Palmer and Totterdell 2001).

TRIFFID, the dynamic global vegetation component of the land surface scheme, MOSES (Cox *et al.* 1999), simulates growth and replacement of vegetation of five functional types: broadleaf tree, needleleaf tree, C_3 grass, C_4 grass, and shrubs. Each modelled grid cell is predicted to contain a combination of functional groups, depending on (1) climatic impacts on photosynthesis, plant respiration, and leaf mortality; (2) competition for light (i.e. taller trees outcompete shorter grasslands); (3) competition for water (i.e. according to differences in rooting depth), and (4) competition for carbon (i.e. shrubs outcompete trees in resource-depleted environments).

An allocation scheme determines the distribution of carbon resources into woody stem, root, and leaf components, and is based on empirical allometric relationships (Cox 2001). A hydrological budget distributes precipitation between canopy interception, throughfall, water percolation through four soil layers, and runoff. The most important terrestrial variables in modifying regional climate processes are leaf surface area (i.e. leaf area index, LAI, $m^2 m^{-2}$: the area of leaf per unit area of ground) and stomatal conductance. LAI is estimated from a percentage of whole-plant carbon balance that is allocated to leaf production. Stomatal conductance and photosynthesis are calculated via a coupled leaf-level model (Cox *et al.* 1998).

Climate scenarios and simulations

The *control* state of the climate system is derived from an experiment with pre-1850 boundary conditions, thus it represents a climate–vegetation system defined by non-anthropogenic influence on atmospheric CO_2 (i.e. 'potential vegetation'). Atmospheric CO_2 concentration was prescribed at 286 ppmv for the pre-industrial simulation, based on ice core CO_2 records. Two experiments with HadCM3LC were conducted to examine vegetation interactions with significantly different climates. The first was a simulation of the last glacial maximum (LGM), and the second, an approximate representation of the climate of the YD cold period.

For the LGM simulation, atmospheric CO_2 concentration was prescribed at 200 ppmv. Boundary conditions such as orbital parameters, changes in land area due to sea level decrease, and the extent of the ice sheets were also prescribed to be consistent with the LGM period (Hewitt *et al.* 2003; Spall *et al.* 2003). Sea surface temperature and salinity values from a previous LGM simulation using HadCM3 (Hewitt *et al.* 2003) were used to derive flux adjustments for the HadCM3LC experiment. Vegetation cover was initialized from Wilson and Henderson-Sellers (1985) and simulations were continued until vegetation reached equilibrium.

The YD is a brief cold climate anomaly occurring mostly in the Northern Hemisphere during the last deglaciation. A pulse of freshwater (glacial meltwater) into the North Atlantic is believed to have caused this anomaly via effects on the thermohaline circulation (THC) (Broecker *et al.* 1988; Hughen *et al.* 2000). The response and stability of the THC in HadCM3 to both freshwater and anthropogenic

forcing have been explored in previous studies (Thorpe *et al.* 2001; Vellinga and Wood 2001; Vellinga *et al.* 2001). These studies show that a single pulse of freshwater in the North Atlantic causes a temporary shut down of the THC and a significant cooling not just within the region of the North Atlantic, but across most of the Northern Hemisphere. Such climate anomalies are characteristic of reconstructed YD climatology.

For our second simulation, therefore, a single pulse of freshwater was applied to the North Atlantic. Climate and vegetation responded dynamically to this forcing (i.e. the simulation experiment followed a trajectory defined by a transient system response rather than be permitted to reach an equilibrium). It should be noted that both atmospheric CO_2 concentration and insolation values were kept at the same level as for our pre-industrial control simulation, thus this should not be viewed as a fully representative YD simulation, but rather a 'YD-like' modelling experiment.

Future climate change simulations using HadCM3LC have been reported and published elsewhere (Cox *et al.* 2000), but have been included in this chapter as a point of comparison and as a focus for discussion. Future simulations were conducted using an arbitrary trajectory of atmospheric CO_2 and temperature change (IS92a: a middle-of-the-range climate change scenario termed 'business-as-usual' in early IPCC reports). At the end of the future climate simulation (2100 AD) CO_2 concentration reaches 980 ppmv (including vegetation–climate feedbacks), with temperatures warmed by about 5.5°C (global mean) or 8.5°C (global mean over land) excluding localized vegetation–climate feedbacks.

Results and discussion

Persistence of Amazonian forest to past changes in climate

Wherever mean values are reported, averages were calculated on an area of the Amazon Basin defined by 1.25° N and 16.25° S and by 69.38° W and 50.63° W. Key parameters describing the climate and biophysical features of each simulated scenario are presented in Table 16.1. For pre-1850, surface temperatures fall within the range of 18°C and 31°C, with

Table 16.1 Key biophysical variables simulated for each of our climate scenarios, pre-1850 (control), the LGM, YD, and Future (future global warming)

Variable	Control	LGM	YD	Future
T	27	25	26	38
P	4.8	4.0	5.0	1.6
H	29.7	28.9	25.4	69.2
LE	98.1	91.5	101.0	34.2

Notes: Variables include temperature (T, °C), precipitation (P, mm per day), sensible heat flux (H, $W m^{-2}$), and latent heat flux (LE, $W m^{-2}$). Numbers represent mean values.

an average basin temperature of 27°C. Precipitation values were simulated at 4.8 mm per day, with surface sensible and latent heat fluxes averaging around 28.9 and 91.5 $W m^{-2}$, respectively.

The LGM surface temperatures in the Amazon Basin were simulated to decrease by an average of 2°C, although some regions experienced temperature cooling in the order of 10°C. YD climate-forcing resulted in an average cooling of 1°C relative to pre-1850. Precipitation averaged 4.0 and 5.0 mm per day for LGM and YD scenarios, respectively, with YD simulations showing strong regional trends. Whereas north-west Amazonia is simulated to be drier during the YD relative to pre-1850, the south-east is wetter, rendering an overall basin average that is slightly wetter than today. LGM climate resulted in a greater proportion of solar radiation converted into sensible relative to latent heat compared to pre-1850 (Bowen ratio = 0.32 relative to 0.30, respectively). In contrast, YD palaeoclimate simulations indicate a greater proportion of solar radiation distributed into latent versus sensible heat production (Bowen ratio = 0.25 relative to 0.30).

Scarcity of Amazonian palaeovegetation proxy data spanning the last glacial is to blame for over three decades of controversy concerning the integrity of the Amazon rainforest during Pleistocene climate cycles (Colinvaux et al. 2000, 2001). In connection with early theories on the origin of high diversity and high endemism in the neotropics, it was once assumed that the Amazonian rainforest disintegrated into small, isolated patches of forest because of (arid-induced) replacement of forest with grassland (Haffer's Forest Refugia Hypothesis, Haffer 1969; Haffer and Prance 2001).

The presence of forest patches ('refugia') in the Amazon Basin during glacial–interglacial cycles is much less an issue of contention today (Colinvaux et al. 2000, 2001). A growing number of pollen data indicate the presence of forest across the majority of the lowland basin (Colinvaux et al. 1996; Haberle and Maslin 1999; Mayle et al. 2000; De Freitas et al. 2001; Bush et al. 2002). Earlier simulations using biophysical dynamic vegetation models, BIOME3 (Cowling et al. 2001) and BIOME4 (Harrison and Prentice 2003) provide further evidence of the likely persistence of Amazonian forest over Pleistocene climate cycles. It is only in marginal areas near the outer boundaries of rainforest (i.e. forest–grassland ecotones) that grassland is simulated to replace tree functional types.

Similarly, our simulations using HadCM3LC strongly support the growing consensus that the Amazonian forest remained relatively intact during past cold and dry periods. Both LGM and YD-like climate simulations result in broadleaf tree-types dominating much of Amazonia (Fig. 16.1). The Hadley earth system model does not distinguish between trees of evergreen and deciduous life histories, and therefore can only identify ecosystems dominated by either forest or grassland. Modelling results indicate an increase in grass dominance mostly in regions near tree–grass ecotones, but widespread proliferation of C4 grassland within the interior of the basin is not simulated (Fig. 16.2).

Understanding the degree to which palaeovegetation of the Amazon Basin responded to past changes in climate is relevant for modern-day ecological theory for several reasons. First, the Amazonian forest seems to be surprisingly resilient to relatively large changes in past climate, confirmed by our simulated LGM and YD-like climate scenarios, as well as other independent modelling (Cowling et al. 2001; Mayle et al., Chapter 15, this volume) and paleodata research (Harrison and Prentice 2003; Maslin, Chapter 14, this volume). Rainforest resilience to environmental change is strongly tied to basic principles of plant carbon allocation, in particular, strategies that promote carbon conservation in the face of carbon-depleting environmental stresses.

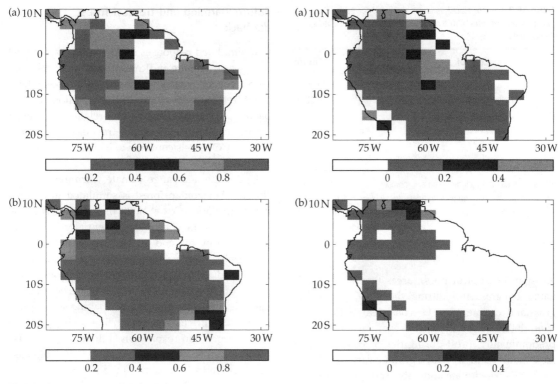

Figure 16.1 Percentage of broadleaf vegetation simulated for (a) LGM, and (b) YD-like palaeoclimate. (See Plate 5(a),(b)).

Figure 16.2 Difference in the percentage of C4-grass simulated for (a) LGM, and (b) YD-like palaeoclimate. Positive numbers represent an increase in grass percentage in the LGM and YD climate scenarios relative to pre-1850. (See Plate 5(c),(d)).

The resilience of Amazonian forest to past changes in climate is relevant for a second reason, relating to the political framework of deforestation and tropical land-use change in the tropics. Many who seek to provide support for continued deforestation in the Amazon Basin cite the apparent ephemeral nature of the rainforest over geological time (see discussion, Bunyard 2002). Misguided rationale follows that the forest has fragmented substantially in the past and recovered, so it can do so again today. There is now little credible scientific evidence supporting such arguments.

The role of temperature cooling in promoting tropical forest resilience

The mechanism underlying tropical forest persistence in past cold periods is not intuitive, and is likely to lie in the combination of biophysical responses of tropical plants to decreasing temperature. Comparison of simulated carbon-based variables between the LGM and YD-like climate supports the role of cooling in maintaining Amazonian forest (Table 16.2). LGM climate differed from pre-1850 in that it was colder, drier, and under the influence of lower atmospheric CO_2 levels (200 ppmv). Although above-ground (vegetation) carbon balance decreases during the LGM, the decline is not sufficient to promote widespread forest deterioration. The YD-like climate scenario is slightly colder than pre-1850. Even though the latter scenario is simulated for only a brief period, YD-like climate actually results in an *enhanced* above-ground carbon budget relative to pre-1850 controls (Table 16.2).

Because the YD-like climate occurs over a relatively short period of time (with respect to the

Table 16.2 Carbon-related variables simulated for each of our climate scenarios, pre-1850 (control), the LGM, YD, and Future (future global warming)

Variable	Control	LGM	YD	Future
GPP	2.06	1.62	2.13	0.88
NPP	0.84	0.73	0.91	0.20
C_v	13.0	8.0	13.5	2.5
R_v	1.22	0.89	1.22	0.68
R_s	0.87	0.69	0.92	0.31
LAI_b	6.5	5.2	6.7	3.5

Notes: Variables include gross primary production (GPP, $kg\,C\,m^{-2}\,yr^{-1}$), net primary productivity (NPP, $kg\,C\,m^{-2}\,yr^{-1}$), vegetation carbon (C_v, $kg\,C\,m^{-2}$), plant respiration (R_v, $kg\,C\,m^{-2}\,yr^{-1}$), soil respiration (R_s, $kg\,C\,m^{-2}\,yr^{-1}$), and leaf area index of broadleaf types (LAI_b, $m^2\,m^{-2}$). Numbers represent mean values.

life-span of tropical trees), areas that were dominated by grassland during the LGM remain as grassland during the YD (consistent with Mayle *et al.* 2000). Those areas, however, that remained predominantly forest during the LGM, experience an enhancement in above-ground carbon following a YD-like climate anomaly. The gross primary productivity GPP (= total photosynthesis), net primary productivity NPP (= GPP – plant respiration), vegetation carbon, and LAI all indicate a trend towards greater carbon retention and growth of forest during the cooler YD-like period relative to pre-1850 (Table 16.2). This positive enhancement of above-ground carbon balance in forest vegetation following climate cooling would have also occurred in the LGM, thus may also be partly responsible for the maintenance of forest cover during glacial climate cycles.

We suggest that the trend towards increasing above-ground plant carbon in tropical regions during the YD-like cold anomaly is in part due to the influence of cooling on biophysical processes such as plant respiration and evapotranspiration. Cooler temperatures cause rates of evapotranspiration to decrease, thus help to conserve carbon and water resources for plant growth and acclimatory processes. A decrease in total plant respiration (R_v) can also lead to conservation of plant carbon resources.

Future warming and tropical forest die-back

Our simulated responses of Amazon forest to past changes in climate indicate a stabilizing relationship between low atmospheric CO_2, cooler temperatures, and above-ground (vegetation) carbon balance. What happens to this relationship when environmental catalysts are reversed, as atmospheric CO_2 levels continue to increase and surface temperatures warm?

Under the mid-range climate warming scenario (IS92a), Amazonian forest is simulated to become highly unstable (Cox *et al.* 2000); the interior of the Amazon Basin becomes essentially void of vegetation by the end of the twenty-first century, with only a very small percentage of broadleaf cover and C4 grass simulated at the outermost edges of the basin (Fig. 16.3). All carbon-related variables indicate symptoms of wide-scale forest die-back: NPP is less than 25% of its pre-1850 value, vegetation respiration is depressed because of dramatic reductions in above-ground biomass (2.5 relative to 13 $kg\,C\,m^{-2}$) (Table 16.2).

A critical mechanism promoting de-stabilization of the Amazon Basin in the future involves vegetation feedbacks on regional climate (Sellers *et al.* 1996; Betts *et al.* 1997; Betts *et al.* 2000, 2003; Cox *et al.* 2000). With progressively decreasing forest cover as a consequence of greater respiration under warming climate (Amthor 1991), recycling of water in the Amazonian watershed begins to break down, contributing to lower basin-wide precipitation (1.6 relative to 4.8 mm per day) (Table 16.1). Coupling between vegetation processes (mostly evapotranspiration) and watershed hydrology can be extremely tight, with some regions of the Amazon Basin returning as much as 74% of annual rainfall to the atmosphere (Sommer *et al.* 2002). Due to the significant increase in the partitioning of incoming radiation into sensible versus latent heat production under future climate scenarios, surface temperatures soar, reaching on average 38°C (Table 16.1).

The contribution of vegetation feedbacks relative to CO_2 climate-forcing can be evaluated by

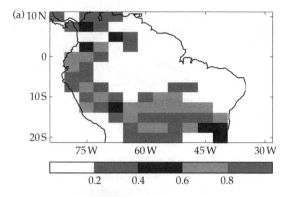

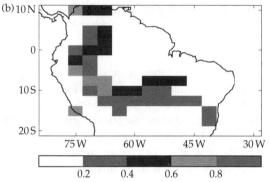

Figure 16.3 Simulations of (a) percentage of broadleaf cover and (b) C4-grass under a mid-range scenario of future climate change (IS92a). (See Plate 5(e),(f)).

comparing predicted Amazonian climate under similar CO_2-warming scenarios, with and without vegetation–climate feedbacks. Two competing processes involving transpiration can occur under scenarios of elevated atmospheric CO_2. A decrease in stomatal conductance reduces transpiration, but increased NPP leads to higher LAI that in turn results in larger surface areas available for transpiration. Betts *et al.*'s (2003) comparison of factorial modelling experiments indicate that the prominent effect on transpiration arises from stomatal responses (not LAI) to rising atmospheric CO_2. The authors predict that the overall physiological effect of vegetation responses to increased CO_2 was to cause further localized drying. The likelihood that LAI effects are relatively less important than stomatal effects under conditions of increased CO_2 is supported by a meta-data analysis of the response of LAI to elevated CO_2 within a variety of tree, crop, and herbaceous experimental species (Cowling and Field, 2003).

Whereas it is likely that stomatal functioning takes prominence over LAI-related processes under elevated CO_2 climates, the same cannot be said for past periods of lower CO_2. For example, LGM simulations by Levis *et al.* (1999) show that LAI has a more dominant control of vegetation–climate feedbacks than physiologically based variables. This seeming contradiction highlights the dynamic and complex nature of biosphere–atmosphere feedbacks.

Tropical forest thresholds and atmospheric change

In contrasting the modelled influence of past and future atmospheric change on features of the Amazonian forest, we conclude the presence of an ecosystem-level resilience threshold that is exceeded under future global change. In other words, the combined strengths of positive versus negative feedbacks in past climate scenarios greatly differ from the future, in that future atmospheric change results in strong positive feedbacks promoting enhanced drying and forest die-back. It is critical that this threshold be identified and fully understood to improve our predictions of future forest integrity.

The effects of low atmospheric CO_2 on ecosystem carbon balance is partially compensated by the influence of decreases in temperature on tropical forests, such that a fairly wide range of low CO_2 and global cooling scenarios can be experienced without the Amazonian forest becoming unstable. On the other hand, despite increases in atmospheric CO_2 (that should improve vegetation carbon balance), global warming results in a progressive decline in forest productivity, that in turn leads to further warming and drying climate feedbacks. Of the potential limiting environmental controls on tropical forest, precipitation (water balance) is often indicated as pre-eminent in defining rainforest structure and function. We speculate, however, that tropical forests like the Amazon

Basin are equally sensitive to high-temperature anomalies, an observation that will become increasingly more apparent in the near future.

Defining the point (threshold) at which tropical ecosystems exceed their capacity for internal/external feedbacks compensating for the deleterious effects of warming on tropical plants (i.e. increased carbon loss due to ecosystem respiration) becomes difficult. We suggest that this threshold exists very near to current climatic (temperature) conditions. A source of supporting evidence can be found in the bulk of eddy-flux data collected to determine whether modern-day tropical forests are sources or sinks of CO_2 to the atmosphere (Malhi and Grace 2000; Falge *et al.* 2002; Malhi 2002).

Tropical forests can switch from source to sink (and *vice versa*) very rapidly (i.e. < 1 yr) and under relatively minor changes in climate (Tian *et al.* 1998). During years of El Niño-induced warming, Amazonian rainforest releases a large flux of CO_2 into the atmosphere, an amount ranging between 0.2 and 0.7 Pg C (Tian *et al.* 1998) or 10–20% of average annual regional productivity (Potter *et al.* 2001). Heightened tree mortality during warmer, drought-prone El Niño years is suggested to contribute to inter-annual fluctuations in biospheric carbon release (Williamson *et al.* 2000; Laurance *et al.* 2001a), along with enhanced respiration (Jones *et al.* 2001) and increased fire activity (Keeling *et al.* 1995; Langenfelds *et al.* 2003).

Conclusion

Our ability to make predictions about the potential fate of the Amazonian rainforest in the near and distant future is strengthened by our understanding of the processes responsible for modifying past tropical forest integrity. Although rising atmospheric CO_2 should enhance tropical ecosystem carbon balance, this gain is more than compensated for by future positive feedback interactions on carbon loss processes. The underlying mechanisms maintaining tropical forest stability in the past (i.e. low-temperature effects on evapotranspiration and plant respiration) will not occur in the future. Further research aimed at determining critical thresholds for tropical forest stability is strongly needed, and could involve a combination of methods including process-based modelling and manipulation-type experiments (Korner, Chapter 6, this volume).

CHAPTER 17

Climate change and speciation in neotropical seasonally dry forest plants

R. Toby Pennington, Matt Lavin, Darién E. Prado, Colin A. Pendry, and Susan K. Pell

Historical climate changes have had a major effect on the distribution and evolution of plant species in the neotropics. What is more controversial is whether relatively recent and rapid Pleistocene climatic changes have driven speciation, or whether neotropical species diversity is more ancient. This question is addressed using evolutionary rates analysis of nuclear ribosomal internal transcribed spacers (ITS) sequence data on diverse taxa occupying neotropical seasonally dry forests: *Ruprechtia* (Polygonaceae), robinioid legumes (Leguminosae), *Chaetocalyx* and *Nissolia* (Leguminosae), and *Loxopterygium* (Anacardiaceae). Species diversifications in these taxa occurred both during and before the Pleistocene in Central America, but were primarily pre-Pleistocene in South America. This indicates plausibility both for models that predict tropical species diversity to be recent and that invoke a role for Pleistocene climatic change, and those that consider it ancient and implicate geological factors such as the Andean orogeny and the closure of the Panama Isthmus.

Introduction

The neotropics have an estimated 90,000 plant species, more than any other continental area (Thomas 1999). The fossil record suggests that the majority of these species must have originated during the past 65 million yrs of the Cenozoic era. While eudicot taxa, which account for 75% of extant angiosperm species, have a fossil record stretching back 125 million yrs to the lower Cretaceous, there is little record of many of the species-rich clades until the upper Cretaceous. For example, the earliest putative fossils of Leguminosae, the family now dominating many neotropical forest ecosystems, appear in Maastrichtian strata of the late Cretaceous (Magallón *et al*. 1999). These are, however, pollen fossils assignable at best to the relatively species-poor subfamily Caesalpinioideae. Pollen and macrofossils of the species-rich subfamilies Mimosoideae and Papilionoideae, as well as macrofossils of Caesalpinioideae are not well documented until the Eocene, ca. 50 Ma (Herendeen *et al*. 1992).

The evolutionary basis for the Cenozoic angiosperm diversification that led to the high species numbers in the neotropics and other tropical areas remains uncertain. Two hypotheses have been proposed. The first held sway for much of the twentieth century, and sees the tropics as a 'museum', where relatively stable climates through the Cenozoic resulted in low extinction rates, allowing species to accumulate over time (e.g. Stebbins 1974). This hypothesis was later challenged by evidence that tropical climates had not been stable, especially over the past 2 million yrs of the Pleistocene. This led to hypotheses of more recent speciation, the most popular of which was a 'refuge' model invoking allopatric differentiation in populations of rainforest species that became isolated from one another by vegetation adapted to more xeric conditions during times of cool dry climate (e.g. Haffer 1969). The refuge hypothesis was popular in the 1970s and 1980s, but has more recently attracted strong criticism, especially for neotropical biota, from three sources. First, from palaeoecologists who see no evidence from pollen cores for reduced cover of rainforest in the Amazon Basin (e.g. Colinvaux *et al.* 2001; Mayle *et al.*, Chapter 15, this volume). Second, dynamic vegetation model simulations also reject the hypothesis of widespread savanna in Amazonia at the last glacial maximum (Cowling *et al.* 2001; Cowling *et al.*, Chapter 16, this volume; Maslin, Chapter 14, this volume; Mayle *et al.*, Chapter 15, this volume). Third, genetic divergence data show speciation in tropical rainforest animals to generally predate the Pleistocene (summarized by Moritz *et al.* 2000; Glor *et al.* 2001). These animal genetic data alone are sufficient for some workers to dismiss the possibility of Pleistocene speciation by any model (Colinvaux *et al.* 2001, p. 611). More recent molecular divergence literature, however, shows some evidence for Pleistocene speciation in neotropical rainforest trees (Richardson *et al.* 2001). Complete dismissal of the specific refuge model of recent speciation might be premature because pollen data are consistent with the presence of seasonally dry tropical forest in the Amazon Basin during the Pleistocene, and this vegetation might have isolated refugial islands of rainforest (Pennington *et al.* 2000; but see Mayle *et al.*, Chapter 15, this volume; Mayle in press). In the context of this volume, understanding whether recent Pleistocene climatic changes did provide a significant evolutionary engine in the neotropics may help us to understand how species might react to future anthropogenic climate changes.

Seasonally dry tropical forests and climatic change

In this chapter we address the question of the age of species in the neotropics, but by using the novel approach of focusing upon seasonally dry tropical forest (SDTF) rather than rainforests. Indirectly, this

allows an assessment of how species occupying this vegetation reacted to climatic change in the Pleistocene. The SDTF has been relatively neglected by scientists, but may offer a better system than rainforests for testing whether Pleistocene speciation occurred. In contrast to wet forests, the disjunct areas of present-day distribution of neotropical SDTF (Fig. 17.1) may represent actual present-day refugia for SDTF species. This is because these forests grow in strongly seasonal areas where rainfall is less than 1600 mm yr^{-1}, with at least 5–6 months receiving less than 100 mm (Gentry 1995). We are currently in a wet interglacial period, but glacial climates in the neotropics were drier (Clapperton 1993; Van der Hammen and Absy 1994; Mayle *et al.* 2000), suggesting that areas suitable for growth of SDTF are likely to be at their minimum, but may have expanded in arid glacial times. Recent fossil pollen data (Mayle *et al.*, Chapter 15, this volume; Mayle in press) show, however, that in Bolivia, the Chiquitano SDTF did not increase in area during the last glacial maximum, but moved northwards. Even if regarding all current areas of SDTF as remnants of more widespread seasonal woodland does not prove accurate, identifying areas of endemism for SDTF for biogeographic analysis is straightforward because of their geographic separation. In contrast, much debate of Amazonian biogeography has not proceeded beyond postulating the locations of ever-wet refugia when past climates were drier (e.g. Nelson *et al.* 1990).

A full discussion of SDTFs, their location, ecology, and floristic composition is given by Pennington *et al.* (2000), and this will only be summarized here. These forests have a smaller stature and lower basal area than tropical rainforests (Murphy and Lugo 1986), and thorny species are often prominent. Net primary productivity is lower than in rainforests because growth only takes place in the wet season. Leaf litter builds up in the dry season because the vegetation is mostly deciduous (Fig. 17.2) and sunlight penetrates to the forest floor and reduces decomposition by lowering the relative humidity. Wetter forests are semi-deciduous, and deciduousness tends to increase as rainfall declines. However, in the driest forests there is an increase in evergreen and succulent species (Mooney *et al.* 1995). Our definition of SDTF is

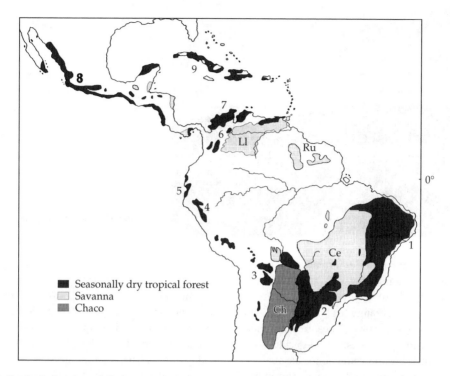

Figure 17.1 The distribution of seasonally dry vegetation in the neotropics highlighting areas of endemism. Seasonally dry forest; 1. Caatingas. 2. Misiones Nucleus. 3. Piedmont Nucleus (including northern Bolivian SDTF). 4. Northern Peruvian InterAndean valleys. 5. Pacific coastal Ecuador and Peru. 6. Colombian and Venezuelan InterAndean valleys. 7. Caribbean coast of Colombia and Venezuela. 8. Mexico and Central America. 9. Caribbean Islands. Savannas: Ce. Cerrado. Ll. Llanos. Ru. Rupununi. Ch: Chaco. Modified after Pennington *et al.* (2000).

Figure 17.2 Seasonally dry tropical forest near Loja, Ecuador showing almost complete deciduousness in the dry season.

Figure 17.3 Seasonally dry tropical forest in Mexico, with abundant cacti.

distinctly general, and includes formations as diverse as tall forest on more moist sites to cactus scrub (Fig. 17.3) on the driest. Many different names are used for SDTF over its range (e.g. tropical and sub-tropical dry forests, caatinga, mesotrophic, meso-philous or mesophytic forest, semi-deciduous or deciduous forest, bosque caducifolio, bosque espi-noso). The Leguminosae and Bignoniaceae dominate the woody flora with the Anacardiaceae, Burseraceae, Myrtaceae, Rubiaceae, Sapindaceae, Euphorbiaceae, Flacourtiaceae, and Capparidaceae also more or less strongly represented (Gentry 1995). The Cactaceae are prominent, particularly at the formation's latitudinal extremes, and are an important element in the diversity of these forests (Fig. 17.3; Gentry 1995). Seasonally dry tropical forests usually have a closed canopy, with a sparse ground flora consisting of rather few grasses, with Bromeliaceae, Compositae, Malvaceae, and Marantaceae also represented.

The largest areas of SDTF in South America are found in north-eastern Brazil (the 'caatingas', extending south to eastern Minas Gerais), in two areas defined by Prado and Gibbs (1993) as the 'Misiones' and 'Piedmont' nuclei (Fig. 17.1) and on the Caribbean coasts of Colombia and Venezuela. Other, smaller and more isolated areas of SDTF occur in dry valleys in the Andes in Bolivia, Peru, Ecuador, and Colombia, coastal Ecuador and northern Peru, the 'Mato Grosso de Goiás' in central Brazil, and scattered throughout the Brazilian cer-rado region on areas of fertile soils (Ratter *et al.* 1997). In Central America, SDTFs are concentrated

along the Pacific coast from Guanacaste in northern Costa Rica, to just north of the Tropic of Cancer in the Mexican state of Sonora. Over much of their neotropical range, SDTFs have been virtually entirely destroyed, with the largest intact block remaining in the Chiquitano region of Bolivia (Mayle *et al.*, Chapter 15, this volume).

Tropical savannas such as the cerrado of central Brazil (Fig. 17.1), and the chaco woodlands of the plains of northern Argentina, western Paraguay, and south-eastern Bolivia (Fig. 17.1), both grow in areas of low precipitation and high seasonality. They are also likely to cover less area currently than in past glacial times. Cerrado and SDTF areas have similar climates, but cerrado is fire-prone and its soils are acid, dystrophic, with low calcium and magnesium availability, and often with high levels of aluminium (Ratter *et al.* 1997). In contrast, SDTF grows on fertile soils with moderate to high pH and nutrient levels, and is less subject to burning, as evidenced by the abundance of cacti, which are generally not adapted to survive fire. Chaco is subjected to low soil moisture and freezing in the dry season and waterlogging and air temperatures up to 49°C in the wet season. Because of the regular frosts that it receives its floristic links are to temperate Monte and Andean Prepuna formations (Cabrera 1976) and it can be considered a sub-tropical extension of a temperate formation.

The debate of what may have replaced moist forest in the neotropics during glacial periods has been strongly influenced by African studies that showed savanna increased in extent in dry ice age climates. For many, there has been an implicit acceptance that the dominant vegetation of ice age Amazonia was a form of cerrado. Tropical savannas such as the cerrado differ from SDTF in a xeromorphic, fire tolerant grass layer. For some palaeoecologists, the failure to find grass pollen in Amazon lake cores (Colinvaux *et al.* 1996), or the offshore sediments at the mouth of the Amazon River (Hoorn 1997) means that refuge theory can be entirely rejected because cerrado must never have been present in the Amazon Basin. This, together with the appearance of cool adapted taxa such as *Alnus, Podocarpus,* and Ericaceae in pollen cores (Liu and Colinvaux 1985; Bush *et al.* 1990; Colinvaux *et al.* 1996), has lead to the idea that climatic cooling was the dominant force in the ice age Amazon, and that aridification of the climate was less significant (Colinvaux *et al.* 1996, 2001). More recent work (Cowling *et al.* 2001; Huang *et al.* 2001; Mayle *et al.*, Chapter 15, this volume) has also

implicated changes in carbon dioxide concentrations in driving vegetation changes. Lowered CO_2 concentrations coupled with less precipitation and higher temperatures are thought to favour grasses such as those that dominate neotropical savannas, and which photosynthesize using the C4 pathway.

Much work, however, neglects the possibility that SDTF, a drought-adapted, tree-dominated ecosystem with a more or less continuous canopy and in which grasses are a minor element (Mooney *et al.* 1995) may have spread into Amazonia in times of drier climate. Study of the contemporary distributions of species inhabiting SDTF has lent support to the notion that these formations were more widespread in cooler and drier periods of the Pleistocene (Prado and Gibbs 1993; Pennington *et al.* 2000). At least 104 plant species from a wide range of families and often with limited dispersal capabilities are each found in two or more of the isolated areas of SDTF scattered across the neotropics (Figs 17.4 and 17.5). It is more

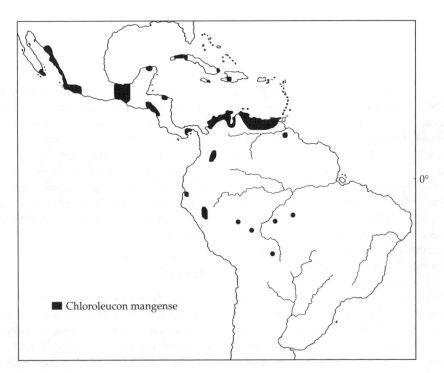

Figure 17.4 Distribution of *Chloroleucon mangense* (Jacquin) Britton & Rose s.l. (Leguminosae: Mimosoideae). Redrawn from Barneby and Grimes (1996) and reproduced with permission from Pennington *et al.* (2000, fig. 2; Blackwell Publishing).

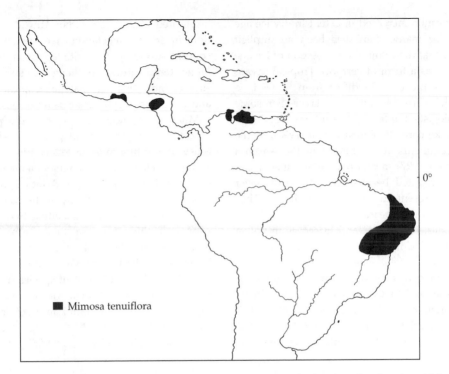

Figure 17.5 Distribution of *Mimosa tenuiflora* (Willdenow) Poiret (Leguminosae: Mimosoideae). Redrawn from Barneby and Grimes (1996) and reproduced with permission from Pennington *et al.* (2000, fig. 3; Blackwell Publishing).

parsimonious to assume that these repeated patterns of distribution imply a former more continuous extent of these forests rather than a series of long-distance dispersal events (Prado and Gibbs 1993; Pennington *et al.* 2000).

In this chapter, we focus on a series of taxonomically unrelated groups (*Ruprechtia* [Polygonaceae], *Coursetia* [Leguminosae–Papilionoideae–Robinieae], *Nissolia* plus *Chaetocalyx* [Leguminosae–Papilionoideae–Aeschynomeneae], and *Loxopterygium* [Anacardiaceae] whose individual species show a different distribution pattern—high levels of endemicity in the separate areas of neotropical SDTF. A classic example of this type of distribution pattern (Pennington *et al.* 2000, 2004) is displayed by species of the leafy cactus *Pereskia* (Fig. 17.6). Pennington *et al.* (2000) suggested that the endemic species in these groups might have been produced by an allopatric process driven by wet interglacial climates fragmenting South American SDTF.

It might be expected that any effect of Pleistocene vicariance of these forests on speciation would be most pronounced in South America, where they are currently distributed in disjunct areas, as opposed to Central America where they are more continuous. We test this hypothesis of Pleistocene speciation using molecular biogeographic approaches and thereby test indirectly whether SDTF did expand and contract in response to Quaternary climatic changes.

Methods

For full details of methods, refer to Pennington *et al.* (2004). This chapter includes voucher specimen information for the sampling of molecular data, details of phylogenetic analyses, and information on how individual phylogenies were calibrated with a dimension of time using fossil information and other criteria. Datasets of aligned nucleotide

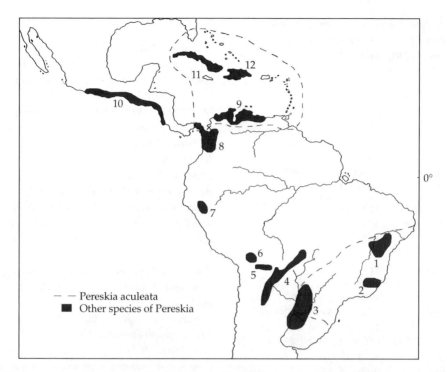

Figure 17.6 Distribution of *Pereskia* (Cactaceae). The dashed line indicates the limits of *P. aculeata* Miller, found in the Caribbean region and eastern South America. The other species are distributed as follows : 1. *P. bahiensis* Gürke, *P. aureifolia* Ritter, *P. stenantha* Ritter; 2. *P. grandifolia* Haworth; 3. *P. nemorosa* Rojas Acosta; 4. *P. sacharosa* Griseb.; 5. *P. diazromeroana* Cárdenas; 6. *P. weberiana* K. Schumann; 7. *P. humboldtii* Britton & Rose; 8. *P. bleo* (Kunth) DC.; 9. *P. guamacho* F.A.C. Weber; 10. *P. lychnidiflora* DC.; 11. *P. zinniiflora* DC.; 12. *P. portulacifolia* (L.) DC., *P. quisqueyana* Liogier. Redrawn from Leuenberger (1986) and reproduced with permission from Pennington *et al.*

sequences and trees are available at TreeBase (study accessions S928) and the website http://gemini. oscs.montana.edu/~mlavin/data/dryforest.htm.

Evolutionary rates analysis

Evolutionary rates analysis places a time-dimension on a molecular phylogeny. A date gained from the fossil record or a geological event is placed on a particular node of a phylogenetic tree, and the distribution of DNA base substitutions along the branches of the rest of the tree allows the ages of all other nodes to be inferred from this single point of calibration. New methods of analysis such as those used in this chapter can allow for variation in the rate of substitution, and therefore do not rely upon the assumption of a constant 'molecular clock'.

In all cases, phylogenies were based on nucleotide sequence data, largely from the nuclear ribosomal internal transcribed spacers (ITS) and intervening 5.8S sequences. For the phylogenetic analysis, branch lengths were estimated during a Bayesian likelihood approach. The rate smoothing program r8s (Sanderson 2001) was used to assess variance in evolutionary substitution rates in and to incorporate such variance into the estimation of ages of lineages (Sanderson 1997, 2001, 2002). We compared the results of a penalized likelihood (PL) approach with those of non-parametric rate smoothing (NPRS; Sanderson 1997), and a rate constant model (Langley and Fitch [LF], 1974). NPRS and LF theoretically define the extremes of a continuum from assuming a molecular clock (LF) to allowing extreme rate variation from ancestral to descendant nodes. We consider PL to provide a reasonable intermediate position between these extremes, and PL results are reported in the text. Values from NPRS and LF are reported in Tables 17.1–17.3.

Table 17.1 Age (Ma) and rate (s/s/Ma) estimates for selected clades of *Coursetia* based on PL, LF, and NPRS analyses

Crown clade	PL (age and rate)	LF (age and rate)	NPRS (age and rate)
A	38.3 ± 3.8	38.4 ± 3.8	37.9 ± 4.1
	0.0031 ± 0.0012	0.0035 ± 0.0002	0.0055 ± 0.0022
B	16.3 ± 4.3	16.3 ± 4.3	26.4 ± 3.9
	0.0034 ± 0.0004	0.0035 ± 0.0002	0.0060 ± 0.0011
C	18.1 ± 2.1	17.7 ± 2.1	25.8 ± 2.2
	0.0034 ± 0.0002	0.0035 ± 0.0002	0.0036 ± 0.0007
D	19.8 ± 2.5	19.6 ± 2.4	28.2 ± 3.0
	0.0034 ± 0.0002	0.0035 ± 0.0002	0.0046 ± 0.0008
D1	15.7 ± 1.8	15.6 ± 1.8	23.8 ± 2.9
	0.0034 ± 0.0002	0.0035 ± 0.0002	0.0037 ± 0.0006
D2	4.1 ± 1.0	4.1 ± 1.0	15.2 ± 3.8
	0.0034 ± 0.0002	0.0035 ± 0.0002	0.0020 ± 0.0003
D3	5.8 ± 1.1	5.8 ± 1.1	11.6 ± 3.0
	0.0034 ± 0.0002	0.0035 ± 0.0002	0.0022 ± 0.0004
D4	7.0 ± 1.2	6.9 ± 1.1	10.0 ± 2.0
	0.0035 ± 0.0002	0.0035 ± 0.0002	0.0034 ± 0.0007
D5	7.9 ± 1.2	7.8 ± 1.2	11.8 ± 2.1
	0.0035 ± 0.0002	0.0035 ± 0.0002	0.0034 ± 0.0006
D6	4.8 ± 0.9	4.8 ± 0.9	8.1 ± 1.9
	0.0035 ± 0.0002	0.0035 ± 0.0002	0.0026 ± 0.0006

Notes: Age constraints for calibration are marked in Fig. 17.7. Crown clades are those labelled in Fig. 17.7.

Taxa

Coursetia and *Poissonia*. *Coursetia* and *Poissonia* (Leguminosae, Papilionoideae, tribe Robinieae) contain 35 and 4 species, respectively, of mostly shrubs that primarily inhabit SDTFs, and other dry or desert environments. The two genera were considered collectively monophyletic, but a recent multiple dataset phylogeny of all robinioid genera (Lavin *et al.* 2003) sampled 33 species from these two genera and demonstrated that they do not form a clade. Four *Coursetia* species from Peru, Argentina, and Bolivia are placed in a separate monophyletic group, *Poissonia*, which is sister to *Robinia* rather than being most closely related to the remaining species of *Coursetia*. We thus have to consider the entire phylogeny of the robinioid legumes for molecular biogeographic analysis.

Ruprechtia. *Ruprechtia* (Polygonaceae, tribe Triplarideae) comprises 38 species of trees, shrubs, and lianas. It is primarily a genus of SDTF, but nine species are confined to more moist conditions in rainforests or gallery forests in seasonally dry areas. The SDTF species show high levels of endemicity in single, disjunct SDTF areas, whereas the moist forest species are more widespread.

Chaeotocalyx/Nissolia. *Chaetocalyx* and *Nissolia* (Leguminosae, Papilionoideae, tribe Aeschynomeneae) are twining herbaceous to woody vines characteristic of neotropical SDTF. *Chaetocalyx* includes ca. 13 species centred in South America, and *Nissolia* ca. 13 species centred in Mesoamerica.

Table 17.2 Age (Ma) and rate (s/s/Ma) estimates for selected clades of *Ruprechtia* based on PL, LF, and NPRS analyses

Crown clade	PL (age and rate)	LF (age and rate)	NPRS (age and rate)
A	6.6 ± 0.8	5.1 ± 1.4	7.4 ± 0.3
	0.00127 ± 0.00059	0.00231 ± 0.00028	0.00265 ± 0.00121
B	4.1 ± 0.5	2.8 ± 0.5	6.0 ± 0.3
	0.00142 ± 0.00016	0.00231 ± 0.00028	0.00291 ± 0.00021
C	3.7 ± 0.5	2.5 ± 0.5	5.6 ± 0.2
	0.00135 ± 0.00016	0.00231 ± 0.00028	0.00176 ± 0.00015
D	1.2 ± 0.2	0.8 ± 0.1	3.1 ± 0.3
	0.00145 ± 0.00018	0.00231 ± 0.00028	0.00212 ± 0.00014

Notes: The age constraint for calibration is marked in Fig. 17.8. Crown clades are those labelled in Fig. 17.8.

Table 17.3 Age (Ma) and rate (s/s/Ma) estimates for selected clades of *Chaetocalyx/Nissolia* based on PL, LF, and NPRS analyses (based on 10 Ma root)

Crown clade	PL (age and rate)	LF (age and rate)	NPRS (age and rate)
A	7.9 ± 1.3	7.9 ± 1.3	8.7 ± 0.9
	0.00479 ± 0.00271	0.00606 ± 0.00054	0.00736 ± 0.00422
B	5.9 ± 1.7	5.9 ± 1.7	7.0 ± 1.4
	0.00547 ± 0.00205	0.00606 ± 0.00054	0.00785 ± 0.00311
C	1.6 ± 0.4	1.6 ± 0.4	5.0 ± 0.8
	0.00605 ± 0.00053	0.00606 ± 0.00054	0.00525 ± 0.00130

Notes: The age constraint for calibration is marked in Fig. 17.9. Crown clades are those labelled in Fig. 17.9.

Loxopterygium. *Loxopterygium* (Anacardiaceae) comprises three species. *Loxopterygium huasango* grows in the SDTF of northern Peru and Ecuador, and *Loxopterygium grisebachii* in the Piedmont nucleus of SDTF in northern Argentina and southern Bolivia. *Loxopterygium sagotii* is found in the moist forests of the Guianas and Venezuelan Guayana.

Results

Coursetia and Poissonia

Two clades confined primarily to Caribbean and Mesoamerican dry forests (clades A and B Fig. 17.7; Table 17.1) involve stem and crown clades that are all Tertiary in age. The estimated age of the diversification of lineages in the entirely South American crown clade *Poissonia* (C), starting at 18 Ma and continuing until 5 Ma, coincides approximately with Alpers and Brimhall's (1988) estimate of 9–15 Ma for the onset of aridification of the Atacama Desert and adjacent Andes where the species of this clade occur. Within the *Coursetia* clade (D), all the subclades of South American species occur within the crown group labelled D1 (Fig. 17.7; Table 17.1), which started to diversify by at least 15 Ma. The crown groups D3 and D4 contain only South American species, whereas the basal lineages of the crown clade D5 include South American taxa (e.g. *Coursetia caribaea* var. *ochroleuca* and *Coursetia hassleri*). Collectively these three clades began diversifying around 6 Ma, although a few constituent species are Pleistocene

in age (e.g. *Coursetia dubia*, *Coursetia grandiflora*; Fig. 17.7). Notably, many species in the two primarily Central American diversifications, the crown clades D2 and D6, originated during the Pleistocene (Table 17.1; Fig. 17.7). These age estimates are probably accurate because the estimated substitution rates for the ITS region of ca. $3–4 \times 10^{-9}$ s/s/y substitutions per site per year (see Table 17.1, especially the PL and LF estimates) for the robinioid legumes are close to the range for other legume groups (Richardson *et al.* 2001).

Ruprechtia

Figure 17.8 shows that all South American SDTF lineages are basal and paraphyletic with the exception of species from the Caribbean coast of Colombia and Venezuela. These are placed in a more apical monophyletic group (clade D in Fig. 17.8) with all Central American species and two subclades of South American rainforest species. The basally divergent South American SDTF lineages of *Ruprechtia* are shown to pre-date the Pleistocene, and the only evidence for Pleistocene speciation is in northern Peru within the *Ruprechtia aperta/albida* clade (Table 17.2). It should, however, be noted that several South American dry forest species were not sampled in this analysis, and Pleistocene origin for these cannot be discounted. The diversification of the Central American species of *Ruprechtia* is clearly post-Pleistocene. Intriguingly, the few sampled species of *Ruprechtia* that are characteristic of rainforests are some of the youngest in the phylogeny.

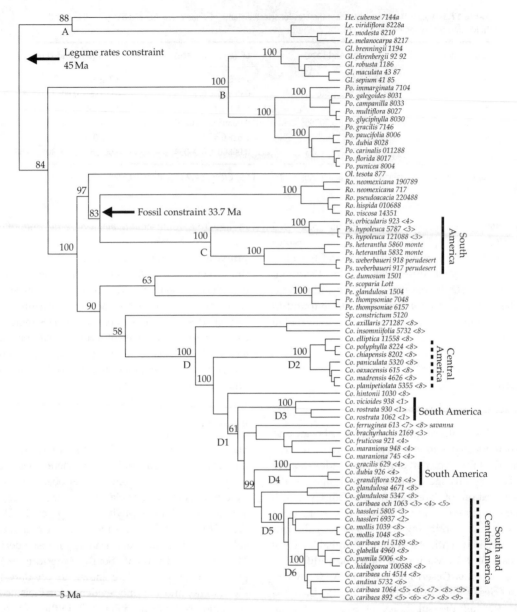

Figure 17.7 Chronogram for robinioid legumes derived from penalized likelihood rate smoothing of a Bayesian likelihood consensus tree, which was estimated with sequences from the ITS region. Codes for crown clades (e.g. A, B, C, and D) are those used in Table 17.1 and the text. Numerical codes in brackets after terminal taxa labels are the SDTF areas of endemism numbered in Fig. 17.1. Non-SDTF areas are savanna = savanna areas in Central America and northern South America; perudesert = Peruvian coastal desert; and monte = arid Andean vegetation. The 45 Ma fixed age constraint at the basal node is derived from a large-scale rates analysis of all legumes (Lavin *et al.* submitted). The 33.7 Ma minimum age constraint at the second lowest node is derived from the fossil wood record (see Pennington *et al.* 2004). See Table 17.1 for the estimated ages and rates of substitution. Numbers above branches are Bayesian posterior probabilities.
He. = *Hebestigma*, *Le.* = *Lennea*, *Gl.* = *Gliricidia*, *Po.* = *Poitea*, *Ol.* = *Olneya*, *Ro.* = *Robinia*, *Ps.* = *Poissonia*, *Ge.* = *Genistidium*, *Pe.* = *Peteria*, *Sp.* = *Sphinctospermum*, and *Co.* = *Coursetia*.

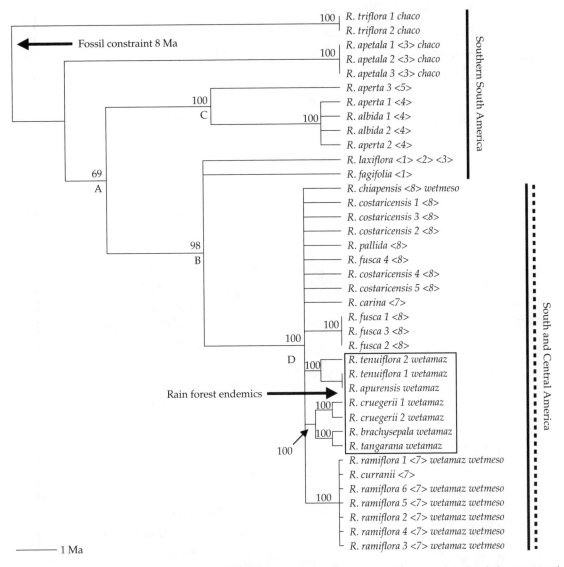

Figure 17.8 Chronogram for *Ruprechtia* derived from penalized likelihood rate smoothing of a Bayesian likelihood tree, which was estimated with sequences from the ITS region. Codes for crown clades (A, B, C, and D) are those used in Table 17.2 and the text. Numerical codes in brackets after terminal taxa labels are the SDTF areas of endemism numbered in Fig. 17.1. Non-SDTF areas are wetmeso = Central American rainforest; wetamaz = Amazonian rainforest; and chaco = chaco woodlands of Argentina, Paraguay, and Bolivia. The 8 Ma fixed age constraint at the basal node is derived from fossil data (see Pennington *et al.* 2004). See Table 17.2 for the estimated ages and rates of substitution. Numbers above branches are Bayesian posterior probabilities. *R.* = *Ruprechtia*.

Chaetocalyx/Nissolia

The *Nissolia* crown group is dated as ca. 1.6 Ma (using either PL or LF; Table 17.3) and the *Chaetocalyx* subclade that is its sister as ca. 5.9 Ma (Fig. 17.9). The age of the *Nissolia* stem lineage is estimated at ca. 7.9 Ma. This suggests that much of the South American *Chaetocalyx* diversification took place well before the Pleistocene, although some individual species (e.g. *Chaetocalyx longiflora*, *Chaetocalyx glaziovii*) are Pleistocene in age.

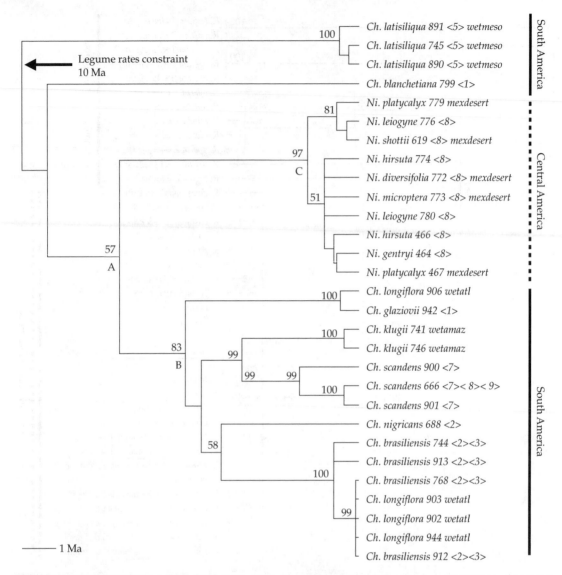

Figure 17.9 Chronogram for *Chaetocalyx/Nissolia* derived from penalized likelihood rate smoothing of a Bayesian likelihood tree, which was estimated with sequences from the ITS region. Codes for crown clades (A, B, and C) are those used in Table 17.3 and the text. Numerical codes in brackets after terminal taxa labels are the SDTF areas of endemism numbered in Fig. 17.1. Non-SDTF areas are wetmeso = Central American rainforest; mexdesert = Mexican deserts; wetamaz = Amazonian rainforest; and wetatl = eastern Brazilian Atlantic coastal rainforest. The 10 Ma fixed age constraint at the basal node is derived from a large-scale rates analysis of all legumes (Lavin *et al.* submitted). See Table 17.3 for the estimated ages and rates of substitution. Numbers above branches are Bayesian posterior probabilities. Ch. = *Chaetocalyx*, Ni. = *Nissolia*.

In contrast, *Nissolia* diversified entirely during Pleistocene, even though it became an isolated lineage well before.

The 10 Ma calibration of the *Chaetocalyx–Nissolia* crown clade derived from a global rates analysis of all legumes (Lavin *et al.* 2005) rather than fossil evidence. This calibration yields a range of substitution rates at ca. 5–8×10^{-9} substitutions per site per year (s/s/y; Table 17.3). This is on the high end of rates reported for the ITS region for leguminous shrubs and herbs by Richardson *et al.* (2001). If we are truly estimating too fast a rate, then this can only mean that our estimated ages are too young and the true ages of the clades reported above are older.

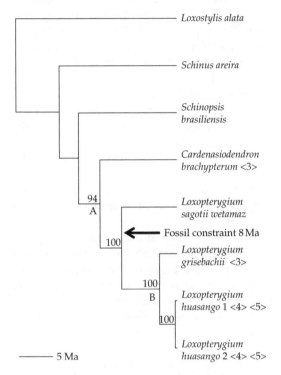

Figure 17.10 Chronogram for *Loxopterygium* derived from penalized likelihood rate smoothing of a Bayesian likelihood tree, which was estimated with sequences from the ITS region. Codes for crown clades (A, B) are those used in Table 17.4 and the text. Numerical codes in brackets after the *Loxopterygium* labels are the SDTF areas of endemism numbered in Fig. 17.1. The non-SDTF area is wetamaz = Amazonian rainforest. The 8 Ma fixed age constraint at the node marking the *Loxopterygium* crown clade is derived from fossil data (Burnham and Carranco 2004). See Table 17.4 for the estimated ages and rates of substitution. Numbers above branches are Bayesian posterior probabilities.

Loxopterygium

The divergence of the two SDTF species of *Loxopterygium*, *L. huasango* (Peru/Ecuador), *L. grisebachii* (Bolivia/Argentina), is dated as 2.6 Ma (clade B in Fig. 17.10; Table 17.4), just prior to the Pleistocene. The nucleotide substitution rates of ca. 1×10^{-9} are somewhat lower than those reported for the ITS region in the woody, neotropical genus *Inga* (Richardson *et al.* 2001), but *Loxopterygium* species are woody and probably slow growing with a long generation time, which would suggest that substitution rates should be less. It therefore seems likely that these substitution rates are a reasonable estimate.

Discussion

Miocene/Pliocene patterns in South American SDTF

The distribution patterns of species of *Ruprechtia*, *Coursetia*, *Loxopterygium*, and *Chaetocalyx–Nissolia* in South American SDTF, where different species are endemic to separate areas, suggested that they might have been produced allopatrically by Pleistocene vicariance of these forests (Pennington *et al.* 2000). This hypothesis is largely rejected by the evolutionary rates analysis presented here. In no case, irrespective of the reconstruction method (LF, NPRS, PL) or the means of calibration, are crown groups of South American species indicated to have started diversifying during the Pleistocene (1.8 Ma or less). Furthermore, few individual species are shown to be Pleistocene in age. This finding is bolstered by estimated substitution rates that approximate those published

Table 17.4 Age (Ma) and rate (s/s/Ma) estimates for selected clades of *Loxopterygium* and outgroups based on PL, LF, and NPRS analyses

Crown clade	PL (age and rate)	LF (age and rate)	NPRS (age and rate)
A	11.4 ± 1.6	10.0 ± 0.9	11.2 ± 1.5
	0.0085 ± 0.0015	0.0092 ± 0.0014	0.0091 ± 0.0016
B	2.6 ± 0.5	2.8 ± 0.5	2.8 ± 0.5
	0.0122 ± 0.0019	0.0092 ± 0.0014	0.0152 ± 0.0031

Notes: The age constraint for calibration is marked in Fig. 17.10. Crown clades are those labelled in Fig. 17.10.

for the ITS region from other taxa (e.g. Richardson *et al.* 2001). Again in no case did we estimate substitution rates on the slow end of the reported range. Thus, our findings are unlikely to be biased towards older age estimates. In summary, these genera diversified during roughly the same time frame in South America, from the mid-Miocene to the Pliocene.

Recent patterns in Central American SDTF and South American rainforests

In contrast to the old species diversity in South American SDTF, Central American dry forest species have in many cases originated more recently. In *Ruprechtia,* all Central American species are more recent than 1.2 Ma. Similarly, the majority of the 12 Central American taxa in two *Coursetia* crown clades (D2 and D6; Fig. 17.7) are Pleistocene in age. Similarly, the primarily Central American *Nissolia* crown clade began to diversify well after *Chaetocalyx* did in South America, and its species probably have a Pleistocene origin.

One intriguing aspect of the results is the recent origin of rainforest species of *Ruprechtia*. The two clades of rainforest species (Fig. 17.8) are dated as ca. 1 Ma or less. This recent origin is corroborated by species such as *Ruprechtia tenuiflora* where individual accessions are resolved as paraphyletic with respect to *Ruprechtia apurensis*, which is consistent with the recent origin of *R. apurensis* from populations of *R. tenuiflora*. Similar recent speciation, evidenced by low sequence divergence for ITS between species, was found in the neotropical rainforest genus *Inga* (Richardson *et al.* 2001).

The clear evidence for Pleistocene speciation in Central American SDTF and rainforests, contradicts most studies of genetic divergence in rainforest animals that demonstrate species to have originated before the Pleistocene in virtually all cases (reviewed by Moritz *et al.* 2000). The conclusion of Colinvaux *et al.* (2001), that 'data from molecular genetics are probably sufficient by themselves to discount the [Pleistocene refuge] hypothesis', is now clearly overstated.

SDTF and Pleistocene climate change

If all the South American SDTF species in our study group had been shown to have Pleistocene origins, this would have represented substantial evidence that ice age climates had strongly influenced this vegetation. However, the finding that these species are primarily more ancient cannot entirely refute the hypothesis that SDTF was more widespread in glacial times. Perhaps the time of isolation in SDTF areas of endemism (Fig. 17.1) has not been long enough to promote allopatric speciation in most cases. Evidence from the distribution of species that are widespread in Neotropical SDTF (Prado and Gibbs 1993; Pennington *et al.* 2000) is still suggestive that these forests were historically more widespread. This is partly supported by the vegetation modelling of (Mayle *et al.* 2004, Chapter 15, this volume), which shows deciduous broad-leaf forests covered the southern half of Amazonia at the last glacial maximum. Elsewhere, however, their model simulates northwestern Amazonia to have retained evergreen broadleaved forest, and the SDTF Chiquitano nucleus in Bolivia to have moved north rather than expanded. If this model is correct, then some of the inferences made from plant distribution patterns by Prado and Gibbs (1993) and Pennington *et al.* (2000) may be erroneous because some species may have become widespread by long-distance dispersal between separated areas of SDTF.

The future of evolution of SDTF plants in the face of anthropogenic climate change

Some authors (e.g. Myers and Knoll 2001; Woodruff 2001) have pointed out that conservation policies fail to consider the 'future of evolution'—the long-term evolutionary aspects of biodiversity loss, habitat alteration, and climatic change. Although predicting the products of evolution is impossible given that it is a stochastic process, these authors argue that we should attempt to allow the processes of evolution to proceed as they would without human intervention. A difficulty of this approach is the disparity between evolutionary and human timescales, which is emphasized by the findings of this chapter. Only a minority of South American SDTF species are 'recent', but even this is 1–2 Ma, 5000 times greater than a timescale considered 'long' by humans (200 yrs). Persuading governments to plan for such 'short' evolutionary timescales would be doomed to failure. Arguments concerning extinction, which will certainly be a more rapid process, will perhaps be more persuasive, especially if they are framed in terms of the extinction of biomes as well as of species. The SDTFs are already so heavily impacted across the neotropics, and in many cases surrounded by agricultural landscapes, that it is hard to envisage SDTF species expanding their ranges in the face of hotter and drier climates. Either adequate species populations in SDTF nuclei do not exist, or migration would be impossible due to lack of habitat in an anthropogenic landscape. If the case demonstrated for the Bolivian Chiquitano region in the face of drier climates (Mayle *et al.*, Chapter 15, this volume; Mayle in press)—that SDTF species merely shift their ranges rather than expand them—is general, then the future is bleaker still because species will die where they stand. Clearly, conservation planning must allow for migration across ecotones, such as that advocated for the forests of the Andean flanks by Bush (2002). In the case of SDTF, this would require the protection of relatively intact, large SDTF areas, plus other intact biomes surrounding them. This ecotonal approach might indirectly address the future evolutionary concerns of authors such as Myers and Knoll (2001) because speciation associated with ecotones is indicated to be widespread in our study. It is clear that most SDTF species come from genera containing species characteristic of several other vegetation types. This indicates that ecotonal speciation may be generally important for the neotropical flora. Furthermore the individual phylogenies often indicate multiple, rather than single diversification events across ecotonal boundaries. For example, in the *Ruprechtia* phylogeny, there are two recent rainforest radiations in this primarily SDTF genus.

Conclusion

The debate of when neotropical species arose and how climatic changes have affected their evolution has focused principally on rainforests, and especially those of the Amazon Basin. Our study differs in its focus on SDTF, especially those in South America. Evolutionary rates analysis demonstrates that in monophyletic radiations of species with high levels of endemicity in these forests, diversification took place beginning in the late Miocene and Pliocene. This contradicts our own earlier assertions (Pennington *et al.* 2000) that Pleistocene climatic changes were a major force driving the speciation of woody, South American SDTF plants.

Many readers may interpret the principal finding of this Chapter—Miocene and Pliocene speciation in South American SDTFs—as confirming the now popular rejection of neotropical Pleistocene speciation, especially that of the specific model of allopatric speciation in rainforest refuges (e.g. Colinvaux *et al.* 2001). We, however, emphasize that our data do show Pleistocene diversification in Central American SDTF in cases where taxa have an ancestral area of South America, and may have reached Central America after the Panama Isthmus closed. Pleistocene climatic change as a factor in neotropical speciation cannot be entirely discounted. Furthermore, there is clear evidence for Pleistocene speciation in South American rainforest species of *Ruprechtia*, just as there is for the species-rich genus *Inga* (Richardson *et al.* 2001). Our data, perhaps unsurprisingly, show that a mixture of ancient and recent diversification explains the extant diversity in five genera of plants centred in neotropical SDTF ecosystems. The museum and recent speciation hypotheses are not mutually exclusive, and an explanation for high tropical diversity may lie in some combination of ancient and recent speciation, with climatic changes playing an important, albeit partial, role.

Acknowledgements

This work was supported largely by: Leverhulme Trust (F/771/A to Royal Botanic Garden

Edinburgh [RBGE]) and US National Science Foundation (DEB-0075202). Darién Prado is a researcher of Consejo Nacional de Investigaciones Científicas y Tecnológicas (CONICET); he received travel grants from Universidad Nacional de Rosario (Programa Viajes al Exterior) to support visits to Montana State University and RBGE. Sequencing of Anacardiaceae was supported by the Lewis B. and Dorothy Cullman Program for Molecular Systematics Studies (New York Botanical Garden). We thank Carlos Reynel, Aniceto Daza, Monica Moraes, Stephan Beck, Mario Saldias, Felipe Ribeiro, Jim Ratter, Sam Bridgewater, Mario Sousa, Alberto Reyes for help with fieldwork. We thank Michelle Hollingsworth, Alex Ponge, Stephen O'Sullivan, Will Goodall-Copestake, and Ben Mosse for assistance in the RBGE laboratory.

CHAPTER 18

The prospects for tropical forests in the twenty-first-century atmosphere

Oliver L. Phillips and Yadvinder Malhi

At the start of the twenty-first century the human race dwells in a radically altered biosphere, and in an atmosphere that is shifting rapidly to conditions with no direct analogue in Earth's prehistory. These atmospheric changes are certain to have impacts on the Earth's ecosystems, but the complexity of these ecosystems and the multiplicity of processes and scales make the exact nature of these impacts difficult to tease out. In this book we have examined the potential impacts on the most complex of terrestrial ecosystems, the great tropical forests. We have drawn together insights from a range of scientific disciplines, including laboratory ecophysiology, forest ecology, palynology, meteorology, climate modelling, and molecular genetics. In this final chapter we present a synthesis and personal perspective on the contributions and insights from the various chapters of this book, with the aim of addressing a specific question that could be encapsulated as: what will (any remaining) mature old-growth tropical forests look like at the end of this century?

Answering this question has implications for conservation, global biodiversity, and global biogeochemical cycling, as well as being a test for our understanding of natural- and human-impacted ecosystems. At the end of this process we cannot claim to be able to answer this question definitively, but there has been rapid recent progress in several relevant scientific fields. By examining this question from a variety of perspectives and through a range of scientific disciplines we hope to have made significant progress.

The age and continuity of tropical rain forests

Some paleoperspectives help put current change into context, and give insight into the resilience of tropical forests to atmospheric perturbations. Two chapters in this book (Chapter 14 by Maslin, and Chapter 15 by Mayle and Bush), place the contemporary tropical forests of Amazonia in their pre-historical context.

The chapter by Maslin demonstrates that megathermic angiosperm-dominated tropical forests have experienced extremes of global climate through their long evolutionary history, ranging from extreme warmth and high carbon dioxide in the Cretaceous and early Cenozoic, to cooler conditions and periodic aridity in the Pleistocene.

Maslin argues that over much of this time some areas such as Amazonia appear to have experienced continuous forest cover, a fact that highlights the potential long-term significance of contemporary tropical deforestation. Other areas such as Africa may be more variable in their continuity of forest cover. However, continuity of forest cover does not imply continuity of forest structure, composition, or dynamics, and the species composition of Amazonian forests appears to have varied with the atmospheric conditions. During the high temperatures of the Cretaceous there was significant aridity at equatorial latitudes, although this may be linked to the continentality of the huge Gondwanaland landmass rather than to any high temperature limit to tropical forest function.

Some insight into the responses of tropical taxa to rising temperatures may come from examination of the early Eocene (about 55 million years ago), when there was a rapid rise of tropical temperatures (by 4–5°C; Zachos *et al.* 2003), possibly at a rate similar to projected climate change over the twenty-first century. This was possibly induced by a rapid rise in methane concentrations (methane is a potent greenhouse gas), due to volcanism or submarine seismicity and perhaps accelerated by a release of methane clathrates from ocean sediments. Many contemporary tropical families had evolved by then, and examination of which of these prospered and declined may give insights into their relative tolerance to high temperatures. Morley (2000) reports a worldwide decline in palms at equatorial latitudes at this thermal maximum, and the appearance (and subsequent disappearance with cooling) of numerous new vegetation types without analogues in modern floras. On the other hand, Wing *et al.* (2003) report little evidence of species turnover at a site in the warm 'paratropical' latitudes of North America. This is clearly a rich field for research and possibly the closest past analogue to 21st century climate change. Incidentally, carbon isotope analyses and model simulations suggest that the terrestrial biosphere was a major carbon sink (1000–3000 Pg C) over this early Eocene warming, because of a combination of CO_2 enrichment and global warming (Beerling 2000).

Moving to the more recent timescales of the Pleistocene (from 2 million yrs ago) and Holocene (from ~11,000 yrs ago), the chapter by Mayle and Bush presents a comprehensive review of changes in Amazonia since the last glacial maximum (LGM). Both fossil pollen data and dynamic vegetation models now suggest that much of Amazonia remained forested through the cool, arid conditions of the LGM, but the forest was of substantially different composition from modern-day forests, with drought-adapted forests covering much of southern Amazonia and cold-adapted Andean taxa migrating into the lowlands. Hence it now seems unlikely that Amazonian forests retreated into refugia surrounded by savanna at the last glacial maximum, although 'within-forest' refugia of moist forest surrounded by seasonally

dry forest remain a distinct possibility. This contrasts with Africa, for example, where the mean state of the forests is drier and there is some evidence of large-scale retreat of the forest during arid periods (e.g. Morley 2000), although the extent of this retreat is still disputed. During the early-mid Holocene (ca. 8500–3600 cal yr BP), much of Amazonia experienced drying and perhaps warming, and Mayle *et al.* suggest that this period may give insights into ecological shifts to be expected in warming, drying tropical forest (although the direction of projected precipitation trends this century is far from certain). The early-mid Holocene saw frequent, widespread fires throughout southern and eastern Amazonia, which led to replacement of lowland evergreen rainforest taxa by drought/fire-tolerant semi-deciduous dry forest and savanna taxa. The paleo-analogue is not exact however, as atmospheric CO_2 concentrations this century will be at least double those in the early-mid Holocene, which would favour species with C3 photosynthetic pathways (such as lianas and dry forest trees) over those with C4 pathways (such as savanna grasses).

The potential importance of the seasonally dry tropical forest (SDTF) as a transition biome is also emphasized by Pennington *et al.*, in Chapter 17, (Fig. 18.1). Until recently this ecosystem has been neglected in a debate largely focussed on moist forests and savannas. The SDTF taxa have frequently shifted or expanded to replace moist forest taxa at times of aridity, and some phylogenetic analyses have suggested that current pockets of SDTF are refugia from a broad arc of SDTF that existed at the last glacial maximum. However, SDTF regions have been more affected by agriculture than moist forests, and it is likely that any future drying would see an expansion of agriculture rather than dry forests.

What stands out from the paleo-analyses is the sensitivity of tropical forest structure and composition to climate variations in the Pleistocene (11,000–2 million yrs BP), and even in the relatively stable Holocene (the last 11,000 yrs). Tropical forests are clearly dynamic ecosystems of constantly shifting composition and structure. Viewed from this perspective, the ecological change witnessed in mature forests during the final quarter of the last

Figure 18.1 Seasonally dry tropical forests may increase in importance in rainforest areas experiencing increased water stress under climate change. The example in this photo is from Mexico, with abundant cacti (photo T. Pennington). (See Plate 6.)

century (see below) is perhaps not surprising, and it seems inevitable that the ecology and structure of intact tropical forests will respond to the atmospheric changes projected for the current century.

The chapter by Cowling *et al.*, Chapter 16 presents a climate–vegetation model analysis of the recent past (the Last Glacial Maximum and the Younger Dryas) and the near future. They suggest that the dominant mechanisms for maintenance of forest in cool periods are reduced evapotranspiration and lowered respiration costs, and that tropical forests may now be near an upper temperature threshold where these physiological mechanisms become positive feedbacks that induce forest die-back. Will most plants have sufficient genetic and phenotypic flexibility to respond to rising temperatures by increasing these temperature thresholds? Recent evolutionary history suggests not, since the Pleistocene climates that tropical taxa have experienced for several million years have been typically much cooler than today's. Those (few?) taxa which are tolerant of higher temperatures may be expected to dominate ecosystem composition in a warming world. Identifying these taxa should be a research priority.

Montane regions have an obvious conservation importance at times of warming, as demonstrated by the flight of many cold-adapted taxa to the Andes at the last glacial transition (Mayle and Bush). Their protection should therefore be a conservation priority, not only to save their current endemic-rich biota, but also as a refuge for low-

land taxa. The projected 4°C warming over the twenty-first century can be expected to lead to an 800 m upward migration of ecotone, over half the current altitudinal extent of cloud forest. Cloud forests may be the tropical ecosystems most sensitive to climate change, and monitoring of changes in composition and structure should give early clues to trends in ecosystem compositions.

Current and projected atmospheric change

Atmospheric carbon dioxide concentrations are currently more than 30% above pre-industrial levels, and are rising by about 2 ppm (ca. 0.5%) yearly. These values are already greater than any experienced in the Pleistocene. By the end of our century they are projected to increase beyond any witnessed in at least the last 20 million yrs, and possibly the last 40 million yrs. This is the dominant chemical change in the atmosphere which is driving climate change as well as directly affecting the biosphere through its impacts on photosynthesis and oceanic acidity.

Momentous as these changes are, climate change can still occasionally sound like a theoretical concept, belonging to intricacies of a global circulation model (GCM) rather than to the real world we experience every day. The chapter by Malhi and Wright (Chapter 1) demonstrates that this is clearly not the case. They show that since the mid-1970s there has been a globally synchronous warming in tropical forest regions of $0.26 \pm 0.05°$C per decade: the tropical forests we see today are already experiencing significant rates of warming. This observed warming is consistent with GCM simulations, as discussed by Cramer et al. (Chapter 2), which suggest a further warming in all tropical forest regions of between 3 and 8°C (typically 4°C) by the end of the century. This suggests that tropical temperatures by 2100 will have moved outside the envelope of natural variability of at least the last 2 million yrs.

Changes in precipitation have perhaps the greatest impact on tropical forest structure and composition, but are much more difficult to predict than changes in temperature as they are sensitive to highly non-linear changes in atmospheric and ocean circulation, as opposed to the moderately linear changes in atmospheric radiation balance that dominate temperature changes. Recent rainfall changes indeed show strong spatial variability, with some evidence of a strong drying trend in Africa but few significant trends elsewhere. Future scenarios of precipitation change are also highly variable between regions and between climate models, with changes of up to $\pm 20\%$ predicted by the end of the century. The only consistent tropical rainfall scenario in the four climate models discussed by Cramer et al. is a rise in precipitation in Asian tropical forest regions.

Other aspects of the tropical atmosphere are also changing, but these changes are poorly quantified. In particular, they may be substantial trends in solar radiation. Much of the tropics may be light-limited (Churkina and Running 1998), and thus shifts in cloudiness may affect forest productivity, although these shifts may be entangled with changes in precipitation. These shifts may be driven by changes in global circulation and radiation balance (such as the possible 'global dimming' described in Chapter 4 by Lewis, Malhi, and Phillips), but regional factors may also be important. For example, biomass burning in Amazonia is already causing rapid increases in the concentrations of aerosols that form cloud-condensing nuclei (see Chapter 9 by Laurance), which leads to suppression of rain-bearing clouds but perhaps extension of cloud lifetimes. More directly, biomass burning leads to increased and prolonged haziness in clear sky conditions. Fig. 18.2 demon-

strates the extent of haze generated by biomass burning in Bolivia in September 2004. This may reduce the total amount of light (potentially reducing photosynthesis) but increase the diffuse fraction of light that penetrates deeper into forest canopies (potentially increasing photosynthesis). Biomass burning and fossil fuel burning can also have biogeochemical affects on forests, ranging from tropospheric ozone to the deposition of nitrogen, and potentially other nutrients in ashfalls.

The possible impacts of atmospheric change

A number of chapters in this book focus on contemporary drivers of change in tropical forests, and discuss a suite of possible responses: in forest ecophysiology, ecology, fire dynamics, and human interactions. There are many factors, potentially acting in opposite directions. Equally, these factors may act on different timescales, so transient responses may be in different directions to long-term responses. Moreover, different factors act at different levels (leaf versus whole-plant versus forest stand), and have different effects at each level. This degree of complexity makes prediction of change (and attribution of documented ecological change) difficult. Not surprisingly, this is an area of active debate.

Several chapters (Chapter 4 by Lewis et al., Chapter 5 by Chambers and Silver, Chapter 6 by Korner) discuss the ecophysiological arguments for various responses to contemporary tropical atmospheric change, often taking opposing viewpoints. Some of the major arguments surrounding two potential drivers, increasing atmospheric carbon dioxide concentrations and increasing temperatures, are summarized in Table 18.1.

Increased fire incidence is likely to be another important factor in driving changes in tropical forests (Barlow and Peres, Chapter 8). To some extent this may be driven by drying trends in some tropical forest regions, but human factors such as desiccation associated with fragmentation (Laurance, Chapter 3) and logging, and the spread of the human colonization frontier are likely to be important factors independent of any local background climate change. While much of the book concentrates on the response of tropical vegetation

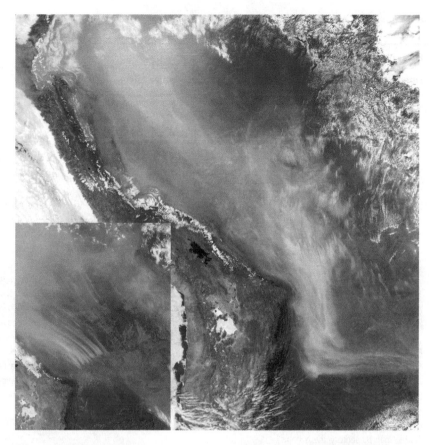

Figure 18.2 Forest fires can modify climate for substantial areas downwind. This MODIS image from 14 September 2004 NASA image by Jesse Allen, Earth Observatory, using data courtesy the MODIS Rapid Response Team shows fires and associated smoke plumes in a forest clearance frontier primarily in Santa Cruz, Bolivia (inset), generating a smoke haze that extends thousands of kilometres across Peru and Brazil to Colombia and Venezuela. (See Plate 7).

to climate change, Barlow and Peres provide an important example of the potential of severe faunal shifts accompanying such vegetation change. Laurance demonstrates how forest deforestation and fragmentation can accelerate climate change, both locally by modifying microclimate and regionally by enhancing surface warming and perhaps suppressing precipitation.

Contemporary observations of change in tropical forests: biomass, dynamics, and structure

Ultimately, model simulations and competing ecophysiological hypotheses are best tested by direct observations of contemporary tropical forests. What is actually being observed in tropical forests today? Can we see any direct evidence of responses to contemporary atmospheric change? Three chapters present such evidence from monitoring plots in old-growth tropical forests across South America, much of these emerging from the RAINFOR project (Malhi *et al.* 2002b), which is supporting the development of a pan-Amazonian monitoring network. These results update and expand substantially upon papers by Phillips and Gentry (1994) and Phillips *et al.* (1998b), which first presented evidence of accelerating forest dynamics, and increasing forest biomass, in old-growth tropical forests.

Baker *et al.*, Chapter 11, present an updated analysis of changes of biomass in old-growth

Table 18.1 Arguments to expect, or not to expect, substantial effects of increasing CO_2 concentrations on tropical forest growth and carbon balance. Direct and indirect effects are considered, including climate change

Scale	Effect on Growth and Carbon Storage	
	Zero or Negative effect	Positive effect
Leaf-level	Increased respiration and photorespiration caused by rising temperatures	Direct fertilization of photosynthesis by high CO_2
	Warming temperatures lead to increased evaporative demand, inducing stomatal closure and reducing photosynthesis.	Reduced photorespiration caused by high CO_2
	Increased emissions of volatile hydrocarbons at higher temperatures consume assimilated carbon.	Improved water-use efficiency caused by high CO_2
		Photosynthetic rates increase with moderate warming.
		Optimum temperature for photosynthesis rises with rising CO_2
	Increased cloudiness or haziness reduces photosynthesis	Reduced cloudiness increases photosynthesis
		Increased diffuse radiation increases photosynthesis
Plant-level		In low light conditions the proportional increase in plant carbon balance (photosynthesis–respiration) may be much greater than any increase in photosynthesis
	Plants are often saturated with respect to non-structural carbohydrates?	Excess carbon may be used preferentially above-ground to acquire rate-limiting resource (light) by investing in wood
	Plant growth limited by nutrients other than carbon (N, P, K, Ca)	Excess carbon may be used preferentially below-ground to acquire rate-limiting resource (nutrients) via fine root development or supporting P-scavenging fungal symbionts
	Rising soil temperatures increase soil acidification and mobilize aluminium, reducing soil nutrient supply.	Rising temperatures increase soil mineralisation rates and improve nutrient supply
	Plant carbon balance is limited by respiration costs rather than by photosynthesis gains?	
	Acclimation (downregulation of photosynthesis) limits any response to increasing CO_2.	
	Local climate drying increases intensity of dry season and reduces plant carbon gain	Local climate wetting ameliorates dry season limitations on carbon gain
Stand-level	Stand biomass ultimately limited by abiotic disturbance (e.g. windthrow risk) or biotic disturbance (e.g. herbivores, termites) rather than by resources.	Faster growth leads to some biomass gains, with mortality gains lagging
	Forest canopies are close to physical limits to forest structure which cannot be increased, (e.g. maximum tree height is limited by hydraulics or mechanics).	Rising CO_2 improves water use efficiency and reduces tension in the water column, allowing an increase in maximum tree height for given cross-sectional area.
	Biomass in forest understorey ultimately limited by intensity of competition for light	Faster growth and turnover may prevent stand dominance by senescent 'over-mature' trees with high respiration costs, creating positive feedback on stand-level growth rates
	Faster growth and turnover may favour disturbance-adapted taxa, with less dense wood	Improved forest water balance leads to reduced drought mortality and fire incidence
	Lianas may benefit from increased CO_2 and disturbance, limiting biomass gains by trees	
	Mortality rates increase because of climatic warming and/or drying, or increased climatic regimes	
	Climatic drying combined with forest fragmentation and degradation lead to increased fire frequency	

Amazonian forests, and conclude that there has been a net increase in biomass in recent decades at a rate of 1.22 ± 0.42 Mg ha^{-1} yr^{-1}, slightly greater than that originally estimated by Phillips *et al.* (1998b). They address a number of methodological issues, including uncertainties in measurement of large trees (Fig. 18.3) and choice of allometric equation to estimate biomass, and find the results robust to these uncertainties. Applying adjustments for the biomass of roots, small trees, and lianas (following Phillips *et al.* 1998b), this is equivalent to a net carbon sink in intact old-growth forests of 0.9 ± 0.2 Mg C ha^{-1} yr^{-1}, or, multiplying by the area of moist forest in Amazonia, about 0.6 Pg C yr^{-1}. This value is consistent in magnitude with expectations from some ecophysiological models of the response of tropical forests to CO_2 fertilization.

Phillips *et al.* in Chapter 10, present an updated analysis of changes in forest turnover in Amazonia (tree recruitment and mortality), using a spatially and temporally extended dataset to explore regional patterns and to separate turnover into tree recruitment and mortality. They confirm that turnover rates have accelerated across Amazonia, with the greatest absolute increases on more fertile soils in western Amazonia. Moreover, the increase in recruitment has been greater than, and in advance of, the increase in mortality, implying that the increases are mostly driven by an acceleration of growth which eventually feeds back on mortality, rather than a direct acceleration of mortality through disturbance.

Lewis *et al.* in Chapter 12, explore the changes in structure and dynamics in greater detail, by examining simultaneous changes in forest biomass, growth, mortality, and stem number in 50 Amazonian forest plots. They demonstrate an acceleration of growth in most of these plots, accompanied by a lagged acceleration of mortality and a general increase in biomass and stem number. This suggests that the observed increase in biomass (Baker *et al.*) is unlikely to be explained by recovery from past disturbance, but instead

Figure 18.3 Large canopy trees account for a substantial fraction of the biomass carbon dynamics of a forest, but can be a challenge to measure with sufficient accuracy. There is some evidence that they may be increasing disproportionately in growth rates: this example is from Noel Kempff National Park, Bolivia (photo Y. Malhi). (See Plate 8).

suggests a direct forcing from carbon dioxide, or possibly solar radiation and/or temperature. One unexpected feature is that the observed acceleration of growth, at $2.55 \pm 1.45\%$ yr^{-1}, is of an order of magnitude higher than that expected from CO_2 fertilisation alone. If CO_2 is fertilization indeed the primary driver of this acceleration, this discrepancy could be explained if trees are disproportionately allocating carbon to stem growth, or there are positive feedbacks via changes in forest structure and composition, or the shift in the compensation point between photosynthesis and respiration is more important than the overall boost in photosynthesis. Alternatively, some other driver such as a possible increase in sunshine may be involved (Nemani *et al.* 2003).

In sum, there is now fairly unequivocal evidence that the biomass of monitored intact forest plots in the neotropics has increased in recent times. There is still room for debate in the interpretation of this increase within the natural disturbance-discovery dynamic of a tropical forest landscape, although the evidence presented by Lewis *et al.* of a simultaneous increase in both biomass and turnover does suggest the presence of an external driver (such as CO_2 or light) accelerating growth. In many ways the observation of increased growth and turnover is perhaps more significant than the observation of increased biomass, although attention has focused more on the latter.

Can the changes in dynamics already documented from plots be verified by satellite observation? The dramatic increases in turnover reported by Phillips *et al.* might have a visible effect on canopy gap fractions or canopy texture. There is some evidence from Landsat imagery of concerted changes in mature forest canopy texture across the biome, consistent with a large-scale shift in forest behaviour (Weishampel *et al.* 2001). Other evidence may lie hidden in radar or lidar estimates of forest height, or the spectral signature of forests canopies, and as the satellite record improves in temporal duration, it may be rewarding to look for trends in these signals.

A key feature of all these spatially extensive analyses, whether based on direct or remotely sensed measurements, is that analysis of the spatial patterns should be able to reveal further clues as to the causal mechanisms behind these observed changes. For example, if the primary driver were increased carbon dioxide levels stimulating growth via improved water-use efficiency, we would expect to see the greatest growth and biomass gains in seasonally dry forests. Alternatively, if the primary driver is a possible increase in solar radiation, this increase would have a distinct geographical pattern which could be compared with forest plot observations.

Contemporary observations of change in tropical forests: species composition and function

An important issue is the extent that we can expect shifts in forest ecology and biodiversity in response to atmospheric change, such that even protected areas will undergo rapid change. Here the observations of changing biomass and dynamics have value as early indicators of the direction of change. The sensitivity of tropical forest composition to climatic change in the Pleistocene and Holocene (Mayle and Bush, Chapter 15) suggest that it is inevitable that tropical forest composition will change in response to atmospheric change, as particular species are favoured by higher CO_2 and changing climate. The increase in tree turnover could also engender floristic change. A change to lighter-wooded, faster growing tree taxa was predicted by Phillips and Gentry (1994) and Korner (Chapter 6). It has not yet been documented on a regional scale, but if lighter-wooded trees benefit across the tropics from accelerated productivity and dynamics, this change alone could potentially wipe out the carbon sink, as well as presenting new threats to diverse plant and animal species.

So far, ground observations have barely begun to address these issues, hampered by the extreme diversity of tropical forests that makes species-level analysis difficult, and the difficulty of assigning coherent functional groups to simplify this diversity in a functionally meaningful manner.

One of the most basic functional divisions within a tropical forest is between trees and lianas. Lianas can account for a significant fraction of canopy photosynthesis, but this is allocated to a low biomass, rapid growth functional form. Variations in liana abundance over time were first investigated by Phillips *et al.* (2002b), and showed a concerted increase in forest plots across western Amazonia since at least the 1980s. Some laboratory studies (e.g. Granados and Korner 2002) have indicated that lianas may have a superior growth response in shaded, high CO_2 conditions. Further evidence has come from the intensely studied plot at Barro Colorado, Panama (Fig. 18.4), one of the few tropical forest studies with sufficient detail, where observations of litterfall rates have shown an increase in liana productivity since the early 1980s (Wright *et al.* 2004).

In this book, Laurance, Chapter 9, takes one particularly well-studied landscape in central Amazonia (the BDFFP project north of Manaus), and examines a number of long-term plots for ecological change among trees since the 1980s. He finds a coherent floristic change in this landscape, with some tree families gaining disproportionately.

The biggest winners appear to be light-demanding canopy trees (Fig. 18.3), with the losers represented by some more slow-growing subcanopy and understorey taxa. Along with the change in species composition, he also finds a trend to faster growth rates and faster turnover, suggesting a shared causal link with the compositional trend. There is little evidence to support a climatic or historical explanation for these changes, and Laurance concludes that enhanced productivity, probably driven by CO_2 fertilization is the most likely explanation. Similar analyses should be possible across the more numerous smaller forest plots scattered across the tropics, and this is an obvious next step.

Tropical forests may be shifting in composition and ecological function regardless of whether they are sources or sinks of carbon. Much greater scientific effort is warranted in understanding this complex and emerging threat, both in terms of the focus of theoretical and modelling effort, and in the monitoring of tropical biodiversity changes on the ground.

Figure 18.4 Atmospheric changes may bring dramatic changes in the relative abundance of different plant functional groups, and hence to the ecological dynamics of tropical forests. There is already evidence that lianas may be increasing in abundance. This example is from Barro Colorado Island, Panama, where liana leaf productivity has increased dramatically over the last two decades (photo Y. Malhi). (See Plate 9).

What can be done?

In the Anthropocene (Crutzen 2002) we have embarked on an unplanned experiment with global ecosystems, mainly through direct action (e.g. deforestation), but also through the indirect effects of climate change. Given the inertia of both the global economy and the ocean–atmosphere system, we have already committed the planet to some further warming even if the world greenhouse gas emissions were eventually to shrink by half or more. Given the more realistic timescales for stabilizing global climate, substantial climate change by the end of the century appears inevitable. Understanding how ecosystems such as tropical forests will respond as they are pushed into environmental regimes unexplored for at least 20 million yrs (Prentice *et al.* 2001) is still riddled with uncertainty, but this book has tried to cast light on possible future trends.

It is of course vital to remember that in most tropical forest regions deforestation and degradation are the most immediate and severe threats, and will still dominate over climate change effects for at least the medium term (see Chapter 2 by Cramer *et al.*). In many regions the debate about the effects of climate change may be rendered irrelevant by the almost complete disappearance of tropical forests, but even if only a small fraction survive in reserves or inaccessible regions, their ecological stability may depend on the degree to which they will be affected by atmospheric change, and the ease with which vulnerable species can migrate to nearby refuges. Fragmented and depleted forest reserves will face the 'double-whammy' of direct human action and global atmospheric change.

Here we offer some recommendations for science and science policy. These can contribute to better understanding, mitigation, and amelioration of the threats facing the world's most diverse and complex ecosystems.

Expect change. Analysis of both contemporary climate records and environmental change since the last glacial maximum shows that tropical climates can vary substantially over time. Both the paleo-record and contemporary observations suggest that the species composition of tropical forests is sensitive to even small shifts in climate and atmospheric composition. Therefore it seems almost inevitable that intact tropical forests are undergoing substantial ecological shifts in response to contemporary global atmospheric change, but our ability to predict what exactly these changes will be in the future remains weak. Some changes may act synergistically to threaten widespread species that are still relatively abundant and appear well-protected. For example, shade-tolerant subcanopy species may suffer from increasing resource supply, increasing turnover, and progressive fragmentation and edge effects, as these are all changes that can be expected to favour faster growing, more disturbance adapted taxa.

Observe and monitor. The changes in tropical forests reported in several chapters in this book would not have been observed if some forest monitoring studies had not been established in the 1980s. For complex ecosystems with the potential of unforeseen emergent changes, 'field-truthing' of ecosystem shifts is an irreplaceable complement to laboratory studies or model projects. This is a much greater challenge in hyperdiverse low income tropical countries than in bio-impoverished but materially wealthy temperate ones. A systematic long-term program for monitoring tropical forests (and other ecosystems) to standard protocols must be a global science priority. Important steps have been made in this direction by networks such as the CTFS (Center for Tropical Forest Science) 50-ha plots network, the RAINFOR network of widespread smaller plots (Malhi *et al.* 2002b), and Conservation International's Tropical Ecosystems and Monitoring (TEAM) program (Figs 18.3 and 18.5), but there is a need for global standardization and integration, and some commitment to long-term funding. A useful analogy is perhaps the global network of meteorological observations that have proved invaluable in confirming and detailing climate change initially predicted by model simulations. A similar monitoring of twenty-first century 'ecosystem weather' would be relatively inexpensive and prove invaluable, both as a test of our understanding and as a 'miner's canary' to provide early warning of unforeseen ecological changes.

Figure 18.5 Atmospheric change may have different effects on canopy trees, gap-specialists, and understorey trees, altering the competitive balance between functional groups. Forest understoreys vary substantially with soil and climate conditions: this example is from Sucusari near Iquitos, Peru (photo Y. Malhi). (See Plate 10).

Bring more ecology into global vegetation models. The current generation of vegetation–climate models represent significant advances in our understanding of the relationship between climate and biome distribution. They are, however, based largely on an understanding of bulk physiology and biogeochemical cycling, and divide potential vegetation into a few 'functional types' such as evergreen broadleaf and evergreen deciduous. This is a substantial degree of complexity and can hardly be considered a failing, but very few explicitly model individual tree dynamics and system-level interactions such as resource competition. It is likely that critical interactions will occur at a fine degree of resolution of functional types. For example, an increase in abundance of lianas (Phillips *et al.* 2002) or shade-intolerant species may have a greater influence on forest biomass and dynamics than any direct ecophysiological response to temperature, but no current model is able to address such possible changes. This largely reflects a current lack of data and understanding, but an important target for a future generation of ecosystem models would be to incorporate a finer degree of resolution of plant functional types.

Do not take a tropical carbon sink for granted. To the extent that intact forests are increasing in biomass, they act as a carbon sink (of order $1 \, \mathrm{Pg\,C\,yr^{-1}}$) and a moderate buffer on the rate of rise of atmospheric CO_2. While the contemporary debate about the magnitude of the carbon sink is clearly of current global scientific and societal interest, in the longer term it may become less significant. Moreover, as pointed out by Cramer *et al.*, deforestation and forest degradation are likely to be more important factors in determining tropical carbon stores than the carbon dynamics of the remaining intact forests. Finally, any change in the terrestrial biosphere is likely to be swamped by expected fossil fuel emissions of CO_2 in the coming decades. For example, the entire live biomass carbon pool of the Amazonian forest is approximately $120 \, \mathrm{Pg\,C}$

(Houghton *et al.* 2000; Wood *et al.* submitted). If we include soil carbon reserves this store increases to about 200 PgC. This is equivalent to about 20 yrs of anthropogenic CO_2 emissions under the most likely future emissions scenarios (IPCC 2001). Hence even a 25% increase in the vegetation biomass of Amazonia (i.e. 30 PgC) would delay anthropogenic global warming by only 3 or 4 yrs. Researchers from different relevant disciplines are in agreement that the sink cannot be relied on, and may even reverse in the coming century, acting as a moderate accelerator rather than a moderate sink. This could be one of several 'tipping points' in the Earth's climate–biosphere system, threatening to send it into an alternative state characterized by uncontrollable and dangerous climate change and substantial modifications in biome distributions and function. Key changes already documented, such as widespread increases in lianas and localized changes in tree species could be early warnings of more significant and irreversible change.

Plan conservation with atmospheric change in mind. Tropical forests face dual pressures this century: direct deforestation and degradation, and atmospheric change. In the face of rapid tropical deforestation, it is essential that the conserved areas that remain are designed to be as robust as possible in the face of atmospheric change. A number of conservation organisations already take climate change into account, focusing more on 'conservation corridors' rather than isolated reserves, and incorporating elevational gradients that allow easier migration to cooler climates. Our understanding of ecosystem responses to atmospheric change is still in its infancy however, and these strategies will need to incorporate new insights as they emerge.

In summary, we are entering unknown environmental territory with the world's most diverse and poorly understood terrestrial ecosystems. Their adaptability and resilience to these changes is open to question. We should tread with care.

References

Absy, M. L., Cleef, A., Fournier, M., Martin, L., Servant, M., Sifeddine, A. *et al.* 1991. Mise en évidence de quatre phases d'ouverture de la forêt dense dans le sud-est de l'Amazonie au cours des 60,000 dernières années. Première comparaison avec d'autres régions tropicales. *Comptes Rendus de l'Academie des Sciences de Paris* **312**(II), 673–78.

Achard, F., Eva, H. D., Stibig, H.-J., Mayaux, P., Gallego, J., Richards, T. *et al.* 2002. Determination of deforestation rates of the world's humid tropical forests. *Science* **297**, 999–1002.

Achard, F., Eva, H. D., Mayaux, P., Stibig, H.-J., and Belward, A. 2004. Improved estimates of net carbon emissions from land cover change in the tropics for the 1990s. *Global Biogeochem. Cycles* **18**, doi: 10.1029/2003GB002142.

Aiba, S.-Y. and Kitayama, K. 2002. The effects of the 1997–98 El Nino drought on rain forests of Mount Kinabalu, Borneo. *J. Trop. Ecol.* **18**, 215–230.

Allen, C. D. and Breshears, D. D. 1998. Drought-induced shift of a forest-woodland ecotone: rapid landscape response to climate variation. *Proc. Natl. Acad. Sci.* **95**, 14839–14842.

Allen, L. H., Jr., Drake, B. G., Rogers, H. H., and Shinn, J. H. 1992. Field techniques for exposure of plants and ecosystems to elevated carbon dioxide and other trace gases. *Crit. Rev. Plant Sci.* **11**, 85–119.

Alpers, C. N. and Brimhall, G. H. 1988. Middle Miocene climatic change in the Atacama Desert, northern Chile: evidence from supergene mineralization at La Escondida. *Bull. Geol. Soc. Am.* **100**, 1640–1656.

Amthor, J. S. 1991. Respiration in a future, higher CO_2 world. *Plant Cell Environ.* **14**, 13–20.

Amthor, J. S. 1999. Increasing atmospheric CO_2 concentration, water use, and water stress: scaling up from the plant to the landscape. In *Carbon dioxide and Environmental Stress* (eds. Y. Luo and H. A. Mooney), pp. 33–55. San Diego, CA: Academic Press.

Amthor, J. S. and Koch, G. W. 1996. Biota growth factor b: stimulation of terrestrial ecosystem net primary production by elevated atmospheric CO_2. In *Carbon dioxide and Terrestrial Ecosystems* (eds. G. W. Koch and H. A. Mooney), pp. 399–414. San Diego, CA: Academic Press.

Andreae, M. 2001. The dark side of aerosols. *Nature* **409**, 671–672.

Andreae, M. O. *et al.* 2002. Biogeochemical cycling of carbon, water, energy, trace gases, and aerosols in Amazonia: the LBA-EUSTACH experiments. *J. Geophys. Res. Atm.* **107** (D20), art. no. 8066.

Anon. 2004. Saving the rainforest. *The Economist: 22 July 2004.* (30 July edn).

Araújo, T. M., Higuchi, N., and Carvalho, J. A., Jr., 1999. Comparison formulae for biomass content determination in a tropical rain forest site in the state of Pará, Brazil. *For. Ecol. Manage.* **117**, 43–52.

Arnone, J. A., III. 1996. Predicting responses of tropical plant communities to elevated CO_2: Lessons from experiments with model ecosystems. In *Carbon dioxide, Populations, and Communities* (eds. Ch. Körner and F. A. Bazzaz), pp. 101–121. San Diego, CA: Academic Press.

Arnone, J. A. I. and Körner, C. 1995. Soil and biomass carbon pools in model communities of tropical plants under elevated CO_2. *Oecologia (Berlin)* **104**, 61–71.

Artaxo, P. *et al.* 2003. Dry and wet deposition in Amazonia: from natural biogenic aerosols to biomass burning impacts. *Int. Glob. Atmos. Chem. Newslett.* **27**, 12–16.

Asner, G. P., Scurlock, J. M. O., and Hicke, J. A. 2003. Global synthesis of leaf area index observations: implications for ecological and remote sensing studies. *Glob. Ecol. Biogeogr.* **12**, 191–205.

Asner, G. P., Nepstad, D. C., Cardinot, G., and Ray, D. 2004. Drought stress and carbon uptake in an Amazon forest measured with spaceborne imaging spectroscopy. *Proc. Natl. Acad. Sci.* **101**, 6039–6044.

Avissar, R. and Liu, Y. 1996. A three-dimensional numerical study of shallow convective clouds and precipitation induced by land-surface forcing. *J. Geophys. Res.* **101**, 7499–7518.

Avissar, R. and Nobre, C. 2002. The large-scale biosphere-atmosphere (LBA) experiment in the Amazon. *J. Geophys. Res.* **107**, 8034, doi: 10.1029/2002JD002507.

Avissar, R., Silva Dias, P. L., Silva Dias, M. A. F., and Nobre, C. 2002. The large-scale biosphere-atmosphere experiment in Amazonia (LBA): insights and future research needs. *J. Geophys. Res.* **107** (D20), 8086, doi: 10.1029/2002JD002704.

Axelrod, D. I. 1970. Mesozoic palaeogeography and early angiosperm history. *Bot. Rev.* **36**, 277–319.

Baidya Roy, S. and Avissar, R. 2000. Scales of response of the convective boundary layer to land-surface heterogeneity. *Geophys. Res. Lett.* **27**, 533–536.

Baker, P. A., Seltzer, G. O., Fritz, S. C., Dunbar, R. B., Grove, M. J., Tapia, P. M. *et al.* 2001. The history of South American tropical precipitation for the past 25,000 years. *Science* **291**, 640–643.

Baker, P. A., Bush, M. B., Fritz, S., Rigsby, C., Seltzer, G., and Silman, M. R. 2003. Last glacial maximum in an Andean cloud forest environment (Eastern Cordillera, Bolivia): comment and reply. *Geology* on-line forum, e26–e27.

Baker, T. R., Phillips, O. L., Malhi, Y., Almeida, S., Arroyo, L., Di Fiore, A. *et al.* 2004a. Variation in wood density determines spatial patterns in Amazonian forest biomass *Glob. Change. Biol.* **10**, 545–562.

Baker, T. R., Phillips, O. L., Malhi, Y., Almeida, S., Arroyo, L., Fiore, A. D. *et al.* 2004b. Increasing biomass in Amazonian forest plots. *Phil. Trans. R. Soc. Lond. B* **359**, 353–356.

Barbosa, R. I. and Fearnside, P. M. 1999. Incêndios na Amazônia Brasileira: estimativa da emissão de gases do efeito estufa pela queima de diferentes ecossistemas de Roraima na passagem do evento 'El Niño' (1997/8). *Acta Amaz.* **29**, 513–534.

Barlow, J. 2003. Ecological effects of wildfires in a central Amazonian forest. In *School of Environmental Science.* Norwich: University of East Anglia.

Barlow, J. and Peres, C. A. 2004. Ecological responses to El Niño-induced surface fires in central Amazonia: management implications for flammable tropical forests. *Phil. Trans. R. Sc. Lond. B* **359**, 367–380.

Barlow, J. and Peres, C. A. 2004. Avifaunal responses to single and recurrent wildfires in Amazonian forests. *Ecol. Appl.* **14**, 1358–1373.

Barlow, J., Haugaasen, T., and Peres, C. A. 2002. Effects of ground fires on understorey bird assemblages in Amazonian forests. *Biol. Conserv.* **105**, 157–169.

Barlow, J., Peres, C. A., Lagan, B. O., and Haugaasen, T. 2003. Large tree mortality and the decline of forest biomass following Amazonian wildfires. *Ecol. Lett.* **6**, 6–8.

Barrett, D. J., Richardson, A. E., and Gifford, R. M. 1998. Elevated atmospheric CO_2 concentrations increase wheat root phosphatase activity when growth is limited by phosphorus. *Aust. J. Plant Physiol.* **25**, 87–93.

Barrett, P. M. and Willis, K. J. 2001. Did dinosaurs invent flowers? Dinosaurs-angiosperm coevolution revisited. *Biol. Rev. Cambridge Phil. Soc.* **76**, 411–447.

Barry, R. G. and Chorley, R. J. 1992. *Atmosphere, Weather and Climate*, 6th edn, p. 392. London: Routledge.

Baskerville, G. L. 1972. Use of logarithmic regression in the estimation of plant biomass. *Can. J. For.* **2**, 49–53.

Bazzaz, F. A. 1998. Tropical forests in a future climate: changes in biological diversity and impact on the global carbon cycle. *Clim. Change* **39**, 317–336.

Beerling, D. J. 2000. Increased terrestrial carbon storage across the Palaeocene–Eocene boundary. *Paleogeogr. Palaeoclimatol. Palaeoecol.* **161**, 395–405.

Behling, H. 1996. First report on new evidence for the occurrence of *Podocarpus* and possible human presence at the mouth of the Amazon during the late-glacial. *Veg. Hist. Archaeobot.* **5**, 241–246.

Behling, H. 2001. Late Quaternary environmental changes in the Lagoa da Curuça region (eastern Amazonia, Brazil) and evidence of *Podocarpus* in the Amazon lowland. *Veg. Hist. Archaeobot.* **10**, 175–183.

Behling, H. and Hooghiemstra, H. 1999. Environmental history of the Colombian savannas of the Llanos Orientales since the last glacial maximum from lake records El Pinal and Carimagua. *J. Paleolim.* **21**, 461–476.

Behling, H. and Hooghiemstra, H. 2000. Holocene Amazon rainforest-savanna dynamics and climatic implications: high-resolution pollen record from Laguna Loma Linda in eastern Colombia. *J. Quatern. Sci.* **15** (7), 687–695.

Behling, H. and Hooghiemstra, H. 2001. Neotropical savanna environments in space and time: late Quaternary interhemispheric comparisons. In *Interhemispheric Climate Linkages* (ed. V. Markgraf), pp. 307–323. Academic Press, San Diego, California, USA.

Behling, H., Keim, G., Irion, G., Junk, W., and Nunes de Mello, J. 2001. Holocene environmental changes in the Central Amazon Basin inferred from Lago Calado (Brazil). *Palaeogeogr. Palaeoclimatol. Palaeoecol.* **173**, 87–101.

Bergen, J. D. 1985. Some estimates of dissipation from the turbulent velocity component gradients over a forest canopy. In *The Forest-Atmosphere Interaction* (eds. B. A. Hutchinson, and B. B. Hicks). Dordrecht: D. Reidel.

Berner, R. A. 1991. A model for atmospheric CO_2 over phanerozoic time. *Am. J. Sci.* **291**, 339–375

Berner, R. A. 1997. The rise of plants and their effect on weathering and atmospheric CO_2. *Science* **276**, 544–546.

Berry, J. A. and Björkman, O. 1980. Photosynthetic response and adaptation to temperature in higher plants. *Annu. Rev. Plant Physiol.* **31**, 491–453.

Betts, R. A., Cox, P. M., Lee, S. E., and Woodward, F. I. 1997. Contrasting physiological and structural vegetation feedbacks in climate change simulations. *Nature* **387**, 796–799.

Betts, R. A., Cox, P. M., and Woodward, F. I. 2000. Simulated responses of potential vegetation to doubled-CO_2 climate change and feedbacks on near-surface temperature. *Glob. Ecol. Biogeogr.* **9**, 171–180.

Betts, R. A., Cox, P. M., Harris, P. P., Huntingford, C., and Jones, C. D. 2004. The role of ecosystem-atmosphere interactions in simulated Amazon forest dieback under global climate warming. *Theoretical and Applied Climatology,* **78**, 157–175.

Bierregaard, R. O. and Lovejoy, T. E. 1989. Effects of forest fragmentation on Amazonian understorey bird communities. *Acta Amaz.* **19**, 215–241.

Boer, C. 1989. Investigations of the steps needed to rehabilitate the areas of East Kalimantan seriously affected by fire: effects of the forest fires of 1982/83 in east Kalimantan towards wildlife. FR Report No. 7. Deutsche Forest Service/ITTO/GTZ, Samarinda.

Bolin, B., Sukumar, R., Ciais, P., Cramer, W., Jarvis, P., Kheshgi, H. *et al.* 2000. *IPCC Special Report on Land Use, Land-Use Change and Forestry*, Chapter 1: Global Perspective. Cambridge University Press, Cambridge, UK.

Botta, A., Ramamkutty, N., and Foley, J. A. 2002. Long-term variations of climate and carbon fluxes over the Amazon basin. *Geophys. Res. Lett.* **29**, art-1319.

Boyd, D. S., Phipps, P. C., Foody, G. M., and Walsh, R. P. D. 2002. Exploring the utility of NOAA A VHRR middle infrared reflectance to monitor the impacts of ENSO-induced drought stress on Sabah rainforests. *Int. J. Remote. Sens.* **23**, 5141–5147.

Broecker, W., Andree, M., Wolfli, W., Oeschger, H., Bonani, G., Kennett, J. *et al.* 1988. The chronology of the last deglaciation: implications to the cause of the Younger Dryas. *Paleoceanography* **3**, 1–19.

Brooks, A. and Farquhar, G. D. 1985. Effects of temperature on the CO_2/O_2 specificity of ribulose-1,5-bisphosphate carboxylase/oxygenase and the rate of respiration in the light. *Planta* **165**, 397–406.

Brovkin, V., Sitch, S., Von Bloh, W., Claussen, M., Bauer, E., and Cramer, W. 2004. Role of land cover changes for atmospheric CO_2 increase and climate change during the last 150 years. *Glob. Change Biol.* **10**, 1–14, doi: 10.1111/j.1365-2486.2004.00812.x.-

Brown, I. F., Martinelli, L. A., Thomas, W. W., Moreira, M. Z., Ferreira, C. A. C., and Victoria, R. A. 1995. Uncertainty in the biomass of Amazonian forests: an example from Rondônia, Brazil. *For. Ecol. Manage.* **75**, 175–189.

Brown, S. 1997. *Estimating Biomass and Biomass Change of Tropical Forests*. Forest Resources Assessment publication, volume 134 of Forestry Papers. Rome, FAO.

Brown, S. 2002. Measuring, monitoring, and verification of carbon benefits for forest-based projects. *Phil. Trans. Roy. Soc. A* **360**, 1669–1683.

Brown, S. and Lugo, A. E. 1982. The storage and production of organic matter in tropical forests and their role in the global carbon cycle. *Biotropica* **14**, 161–187.

Brown, S., Gillespie, A., and Lugo, A. 1989. Biomass estimation methods for tropical forests with applications to forest inventory data. *For. Sci.* **35**, 881–902.

Bruna, E. M. 1999. Seed germination in rainforest fragments. *Nature* **402**, 139.

Bruna, E. M. 2002. Experimental assessment of *Heliconia acuminata* growth in a fragmented Amazonian landscape. *J. Ecol.* **90**, 639–649.

Bucci, S. J., Goldstein, G., Meinzer, F. C., Scholz, F. G., Franco, A. C., and Bustamante, M. 2004. Functional convergence in hydraulic architecture and water relations of tropical savanna trees: from leaf to whole plant. *Tree Phys.* **24**, 891–899.

Bull, G. A. D. and Reynolds, E. 1968. Wind turbulence generated by vegetation and its implications. *Forestry* **41**(Suppl.), 28–37.

Bunyard, P. 2002. Climate and the Amazon: consequences for our planet. *Ecologist* October, 1–11.

Burbridge, R. E., Mayle, F. E., and Killeen, T. J. 2004. Fifty-thousand-year vegetation and climate history of Noel Kempff Mercado National Park, Bolivian Amazon. *Quatern. Res.* **61**, 215–230.

Burnham, R. and Carranco, N. In press. Miocene winged fruits of Loxopterygium (Anacardiaceae) from the Ecuadorian Andes. *Am. J. Bot.* **91**, 1767–1773.

Burnham, R. J. and Johnson, K. R. 2004. South American palaeobotany and origins of neotropical rainforests. *Phil. Trans. R. Soc. Lond. B* **359**, 1595–1610.

Bush, M. B. 2002. Distributional change and conservation on the Andean flank: a palaeoecological perspective. *Glob. Ecol. Biogeogr.* **11**, 463–473.

Bush, M. B. and Silman, M. R. 2004. Observations on late Pleistocene cooling and precipitation in the lowland Neotropics. *J. Quatern. Sci.* **19**(7), 677–684.

Bush, M. B., Piperno, D. R., and Colinvaux, P. A. 1989. A 6,000 year history of Amazonian maize cultivation. *Nature* **340**, 303–305.

Bush, M. B., Colinvaux, P. A., Weinmann, M. C., Piperno, D. R., and Liu, K.-B. 1990. Late Pleistocene temperature depression and vegetation change in Ecuadorian Amazon. *Quatern. Res.* **34**, 330–345.

Bush, M. B., Miller, M. C., De Oliveira, P. E. and Colinvaux, P. A. 2002. Orbital-forcing signal in sediments of two Amazonian lakes. *J. Paleolim.* **27**, 341–352.

Bush, M. B., Silman, M. R., and Urrego, D. H. 2004. 48,000 years of climate and forest change in a biodiversity hot spot. *Science* **303**, 827–829.

Buzas, M. A., Collins, L. S., and Culver, S. J. 2002. Latitudinal difference in biodiversity caused by higher tropical rate. *PNAS* **99**, 7841–7843.

Cabrera, A. L. 1976. *Regiones Fitogeográficas Argentinas*, 2nd edn. Buenos Aires: ACME S.A.C.I., Enciclopedia Argentina de Agricultura y Jardinería.

Cairns, M. A., Brown, S., Helmer, E. H., and Baumgardner, G. A. 1997. Root biomass allocation in the world's upland forests. *Oecologia* **111**, 1–11.

Camargo, J. L. C. and Kapos, V. 1995. Complex edge effects on soil moisture and microclimate in central Amazonian forest. *J. Trop. Ecol.* **11**, 205–211.

Canadell J. G. and Pataki, D. 2002. New advances in carbon cycle research. *Trends Ecol. Evol.* **17**, 156–158.

Cao, M., Prince, S. D., Small, J., and Goetz, S. J. 2004. Remotely-sensed interannual variations and trends in terrestrial net primary productivity 1981–2000. *Ecosystems* **7**, 233–242.

Carswell, F. E. *et al.* 2002. Seasonality in CO_2 and H_2O flux at an eastern Amazonian rainforest. *J. Geophys. Res. Atm.* **107**(D20), art 8076.

Carvalho, G., Barros, A. C., Moutinho, P., and Nepstad, D. 2001. Sensitive development could protect Amazonia instead of destroying it. *Nature* **409**, 131–131.

Carvalho, G., Nepstad, D., McGrath, D., Diaz, M. D. V., Santilli, M., and Barros, A. C. 2002. Frontier expansion in the Amazon: balancing development and sustainability. *Environment* **44**, 34–45.

Casper, B. B., Heard, S. B., and Apanius, V. 1992. Ecological correlates of single-seededness in a woody tropical flora. *Oecologia* **90**, 212–217.

Cerling, T. E., Harris, J. M., MacFadden, B. J., Leakey, M. G., Quade, J., Eisenmann, V., Ehleringer, J. R. 1997. Global vegetation change through the Miocene/Pliocene boundary. *Nature,* **389**, 153–158.

Chambers J. Q. and Silver W. L. 2004. Some aspects of ecophysiological and biogeochemical responses of tropical forests to atmospheric change. *Phil. Trans. R Soc. Lond. B,* **359**, 463–476.

Chambers, J. Q., Higuchi, N., and Schimel, J. P. 1998. Ancient trees in Amazonia. *Nature* **391**, 135–136.

Chambers, J. Q., Higuchi, N., Ferreira, L. V., Melack, J. M., and Schimel, J. P. 2000. Decomposition and carbon cycling of dead trees in tropical forests of the central Amazon. *Oecologia* **122**, 380–388.

Chambers, J. Q., dos Santos, J., Ribeiro, R. J., and Higuchi, N. 2001a. Tree damage, allometric relationships, and above-ground net primary production in a tropical forest. *Ecol. Appl.* **11**, 371–384.

Chambers, J. Q., dos Santos, J., Ribeiro, R. J., and Higuchi, N. 2001b. Tree damage, allometric relationships, and above-ground net primary production in central Amazon forest. *For. Ecol. Manage.,* **152**, 73–84.

Chambers, J. Q., Higuchi, N., Tribuzy, E. S., and Trumbore, S. E. 2001c. Carbon sink for a century. *Nature* **410**, 429.

Chambers, J. Q., Schimel, J. P., and Nobre, A. D. 2001d. Respiration from coarse wood litter in central Amazon forests. *Biogeochemistryx* **52**, 115–131.

Chambers, J. Q., Tribuzy, E. S., Toledo, L. C., Crispim, B. F., Higuchi, N., dos Santos, J., Araújo, A. C., Kruijt, B., Nobre, A. D. and Trumbore, S. E. 2004a. Respiration from a tropical forest ecosystem: Partitioning of sources and low carbon use efficiency. *Ecological Applications,* **14**, S72–S88.

Chambers, J. Q., Higuchi, N., Teixeira, L. M., Santos, J. D., Laurance, S. G., and Trumbore, S. E. 2004b. Response of tree biomass and wood litter to disturbance in a Central Amazon forest. *Oecologia,* **141**, 596–614.

Chapin, F. S. 1980. The mineral nutrition of wild plants. *Ann. Rev. Eco. Syst.,* **11**, 233–260.

Chave, J., Andalo, C., Brown, S., Cairns, M. A., Chambers, J. Q., Eamus, D., Fölster, H., Fromard, F., Higuchi, N., Kira, T., Lescure, J.-P., Nelson, B. W., Ogawa, H., Puig, H., Riéra, B., Yamakura, T. Tree allometry and improved estimation of carbon stocks and balance in tropical forests. *Oecologia.* In press.

Chave, J., Riera, B., and Dubois, M-A. 2001. Estimation of biomass in a neotropical forest of French Guiana: spatial and temporal variability. *J. Trop. Ecol.* **17**, 79–96.

Chave, J., Condit, R., Lao, S., Caspersen, J. P., Foster, R. B., and Hubbell, S. P. 2003. Spatial and temporal variation in biomass of a tropical forest: results from a large census plot in Panama. *J. Ecol.* **91**, 240–252.

Chave, J., Condit, R., Aguilar, S., Hernandez, A., Lao, S., and Perez, R. 2004. Error propagation and scaling for tropical forest biomass estimates. *Phil. Trans. R. Soc. B* **359**, 409–420.

Churkina G. and Running, S. W. 1998. Contrasting environmental controls on the estimated productivity of different biomes. *Ecosystems* **1**, 206–215.

Clapperton, C. M. 1993. Nature of environmental changes in South America at the Last Glacial Maximum. *Palaeogeogr. Palaeoclimatol Palaeoecol.* **101**, 189–208.

Clark, D. A. 2002a. Are tropical forests an important carbon sink? Reanalysis of the long-term plot data. *Ecol. Applic.* **12**, 3–7.

Clark, D. A. 2002b. Are tropical forests an important global carbon sink?: revisiting the evidence from long-term inventory plots. *J. Trop. Ecol.* **17**, 79–96.

Clark, D. A. 2004. Sources or sinks? The responses of tropical forests to current and future climate and atmospheric composition. *Phil. Trans. R. Soc. Lond. B* **359**, 477–491.

Clark, D. A. and Clark, D. B. 1994. Climate-induced annual variation in canopy tree growth in a Costa Rican tropical rainforest. *J. Ecol.* **82**, 865–872.

Clark, D. A., Brown, S., Kicklighter, D. W., Chambers, J. Q., Thomlinson, J. R., and Ni, J. 2001a. Measuring net primary production in forests: concepts and field methods. *Ecol. Appl.* **11**, 356–370.

Clark, D. A., Brown, S., Kicklighter, D. W., Chambers, J. Q., Thomlinson, J. R., Ni, J. *et al.* 2001b. Net primary production in tropical forests: an evaluation and synthesis of existing field data. *Ecol. Appl.* **11**, 371–384.

Clark, D. A., Piper, S. C., Keeling, C. D., and Clark, D. B. 2003. Tropical rain forest tree growth and atmospheric carbon dynamics linked to interannual temperature variation during 1984–2000. *Proc. Natl. Acad. Sci. USA* **100**, 5852–5857.

Clark, D. B. and Clark, D. A. 2000. Landscape-scale variation in forest structure and biomass in a tropical rain forest. *For. Ecol. Manage.* **137**, 185–198.

Clark, D. B., Clark, D. A., Brown, S., Oberbauer, S. F., and Veldkamp, E. 2002. Stocks and flows of coarse woody debris across a tropical rain forest nutrient and topography gradient. *For. Ecol. Manage.* **164**, 237–248.

Clark, D. B., Castro, C. S., Alvarado, L. D. A., and Read, J. M. 2004. Quantifying mortality of tropical rain forest trees using high-spatial-resolution satellite data. *Ecol. Lett.* **7**, 52–59.

Cleary, D. F. R. and Genner, M. J. 2004. Changes in rain forest butterfly diversity following major ENSO-induced fires in Borneo. *Glob. Ecol. Biogeogr.* **13**, 129–140.

CLIMAP Project Members. 1981. Seasonal reconstruction of the Earth's surface at the last glacial maximum. Geological Society of America Map and Chart Series MC-36.

Cochrane, M. A. 2001. In the line of fire: understanding the dynamics of tropical forest fires. *Environment* **43**, 28–38.

Cochrane, M. A. 2003. Fire science for rainforests. *Nature* **421**, 913–919.

Cochrane, M. A. and Laurance, W. F. 2002. Fire as a large-scale edge effect in Amazonian forests. *J. Trop. Ecol.* **18**, 311–325.

Cochrane, M. A. and Schulze, M. D. 1999. Fire as a recurrent event in tropical forests of the eastern Amazon: effects on forest structure, biomass, and species composition. *Biotropica* **31**, 2–16.

Cochrane, M. A., Alencar, A., Schulze, M. D., Souza, C. M., Nepstad, D. C., Lefebvre, P. *et al.* 1999. Positive feedbacks in the fire dynamic of closed canopy tropical forests. *Science* **284**, 1832–1835.

Colinvaux, P. A. and de Oliveira, P. E. 2000. Amazon plant diversity and climate through the Cenozoic. *Palaeogeogr. Palaeoclimatol. Palaeoecol.* **166**, 51–63.

Colinvaux, P. A., de Oliveira, P. E., Moreno, J. E, Miller, M. C., and Bush, M. B. 1996. A long pollen record from lowland Amazonia: forest and cooling in glacial times. *Science* **274**, 85–87.

Colinvaux, P. A., de Oliveira, P. E., and Bush, M. B. 2000. Amazonian and neotropical plant communities on glacial time-scales: the failure of the aridity and refuge hypotheses. *Quatern. Sci. Rev.* **19**, 141–169.

Colinvaux, P. A., Irion, G., Räsänen, M. E., and Bush, M. B. 2001. A paradigm to be discarded: Geological and paleoecological data falsify the Haffer and Prance refuge hypothesis of Amazonian speciation. *Amazoniana* **16**, 609–646.

Condit, R. 1997. Forest turnover, density, and CO_2. *Trends Ecol. Evol.* **12**, 249–250.

Condit, R. 1998a. Ecological implications of changes in drought patterns: shifts in forest composition in Panama. *Clim. Change.* **39**, 413–427.

Condit, R. 1998b. *Tropical Forest Census Plots*, Berlin, Georgetown, T: p 211. Springer-Verlag and R. G. Landes Company.

Condit, R., Hubbell, S. P., and Foster, R. B. 1993. Identifying fast-growing native trees from the neotropics using data from a large, permanent census plot. *For. Ecol. Manage.* **62**, 123–143.

Condit, R., Hubbell, S. P., and Foster, R. B. 1995. Mortality rates of 205 Neotropical tree and shrub species and the impact of servere drought. *Ecol. Monogr.* **65**, 419–439.

Condit, R., Hubbell, S. P., and Foster, R. B. 1996a. Changes in tree species abundance in a Neotropical forest: impact of climate change. *J. Trop. Ecol.* **12**, 231–256.

Condit, R., Hubbell, S. P., and Foster, R. B. 1996b. Assessing the response of plant functional types to climatic change in tropical forests. *J. Veg. Sci.* **7**, 405–416.

Condit, R., Ashton, P. S., Manokaran, N., LaFrankie, J. V., Hubbell, S. P., and Foster, R. B. 1999. Dynamics of the forest communities at Pasoh and Barro Colorado: comparing two 50-ha plots. *Phil. Trans. R. Soc. Lond. B* **354**, 1739–1748.

Condit, R., Watts, K., Bohlman, S. A., Perez, R., Foster, R. B., and Hubbell, S. P. 2000. Quantifying the deciduousness of tropical forest canopies under varying climates. *J. Veg. Sci.* **11**, 649–658.

Condit, R., Pitman, N., Leigh, E. G., Chave, J., Terborgh, J., Foster, R. B., *et al.* 2002. Beta-diversity in tropical forest trees. *Science* **295**, 666–669.

Condit, R., Aguilar, S., Hernandez, A., Perez, R., Lao, S., Angehr, G., *et al.* 2004. Tropical forest dynamics across a rainfall gradient and the impact of an El Niño dry season. *J. Trop. Ecol.* **20**, 51–72.

Connell, J. H. 1978. Diversity in tropical rain forests and coral reefs. *Science* **199**, 1302–1310.

Costa, M. and Foley, J. 2000. Combined effects of deforestation and doubled atmospheric CO_2 concentrations on the climate of Amazonia. *J. Clim.* **13**, 18–34.

Cowling, S. A. 1999a. Plants and temperature-CO_2 uncoupling. *Science* **285**, 1500–1501.

Cowling, S. A. 1999b. Simulated effects of low atmospheric CO_2 on structure and composition of North American vegetation at the Last Glacial Maximum. *Glob. Ecol. Biogeogr.* **8**, 81–93.

Cowling, S. A. and Field, C. B. 2003. Environmental controls on leaf area production: implications for vegetation and land surface modelling. *Glob. Biogeochem. Cyc.* **17**, 1007, doi: 10.1029/2002GB001915.

Cowling, S. A., and Sykes, M. T. 1999. Physiological significance of low atmospheric CO_2 for plant-climate interactions. *Quat. Res.* **52**, 237–242.

Cowling, S. A., Maslin, M. A., and Sykes, M. T. 2001. Paleovegetation simulations of lowland Amazonia and implications for neotropical allopatry and speciation. *Quatern. Res.* **55**, 140–149.

Cowling, S. A., Betts, R. A., Cox, P. M., Ettwein, V. J., Jones, C. D., Maslin, M. A. *et al.* 2004. Contrasting simulated past and future responses of the Amazon rainforest to atmospheric change. *Phil. Trans. R Soc.: Biol. Sci.* **359**, 539–547.

Cox, P. M. 2001. Description of the TRIFFID dynamic global vegetation model. Technical Note 24, Hadley Centre, UK Meteorological Office.

Cox, P. M., Huntingford, C., and Harding, R. J. 1998. A canopy conductance and photosynthesis model for use in a GCM land surface scheme. *J. Hydrol.* **213**, 79–94.

Cox, P. M., Betts, R. A., Bunton, C. B., Essery, R. L. H., Rowntree, P. R., and Smith, J. 1999. The impact of new land surface physics on the GCM simulation of climate and climate sensitivity. *Clim. Dyn.* **15**, 183–203.

Cox, P. M., Betts, R. A., and Jones, C. D., Spall, S. A. and Totterdell, I. J. 2000. Acceleration of global warming due to carbon-cycle feedbacks in a coupled climate model. *Nature* **408**, 184–187.

Cox, P. M., Betts, R. A., Collins, M., Harris, P. P., Huntingford, C., and Jones, C. D. 2004. Amazonian forest dieback under climate-carbon cycle projections for the 21st century. *Theor. Appl. Clim.* **78**, 137–156.

Cramer, W., Bondeau, A., Woodward, F. I., Prentice, I. C., Betts, R. A., Brovkin, V. *et al.* 2001. Global response of terrestrial ecosystem structure and function to CO_2 and climate change: results from six dynamic global vegetation models. *Glob. Change Biol.* **7**(4), 357–373.

Cramer, W., Bondeau, A., Schaphoff, S., Lucht, W., Smith, B., and Sitch, S. 2004. Tropical forests and the global carbon cycle: impacts of atmospheric CO_2, climate change and rate of deforestation. *Phil. Trans. R Soc. B* **359**, 331–343, doi: 10.1098/rstb.2003.1428.

Crane, P. R., Friis, E. M., and Pedersen, K. R. 1995. The origin and early diversification of angiosperms. *Nature* **374**, 27–34.

Cross, S. L., Baker, P. A., Seltzer, G. O., Fritz, S. C., and Dunbar, R. B. 2000. A new estimate of the Holocene lowstand level of Lake Titicaca, central Andes, and implications for tropical palaeohydrology. *Holocene* **10**(1), 21–32.

Cross, S. L., Baker, P. A., Seltzer, G. O., Fritz, S. C., and Dunbar, R. B. 2001. Late Quaternary climate and hydrology of tropical South America inferred from an isotopic and chemical model of Lake Titicaca, Bolivia and Peru. *Quatern. Res.* **56**, 1–9.

Crutzen P. J. 2002. The geology of mankind. *Nature* **415**, 23.

Curran, L. M., Caniago, I., Paoli, G., Astianti, D., Kusneti, M., Leighton, M. *et al.* 1999. Impact of El Niño and logging on canopy tree recruitment in Borneo. *Science* **286**, 2184–2188.

Curtis, P. S. and Wang, X. Z. 1998. A meta-analysis of elevated CO_2 effects on woody plant mass, form, and physiology. *Oecologia* **113**, 299–313.

Curtis, S. and Hastenrath, S. 1999. Trends of upper-air circulation and water vapour over equatorial South

America and adjacent oceans. *Int. J. Clim.* **19**, 863–876.

D'Agostino, K., Seltzer, G., Baker, P., Fritz, S., and Dunbar, R. 2002. Late-Quaternary lowstands of Lake Titicaca: evidence from high-resolution seismic data. *Palaeogeogr. Palaeoclimatol. Palaeoecol.* **179**, 97–111.

D'Angelo, S., Andrade, A. G., Laurance, S. G., Laurance, W. F., and Mesquita, R. 2004. Inferred causes of tree mortality in fragmented and intact Amazonian forests. *J. Trop. Ecol.* **20**, 243–246.

Dargaville, R. J., Heimann, M., McGuire, A. D., Prentice, I. C., Kicklighter, D. W., Joos, F. *et al.* 2002. Evaluation of terrestrial carbon cycle models with atmospheric CO_2 measurements: results from transient simulations considering increasing CO_2, climate, and land-use effects. *Glob. Biogeochem. Cycles* **16**, 1092.

Davidson, E. A. and Artaxo, P. 2004. Globally significant changes in biological processes of the Amazon Basin: results of the large-scale biosphere-atmosphere experiment. *Glob. Changes Biol.* **10**, 519–529.

Davidson, E. A. and Trumbore, S. E. 1995. Gas diffusivity and the production of CO_2 in deep soils of the eastern Amazon. *Tellus* **47B**, 550–565.

Davidson, E. A., Verchot, L. V., Cattanio, J. H., Ackerman, I. L., and Carvalho, J. E. M. 1999. Effects of soil water content on soil respiration in forests and cattle pastures of eastern Amazonia. *Biogeochemistry* **48**, 53–69.

Davidson, E. A., Ishida, F. Y., and Nepstad, D. C. 2004. Effects of an experimental drought on soil emissions of carbon dioxide, methane, nitrous oxide, and nitric oxide in a moist tropical forest. *Glob. Change Biol.* **10**, 718–730.

De Freitas, H. A., Pessenda, L. C. R., Aravena, R., Gouveia, S. E. M., De Souza Ribeiro, A., and Boulet, R. 2001. Late Quaternary vegetation dynamics in the Southern Amazon basin inferred from carbon isotopes in soil organic matter. *Quatern. Res.* **55**, 39–46.

de Noblet-Ducoudri, N., Claussen, M., and Prentice, I. C. 2000. Mid-Holocene greening of the Sahara: first results of the GAIM 6000 yr BP experiment with two asynchronously coupled atmosphere/biosphere models. *Clim. Dyn.* **16**, 643–659.

De Saussure, T. 1804. *Recherches chimiques sur la végétation.* Paris: chez la Veuve Nyon, an XII.

De Saussure, T. 1890. *Chemische Untersuchungen über die Vegetation.* Übersetzt von A. Wieler. Leipzig: Engelmann.

DeConto, R. M. and Pollard, D. 2003. Rapid Cenozoic glaciation of Antarctica induced by declining atmospheric CO_2. *Nature* **421**, 245–249

DeFries, R. S., Houghton, R. A., Hansen, M. C., Field, C. B., Skole, D., and Townshend, J. 2002. Carbon emissions from tropical deforestation and regrowth based on satellite observations for the 1980s. and 1990s. *Proc. Natl. Acad. Sci.* **99**, 14256–14261.

Delire, C., Behling, P., Coe, M. T., Foley, J. A., Jacob, R., Kutzbach, J. *et al.* 2001. Simulated response of the atmosphere-ocean system to deforestation in the Indonesian Archipelago. *Geophys. Res. Lett.* **28**, 2081–2084.

DeLucia, E. H., Hamilton, J. G., Naidu, S. L., Thomas, R. B., Andrews, J. A., Finzi, A. *et al.* 1999. Net primary production of a forest ecosystem with experimental CO_2 enrichment. *Science* **284**, 1177–1179.

Denevan, W. M. 1966. The aboriginal cultural geography of the Llanos de Mojos of Bolivia. *Ibero-Americana* **48**, 1–60.

Denslow, J. S., Schulz, J. C., Vitousek, P. M., and Strain, B. R. 1990. Growth response of tropical tree shrubs to treefall gap environment. *Ecology* **71**, 165–179.

Dewar, R. C., Medlyn, B. E., and McMurtrie, R. 1999. Acclimation of the respiration–photosynthesis ratio to temperature: insights from a model. *Glob. Change Biol.* **5**, 615–622.

Diaz, M. C. V., Nepstad, D., Mendonça, M. J. C., Seroa da Motta, R., Alencar, A., Gomes, J. C. *et al.* 2002. O Prejuízo Oculto do Fogo: Custos Econômicos das Queimadas e Incêndios Florestais na Amazônia (unpublished manuscript).

Dickinson, R. and Kennedy, P. 1992. Impacts on regional climate of Amazon deforestation. *Geophys. Res. Lett.* **19**, 1947–1950.

Didham, R. K. and Lawton, J. H. 1999. Edge structure determines the magnitude of changes in microclimate and vegetation structure in tropical forest fragments. *Biotropica* **31**, 17–30.

Doi, T. 1988. Present status of large mammals in Kutai National Park after a large scale fire in East Kalimantan. In *A Research on the Process of Earlier Recovery of Tropical Rain Forest after a Large Scale Fire in Kalimantan Timur, Indonesia* (eds. H. Tagawa and N. Wirawan). University of Kagoshima, Research Center for the South Pacific, Japan.

Douville, H., Planton, S., Royer, J. F., Stephenson, D., Tyteca, L., Kergoat, L. *et al.* 2000. Importance of vegetation feedbacks in doubled-CO_2 climate experiments. *J. Geophys. Res.* **105**(D), 14841–14861.

Dufresne, J. L., Friedlingstein, P., Berthelot, M., Bopp, L., Ciais, P., Fairhead, L. *et al.* 2002. On the magnitude of positive feedback between future climate change and the carbon cycle. *Geophys. Res. Lett.* **29**, art. no. 1405.

Dunbar, R. B. 2000. El Niño—clues from corals. *Nature* **407**, 956–959.

Ehleringer, J. R., Cerling, T. E., and Helliker, B. R. 1997. C_4 photosynthesis, atmospheric CO_2 and climate. *Oecologia* **112**, 285–299.

Elias, M. and Potvin, C. 2003. Assessing inter- and intra-specific variation in trunk carbon concentration for 32 neotropical tree species. *Can. J. For. Res.* **33**, 1039–1045.

Ellsworth, D. S., Oren, R., Huang, C., Phillips, N., and Hendrey, G. R. 1995. Leaf and canopy responses to elevated CO_2 in a pine forest under free-air CO_2 enrichment. *Oecologia* **104**, 139–146.

Enquist, B. J. and Niklas, K. J. 2001. Invariant scaling relations across tree-dominated communities. *Nature* **410**, 655–650.

Erickson, C. L. 1995. Archaeological methods for the study of ancient landscapes of the Llanos de Mojos in the Bolivian Amazon. In *Archaeology in the American Tropics: Current Analytical Methods and Applications* (ed. P. Stahl). Cambridge: Cambridge University Press.

Erickson, C. L. 2000. An artificial landscape-scale fishery in the Bolivian Amazon. *Nature* **408**, 190–193.

Eva, H. D., Achard, F., Stibig, H.-J., and Mayaux, P. 2003. Response to comment on Determination of deforestation rates of the World's humid tropical forests. *Science*, **299**, 1015b.

Evans, G. C. 1972. The Quantitative Analysis of Plant Growth. Blackwell Scientific Publicationis, Oxford.

Evans, L. T. and Dunstone, R. L. 1970. Some physiological aspects of evolution in wheat. *Aust. J. Biol. Sci.* **23**, 725–741.

Falge, E., Baldocchi, D., Tenhunen, J. *et al.* 2002. Seasonality of ecosystem respiration and gross primary production as derived from FLUXNET measurements. *Agric. For. Meteor.* **113**, 53–74.

FAO 1991. The digitized soil map of the world (release 1.0). Technical Report 67/1, Food and Agriculture Organization of the United Nations.

FAO 2000. *Forest resources assessment 2000. FAO Forestry papers 140*. Rome: FAO.

FAO 2003. State of the world's forests. Technical report, United Nations Food and Agriculture Organisation. www.fao.org/DOCREP/005/Y7581E/Y7581E00.HTM.

Farquhar, G. D. and von Caemmerer, S. 1982. Modelling of photosynthetic response to environmental conditions. In *Encyclopedia of Plant Physiology* (eds. O. L. Lange, P. S. Nobel, and C. B. Osmond, and H. Ziegler), pp. 550–587. Physiological Plant Ecology II. Berlin: Springer-Verlag.

Farquhar, G. D., von Caemmerer, S., and Berry, J. A. 1980. A biochemical model of photosynthetic CO_2 assimilation in leaves of C_3 species. *Planta* **149**, 78–90.

Fearnside, P. M. 1996. Amazonian deforestation and global warming: carbon stocks in vegetation replacing Brazil's Amazon forest. *For Ecol. Manage.* **80**, 21–34.

Fearnside, P. M. 2000. Global warming and tropical land-use change: greenhouse gas emissions from biomass burning, decomposition and soils in forest conversion, shifting cultivation and secondary vegetation. *Clim. Change* **46**, 115–158.

Fearnside, P. M. and Laurance, W. F. 2003. Comment on Determination of deforestation rates of the World's humid tropical forests. *Science* **299**, 1015a.

Fedorov, A. V. and Philander, S. G. 2000. Is El Niño changing? *Science* **288**, 1997–2002.

Field, C. B. 1999. Diverse controls on carbon storage under elevated CO_2: toward a synthesis. In *Carbon dioxide and Environmental Stress* (eds. Y. Luo and H. A. Mooney), pp. 389–408. San Diego, CA: Academic Press.

Field, C. B., Jackson, R. B., and Mooney, H. A. 1995. Stomatal responses to increased CO_2: Implications from the Plant to the Global Scale. *Plant Cell Environ.* **18**, 1214–1225.

Finzi, A. C., DeLucia, E. H., Hamilton, J. G., Richter, D. D., and Schlesinger, W. H. 2002. The nitrogen budget of a pine forest under free air CO_2 enrichment. *Oecologia* **132**, 567–578.

Fisher, R. A., Williams, M., De Lourdes Ruivo, M. R. L., Somebroek, W., Ferreira da Costa, R., Lola da Costa, A. *et al.* 2005. Hydraulic properties of a sandy Eastern Amazonian oxisol: estimation of unsaturated hydraulic conductance using tension infiltrometry and instantaneous profiling (in review).

Foley, J. A., Botta, A., Coe, M. T., and Costa, M. H. 2002. El Nino-southern oscillation and the climate, ecosystems and rivers of Amazonia. *Glob. Biogeochem. Cycle.* **16**, art. no. 1132.

Folland C. K. *et al.* 2002. Observed climate variability and change. In *Climate Change 2001: The Scientific Basis* (eds. J. T. Houghton *et al.*), pp. 99–181. Cambridge: Cambridge University Press.

Foster, M. S. and Terborgh, J. 1998. Impact of a rare storm event on an Amazonian forest. *Biotropica* **30**, 470–474.

Fredericksen, N. J. and Fredericksen, T. S. 2002. Terrestrial wildlife responses to logging and fire in a Bolivian tropical humid forest. *Biodivers. Conserv.* **11**, 27–38.

Freitas, S. R., Silva Dias, M. A. F., and Silva Dias, P. L. 2000. Modeling the convective transport of trace gases by deep and moist convection. *Hybrid Meth. Eng.* **3**, 317–330.

Friedlingstein, P., Bopp, L., Ciais, P., Dufresne, J., Fairhead, L., LeTreut, H. *et al.* 2001. Positive feedback between future climate change and the carbon cycle. *Geophys. Res. Lett.* **28**, 1543–1546.

Galloway, J. N. and Cowling, E. B. 2002. Reactive nitrogen and the world: 200 years of change. *Ambio* **31**, 64–71.

Gandú, A. W. and Silva Dias, P. L. 1998. Impact of tropical heat sources on the South American tropospheric upper circulation and subsidence. *J. Geophys. Res.* **103**, 6001–6015.

Ganopolski, A., Rahmstorf, S., Petoukhov, V., and Claussen, M. 1998. Simulation of modern and glacial climates with a coupled global model of intermediate complexity. *Nature* **391**, 351–356.

Gascon, C., Williamson, G. B., and Fonseca, G. A. B. 2000. Receding edges and vanishing reserves. *Science* **288**, 1356–1358.

Gash, J. H. C. and Nobre, C. A. 1997. Climatic effects of Amazonian deforestation: some results from ABRACOS. *Bull. Am. Meteor. Soc.* **78**, 823–830.

Gedney, N. and Valdes, P. J. 2000. The effect of Amazonian deforestation on the northern hemisphere circulation and climate. *Geophys. Res. Lett.* **27**, 3053–3056.

Gentry, A. H. 1995. Diversity and floristic composition of Neotropical dry forests. In *Seasonally Dry Tropical Forests* (eds. S. H. Bullock, H. A. Mooney, and E. Medina), pp. 146–194. Cambridge: Cambridge University Press.

Gerten, D., Schaphoff, S., Haberlandt, U., Lucht, W., and Sitch, S. 2004. Terrestrial vegetation and water balance—hydrological evaluation of a dynamic global vegetation model. *J. Hydrol.* **286**, 249–270.

Gerwing, J. J. 2002. Degradation of forests through logging and fire in the eastern Brazilian Amazon. *For. Ecol. Manage.* **157**, 131–141.

Gibbons, J. M. and Newbery, D. M. 2003. Drought avoidance and the effect of local topography on trees in the understorey of Bornean lowland rain forest. *Plant Ecol.* **164**, 1–18.

Gifford, R. M. and Evans, L. T. 1981. Photosynthesis, carbon partitioning, and yield. *Annu. Rev. Plant Physiol.* **32**, 485–509.

Giorgi F. 2002. Variability and trends of sub-continental scale surface climate in the twentieth century. Part I: observations. *Clim. Dyn.* **18**, 675–691.

Glor, R. E., Vitt, L. J., and Larson, A. 2001. A molecular phylogenetic analysis of diversification in Amazonian *Anolis* lizards. *Mol. Ecol.* **10**, 2661–2668.

Gonzalez-Meler M. A. and Siedow, J. N. 1999. Direct inhibition of mitochondrial respiratory enzymes by elevated CO_2: does it matter at the tissue or whole-plant level? *Tree Physiol.* **19**, 253–259.

Gonzalez-Meler M. A., Ribascarbo M., Siedow J. N., and Drake B. G. 1996. Direct inhibition of plant mitochondrial respiration by elevated CO_2. *Plant Physiol.* **112**, 1349–1355.

Gordon, C., Cooper, C., Senior, C. A., Banks, H., Gregory, J. M. Johns, T. C. *et al.* 2000. The simulation of SST, sea ice extents and ocean heat transports in a version of the Hadley Centre coupled model without flux adjustments. *Clim. Dyn.* **16**, 147–168.

Goulden M. L., Miller S. D., Rocha H. R. d., Menton M., Freitas H. C., Figueira A. M. S. d. *et al.* 2004. Diel and seasonal patterns of tropical forest CO_2 exchange. *Ecol. Appl.* **14**, 542–554.

Grace, J. and Malhi, Y. (2002). Carbon dioxide goes with the flow. *Nature*, **416**, 594–595.

Grace, J. *et al.* 1995a. Carbon dioxide uptake by an undisturbed tropical rain forest in Southwest Amazonia, 1992 to 1993. *Science* **270**, 778–780.

Grace, J. *et al.* 1995b. Fluxes of carbon dioxide and water vapour over an undisturbed tropical rainforest in south-west Amazonia. *Glob. Change Biol.* **1**, 1–12.

Gradstein, F. M. *et al.* 2004. *A Geologic Time Scale 2004.* New York: Columbia University Press.

Graham, E. A., Mulkey, S. S., Kitajima, K., Phillips, N. G., and Wright, S. J. 2003. Cloud cover limits net CO_2 uptake and growth of a rainforest tree during tropical rainy seasons. *Proc. Natl. Acad. Sci. USA* **100**, 572–576.

Granados, J. and Körner, C. 2002. In deep shade, elevated CO_2 increases the vigor of tropical climbing plants. *Glob. Change Biol.* **8**, 1109–1117.

Grime, J. P. 1988. The C-S-R model of primary plant strategies—origins, implications and tests. In *Plant Evolutionary Biology* (eds. L. D. Gottlieb and S. K. Jain), pp. 371–393. London: Chapman and Hall.

Groombridge, B. and Jenkins, M. D. 2003. *World Atlas of Biodiversity.* University of California Press

Gu, L., Baldocchi, D. D., Wofsy, S. C., Munger, J. W., Michalsky, J. J., Urbanski, S. P. *et al.* 2003. Response of a deciduous forest to the Mount Pinatubo eruption: enhanced photosynthesis. *Science* **299**, 2035–2038.

Guariguata, M. R. and Ostertag, R. 2001. Neotropical secondary succession: changes in structural and functional characteristics. *For. Ecol. Manage.* **148**, 185–206.

Gudhardja, E., Fatawi, M., Sutisna, M., Mori, T., and Ohta, S. 2000. Rainforest ecosystems of East Kalimantan: Él Niño, drought, fire and human impacts. *Ecol. Stud.* **140**, 1–330.

Gurney, K. R. *et al.* 2002. Towards robust regional estimates of CO_2 sources and sinks using atmospheric transport models. *Nature* **415**, 626–630.

Haberle, S. G. 1997. Upper Quaternary vegetation and climate history of the Amazon basin: correlating marine and terrestrial pollen records. In *Proceedings of the Ocean Drilling Program, Scientific Results*, Vol. 155 (eds. R. D. Flood, D. J. W. Piper, A. Klaus, and

L. C. Peterson), pp. 381–396. Ocean Drilling Program, College Station, TX.

Haberle, S. G. and Maslin, M. A. 1999. Late Quaternary vegetation and climate change in the Amazon basin based on a 50,000 year pollen record from the Amazon fan, ODP site 932. *Quatern. Res.* **51**, 27–38.

Haffer, J. 1969. Speciation in Amazonian forest birds. *Science* **165**, 131–137.

Haffer, J. and Prance, G. T. 2001. Climatic forcing of evolution in Amazonia during the Cenozoic: on the refuge theory of biotic differentiation. *Amazoniana* **16**, (3/4), 579–607.

Hamilton, J. G., Thomas, R. B., and Delucia, E. H. 2001. Direct and indirect effects of elevated CO_2 on leaf respiration in a forest ecosystem. *Plant Cell Environ.* **24**, 975–982.

Hamilton, J. G., DeLucia, E. H., George, K., Naidu, S. L., Finzi, A. C., and Schlesinger, W. H. 2002. Forest carbon balance under elevated CO_2. *Oecologia* **131**, 250–260.

Hansen, J. E., Sato, M., Lacis, A., Ruedy, R., Tegen, I., and Matthews, E. 1998. Climate forcings in the Industrial era. *Proc. Nat. Acad. Sci. USA* **95**, 12753–12758.

Harrison, S. P. and Prentice, I. C. 2003. Climate and CO_2 controls on global vegetation distribution at the last glacial maximum: analysis based on palaeovegetation data, biome modeling and palaeoclimate simulations. *Glob. Change Biol.* (in press).

Hättenschwiler, S. 2002. Liana seedling growth in response to fertilisation in a neotropical forest understorey. *Basic Appl. Ecol.* **3**, 135–143.1

Hättenschwiler, S., Miglietta, F., Raschi, A., and Korner, C. 1997. Thirty years of in situ tree growth under elevated CO_2: a model for future forest responses? *Glob. Change Biol.* **3**, 463–471.

Haug, G. H., and Tiedemann, R. 1998. Effect of the formation of the Isthmus of Panama on Atlantic Ocean thermohaline circulation. *Nature* **393**, 673–675.

Haugaasen, T., Barlow, J., and Peres, C. A. 2003. Surface wildfires in central Amazonia: Short-term impact on forest structure and carbon loss. *For. Ecol. Manage.* **179**, 321–331.

Haxeltine, A. and Prentice, I. C. 1996a. BIOME3: an equilibrium terrestrial biosphere model based on eco-physiological constraints, resource availability, and competition among plant functional types. *Glob. Biogeochem. Cycles*, **10**(4), 693–709.

Haxeltine, A. and Prentice, I. C. 1996b. A general model for the light-use efficiency of primary production. *Funct. Ecol.* **10**, 551–561.

Hay, W. W. 1996. Tectonic and climate. *Geol. Reundsch.* **85**, 409–437.

Henderson-Sellers, A., Dickinson, R., Durbidge, T., Kennedy, P., McGuffie, K., and Pitman, A. 1993. Tropical deforestation: modeling local to regional scale climate change. *J. Geophys. Res.* **98**, 7289–7315.

Herendeen, P. S., Crepet, W. L., and Dilcher, D. L. 1992. The fossil history of the Leguminosae: phylogenetic and biogeographic implications. In *Advances in Legume Systematics, Part 4: The Fossil Record* (eds. P. S. Herendeen and D. L. Dilcher), pp. 303–316. Royal Botanic Gardens, Kew, UK.

Hewitt, C. D., Broccoli, A. J., Mitchell, J. F. B., and Stouffer, R. J. 2003. A coupled model study of the last glacial maximum: was part of the North Atlantic relatively warm? *Clim. Dyn.* **20**, 203–218.

Hickey, L. J. and Doyle, J. A. 1977. Early Cretaceous fossil evidence for angiosperm evolution. *Bot. Rev.* **43**, 1–183.

Hillebrand, H. 2004. On the generality of the latitudinal diversity gradient. *Am. Nat.* **163**, 192–211.

Högberg, P. *et al.* 2001. Large-scale forest girdling shows that current photosynthesis drives soil respiration. *Nature* **411**, 789–792.

Holdsworth, A. R. and Uhl, C. 1997. Fire in Amazonian selectively logged rain forest and the potential for fire reduction. *Ecol. Appl.* **7**, 713–725.

Hoorn, C. 1997. Palynology of the Pleistocene glacial/interglacial cycles of the Amazon fan (holes 940A, 944A, and 946A). In *Proceedings of the Ocean Drilling Program, Scientific Results*, Vol. 155 (eds. R. D. Flood, D. J. W. Piper, A. Klaus, and L. C. Peterson), pp. 397–409, Ocean Drilling Program, College Station, TX.

Hoorn, C., Guerrero, J., Sarmiento, G., and Lorente, M. 1995. Andean tectonic as a cause for changing drainage patterns in Miocene northern South America. *Geology* **23**, 237–240.

Hostetler, S. W. and Mix, A. C. 1999. Reassessment of ice-age cooling of the tropical ocean and atmosphere. *Nature* **399**, 673–676.

Houghton, J. T., Callander, B. A. and Varney, S. K. (eds.) 1992. *Climate Change 1992: The Supplementary Report to the IPCC Scientific Assessment*. Cambridge, UK: Cambridge University Press.

Houghton, J. T., Ding, Y., Griggs, D. J., Noguer, M., van der Linden, P. J., Dai, X. *et al.* (eds.) 2001. *Climate Change 2001. The Scientific Basis*. Contribution of Working Group 1 to the Third Assessment Report of the IPCC. Cambridge: Cambridge University Press.

Houghton, R. A. 1991. Tropical deforestation and atmospheric carbon dioxide. *Clim. Change* **19**, 99–118.

Houghton, R. A. 1999. The annual net flux of carbon to the atmosphere from changes in land use 1850–1990. *Tellus*, **51B**, 298–313.

Houghton, R. A., Skole, D. L., Nobre, C. A., Hackler, J. L., Lawrence, K. T., and Chomentowski, W. H. 2000. Annual fluxes of carbon from deforestation and regrowth in the Brazilian Amazon. *Nature* **403**, 301–304.

Houghton, R. A, Lawrence, K. T, Hackler, J. L., and Brown, S. 2001. The spatial distribution of forest biomass in the Brazilian Amazon: a comparison of estimates. *Glob. Change Biol.* **7**, 731–746.

House, J. I., Prentice, I. C., and Le Quéré, C. 2002. Maximum impacts of future reforestation or deforestation on atmospheric CO_2. *Glob. Change Biol.* **8**, 1047–1052.

Howard, P. J. A. and Howard, D. M. 1979. Respiration of decomposing litter in relation to temperature and moisture. *Oikos* **33**, 457–465.

Hsiao, T. C. and Jackson, R. B. 1999. Interactive effects of water stress and elevated CO_2 on growth, photosynthesis, and water use efficiency. In *Carbon dioxide and Environmental Stress* (eds. Y. Luo and H. A. Mooney), pp. 3–26. San Diego, CA: Academic Press.

Huang, Y., Street-Perrott, F. A., Metcalfe, S. E., Brenner, M., Moreland, M., and Freeman, K. H. 2001. Climate change as the dominant control on glacial–interglacial variations in C_3 and C_4 plant abundance. *Science* **293**, 1647–1651.

Hubbell, S. P., Foster, R. B., O'Brien, S. T., Harms, K. E., Condit, R., Wechsler, B., *et al.* 1999. Light-cap disturbances, recruitment limitation, and tree diversity in a neotropical forest. *Science* **283**, 554–557.

Hughen, K. A., Southon, J. R., Lehman, S. J., and Overpeck, J. T. 2000. Synchronous radiocarbon and climate shifts during the last deglaciation. *Science* **290**, 1951–1954.

Hughes, R. F., Kauffman, J. B., and Jaramillo, V. J. 1999. Biomass, carbon, and nutrient dynamics of secondary forests in a humid tropical region of México. *Ecology* **80**, 1892–1907.

Hughes, R. F., Kauffman, J. B., and Jaramillo, V. J. 2000. Ecosystem-scale impacts of deforestation and land use in a humid tropical region of Mexico. *Ecol. Appl.* **10**, 515–527.

Hulme, M., Doherty, R., Ngara, T., New, M., and Lister, D. 2001. African climate change: 1900–2100. *Clim. Res.*, **17**, 145–168.

Huntingford, C., Harris, P. P., Gedney, N., Cox, P. M., Betts, R. A., Marengo, J. A. *et al.* 2004. Using a GCM analogue model to investigate the potential for Amazonian forest dieback. *Theor. App. Clim.* **78**, 177–185.

IPCC. 1996. *Climate Change 1995: Impacts, Adaptations and Mitigation of Climate Change: Scientific Technical Analyses.* Intergovernmental Panel on Climate Change. Cambridge: Cambridge University Press.

IPCC. 2001. *Climate change 2001: The Scientific Basis. Contribution of Working Group I to the Third Assessment Report of the Intergovernmental Panel on Climate Change.* Cambridge, UK; New York. Cambridge University Press, p. 881.

IPCC. 2002. *Climate Change 2001: The Scientific Basis,* pp. 1–881. Cambridge: Cambridge University Press.

Irion, G. 1978. Soil infertility in the Amazonian rainforest. *Naturwissenschaften* **65**, 515–519.

Jackson, R. B., Canadell, J., Ehleringer, J. R., Mooney, H. A., Sala, O. E., and Schulze, E. D. 1996. A global analysis of root distributions for terrestrial biomes. *Oecologia* **108**, 389–411.

Jarvis, P. G. 1998. *European Forests and Global Change.* Cambridge: Cambridge University Press, pp. 380.

Johns, T. C., Carnell, R. E., Crossley, J. F., Gregory, J. M., Mitchell, J. F. B., Senior, C. A. *et al.* 1997. The second Hadley Centre coupled ocean-atmosphere GCM: model description, spin-up and validation. *Clim. Dyn.* **13**, 103–134.

Jones, C. D. and Cox, P. M. 2003. The carbon cycle response to a Younger Dryas climate anomaly (in preparation).

Jones, C. D., Collins, M., Cox, P. M., and Spall, S. A. 2001. The carbon cycle response to ENSO: a coupled climate-carbon cycle model study. *J. Clim.* **14**, 4113–4129.

Jones, P. D., New, M., Parker, D. E., Martin, S., and Rigor, I. G. 1999. Surface air temperature and its changes over the past 150 years. *Rev. Geophys.* **37**, 173–199.

Jones, P. D., Osborn, T. J., Briffa, K. R., Folland, C. K., Horton, E. B., Alexander, L. V. *et al.* 2003. Adjusting for sampling density in grid box land and ocean surface temperature time series. *J. Geophys. Res.-Atmos.* **106**, 3371–3380.

Joos, F., Prentice, I. C., Sitch, S., Meyer, R., Hooss, G., Plattner, G. K. *et al.* 2001. Global warming feedbacks on terrestrial carbon uptake under the Intergovernmental Panel on Climate Change (IPCC) emission scenarios. *Glob. Biogeochem. Cycles* **15**, 891–907.

Kapos, V. 1989. Effects of isolation on the water status of forest patches in the Brazilian Amazon. *J. Trop. Ecol.* **5**, 173–185.

Kapos, V., Ganade, G., Matsui, E., and Victoria, R. L. 1993. $\alpha^{13}C$ as an indicator of edge effects in tropical rainforest reserves. *J. Ecol.* **81**, 425–432.

Kastner, T. P. and Goñi, M. A. 2003. Constancy in the vegetation of the Amazon basin during the late Pleistocene. *Geology* **31**, 291–294.

Kauffman, J. B. 1991. Survival by sprouting following fire in tropical forests of the Eastern amazon. *Biotropica* **23**, 219–224.

Kauffman, J. B., Uhl, C., and Cummings, D. L. 1988. Fire in the Venezuelan Amazon 1: fuel biomass and fire chemistry in the evergreen rainforest of Venezuela. *Oikos* **53**, 167–175.

Kauffman, J. B., Cummings, D. L., Ward, D. E., and Babbitt, R. 1995. Fire in the Brazilian Amazon: biomass, nutrient pools, and losses and slashed primary forests. *Oecologia* **104**, 397–408.

Keeling, C. D., Whorf, T. P., Whalen, M., and Van der Plicht, J. 1995. Interannual extremes in the rate of rise of atmospheric carbon dioxide since 1980. *Nature* **375**, 666–670.

Keller, M., Palace, M., and Hurtt, G. 2001. Biomass estimation in the Tapajos National forest, Brazil; Examination of sampling and allometric uncertainties. *For. Ecol. Manage.* **154**, 371–382.

Kessler, M. and Helme, N. 1999. Floristic diversity and phytogeography of the central Tuichi Valley, an isolated dry forest locality in the Bolivian Andes. *Candollea* **54**, 341–366.

Kinnaird, M. F. and O'Brien, T. G. 1998. Ecological effects of wildfire on lowland rainforest in Sumatra. *Conserv. Biol.* **12**, 954–956.

Kirschbaum, M. U. F., King, D. A., Comins, H. N., McMurtrie, R. E., Medlyn, B. E., Pongracic, S. *et al.* 1994. Modelling forest response to increasing CO_2 concentration under nutrient-limited conditions. *Plant Cell Environ.* **17**, 1081–1099.

Killeen, T. J. 1998. Vegetation and flora of Parque Nacional Kempff Mercado. Pages 61–85 in T. J. Killeen and T. S. Schulenberg, editors. *A biological assessment of Parque Nacional Noel Kempff Mercado and adjacent areas, Santa Cruz, Bolivia.* Conservation International, Washington, D. C.

Klein Goldewijk, K. 2001. Estimating global land use change over the past 300 years: the HYDE database. *Glob. Biogeochem. Cycles,* **15**(2), 417–434.

Klinge, H. and Rodrigues, W. A. 1973. Biomass estimation in a central Amazonian rain forest. *Acta Cientifica Venezoalana* **24**, 225–237.

Koch, G. W. and Mooney, H. A. 1996. Response of terrestrial ecosystems to elevated CO_2: A synthesis and summary. In *Carbon Dioxide and Terrestrial Ecosystems* (eds. G. W. Koch and H. A. Mooney), pp. 415–431. San Diego, CA: Academic Press.

Koch, G. W., Amthor, J. S., and Goulden, M. L. 1994. Diurnal patterns of leaf photosynthesis, conductance and water potential at the top of a lowland rain forest canopy in Cameroon: Measurements from the Radeau des Cimes. *Tree Physiol.* **14**, 347–360.

Körner, C. 1998. Tropical forests in a CO_2-rich world. *Clim. Change* **39**, 297–315.

Körner, C. 2000. Biosphere responses to CO_2 enrichment. *Ecol. Appl.* **10**, 1590–1619.

Körner, C. 2003a. Nutrients and sink activity drive CO_2 responses—a note of caution with literature based analysis. *New Phytol.* **159**, 537–538.

Körner, C. 2003b. Slow in, rapid out: carbon flux studies and Kyoto targets. *Science* **300**, 1242–1243.

Körner, C. 2003c. Carbon limitation in trees. *J. Ecol.* **91**, 4–17.

Körner, C. 2004. Through enhanced tree dynamics carbon dioxide enrichment may cuase tropical forests to lose carbon. *Phil. Trans. R. Soc. Lond. B* **359**, 493–498.

Körner, Ch. 1995. Towards a better experimental basis for upscaling plant responses to elevated CO_2 and climate warming. *Plant, Cell Environ.* **18**, 1101–1110.

Körner, Ch. 2004. CO_2 enrichment may cause tropical forests to become carbon sources. *Phil. Trans. R. Soc., B* **359**, 493–498.

Körner, C. and Arnone, J. A., III. 1992. Responses to elevated carbon dioxide in artificial tropical ecosystems. *Science* **257**, 1672–1675.

Körner, Ch. and Würth, M. 1996. A simple method for testing leaf responses of tall tropical forest trees to elevated CO_2. *Oecologia* **107**, 421–425.

Kutzbach, J., Gallimore, R., Harrison, S., Behling, P., Selin, R. and Laarif, F. 1998. Climate and biome simulations for the past 21,000 years. *Quatern. Sci. Rev.* **17**, 473–506.

Kyllo, D. A., Velez, V., and Tyree, M. T. 2003. Combined effects of arbuscular mycorrhizas and light on water uptake of the neotropical understory shrubs, Piper and Psychotria. *New Phytol.* **160**, 443–454.

Lambers, H. 1997. Respiration and the alternative oxidase. In *A molecular approach to primary metabolism in plants* (eds. C. H. Foyer and P. Quick), pp. 295–309. London: Taylor and Francis.

Lambers, H., Chapin, F. S. I., and Pons, T. L. 1998. *Plant Physiological Ecology.* New York: Springer.

Lambert, F. V. and Collar, N. J. 2002. The future for Sundaic lowland forest birds: long-term effects of commercial logging and fragmentation. *Forktail* **18**, 127–146.

Langenfelds, R. L., Francey, R. J., Pak, B. C., Steele, L. P., Lloyd, J., Trudinger, C. M. *et al.* 2003. Interannual growth rate variations of atmospheric CO_2 and its $d^{13}C$, H_2, CH_4 and CO between 1992 and 1999. linked to biomass burning. *Glob. Biogeochem. Cycles* **16**.

Langley, C. H. and Fitch, W. 1974. An estimation of the constancy of the rate of molecular evolution. *J. Mol. Evol.* **3**, 161–177.

Langstroth, R. P. 1999. Forest islands in an Amazonian savanna of northeastern Bolivia. PhD thesis, University of Washington-Madison.

Lanly, J. P. 1982. Tropical forest resources. FAO Forestry Paper 30. Rome: United Nations Food and Agricultural Organization.

Laurance, W. F. 1997. Hyper-disturbed parks: edge effects and the ecology of isolated rainforest reserves in tropical Australia. In *Tropical Forest Remnants: Ecology, Management, and Conservation of Fragmented Communities* (eds. Laurance, W. F. and R. O. Bierregaard) pp. 71–83. Chicago, IL: University of Chicago Press.

Laurance, W. F. 1999. Reflections on the tropical deforestation crisis. *Biol. Conserv.* **91**, 109–117.

Laurance, W. F. 2001a. Tropical logging and human invasions. *Conserv. Biol.* **15**, 4–5.

Laurance, W. F. 2001b. The hyper-diverse flora of the central Amazon: an overview. In *Lessons from Amazonia: Ecology and Conservation of a Fragmented Forest* (eds. R. O. Bierregaard, C. Gascon, T. E. Lovejoy, and R. Mesquita) pp. 47–53. Yale University Press.

Laurance, W. F. 2004. Forest–climate interactions in fragmented tropical landscapes. *Phil. Trans. R. Soc. Lond. B* **359**, 345–352.

Laurance, W. F., and Bierregaard, R. O. 1997. *Tropical Forest Remnants: Ecology, Management, and Conservation of Fragmented Communities*. Chicago, IL: University of Chicago Press.

Laurance, W. F. and Cochrane, M. A. 2001. Synergistic effects in fragmented landscapes. *Conserv. Biol.* **15**, 1488–1489.

Laurance, W. F. and Fearnside, P. M. 2002. Issues in Amazonian development. *Science* **295**, 1643–1643.

Laurance, W. F. and Williamson, G. B. 2001. Positive feedbacks among forest fragmentation, drought, and climate change in the Amazon. *Conserv. Biol.* **15**, 1529–1535.

Laurance, W. F., Laurance, S. G., Ferreira, L. V., Rankin-de Merona, J. M., Gascon, C., and Lovejoy, T. E. 1997. Biomass collapse in Amazonian forest fragments. *Science* **278**, 1117–1118.

Laurance, W. F., Ferreira, L. V., Rankin-de Merona, J., and Laurance, S. G. 1998a. Rain forest fragmentation and the dynamics of Amazonian tree communities. *Ecology* **79**, 2032–2040.

Laurance, W. F., Ferreira, L. V., Rankin-de Merona, J., Laurance, S. G., Hutchings, R., and Lovejoy, T. E. 1998b. Effects of forest fragmentation on recruitment patterns in Amazonian tree communities. *Conserv. Biol.* **12**, 460–464.

Laurance, W. F., Laurance, S. G., and Delamonica, P. 1998c. Tropical forest fragmentation and greenhouse gas emissions. *For. Ecol. Manage.* **110**, 173–180.

Laurance, W. F., Fearnside, P. M., Laurance, S. G., Delamonica, P., Lovejoy, T. E., Rankin-de Merona, J. M., et al. 1999. Relationship between soils and Amazon forest biomass: a landscape-scale study. *For. Ecol. Manage.* **118**, 127–138.

Laurance, W. F., Delamonica, P., Laurance, S., Vasconcelos, H., and Lovejoy, T. E. 2000. Rainforest fragmentation kills big trees. *Nature* **404**, 836.

Laurance, W. F., Cochrane, M. A., Bergen, S., Fearnside, P. M., Dolmônica, P., Barber, C. et al. 2001a. The future of the Brazilian Amazon. *Science* **291**, 438–439.

Laurance, W. F., Perez-Salicrup, D., Delamonica, P., Fearnside, P., D' Angelo, S., Jerozolinski, et al. 2001b. Rain forest fragmentation and the structure of Amazonian liana communities. *Ecology* **82**, 105–116.

Laurance, W. F., Williamson, G. B., Delamonica, P., Olivera, A., Gascon, C., Lovejoy, T. et al. 2001c. Effects of a strong drought on Amazonian forest fragments and edges. *J. Trop. Ecol.* **17**, 771–785.

Laurance, W. F., Albernaz, A. K. M., Schroth, G., Fearnside, P. M., Ventincinque, E., and Da Costa, C. 2002a. Predictors of deforestation in the Brazilian Amazon. *J. Biogeogr.* **29**, 737–748.

Laurance, W. F., Lovejoy, T. E., Vasconcelos, H. L., Bruna, E., Didham, R., Stouffer, P. et al. 2002b. Ecosystem decay of Amazonian forest fragments: a 22-year investigation. *Conserv. Biol.* **16**, 605–618.

Laurance, W. F., Oliveira, A. A. Laurance, S. G., Condit, R., Nascimento, H. E. M., Sanchez-Thorin, A. C. et al. 2004a. Pervasive alteration of tree communities in undisturbed Amazonian forests. *Nature* **428**, 171–175.

Laurance, W. F., Nascimento, H., Laurance, S. G., Condit, R., D'Angelo, S., and Andrade, A. 2004b. Inferred longevity of Amazonian rainforest trees based on a long- term demographic study. *For. Ecol. Manage.* **190**, 131–143.

Laurance, W. F., Oliveira, A. A., Laurance, S. G., Condit, R., Dick, C. W., Andrade, A. et al. 2005. Altered tree communities in undisturbed Amazonian forests: A consequence of global change? *Biotropica*, **37**.

Lavin, M., Herendeen, P. S., and Wojciechowski, M. F. In press. Evolutionary rates analysis of Leguminosae implicates a rapid diversification of the major family lineages immediately following an Early Tertiary emergence. *Syst. Biol.* **54**.

Lavin, M., Wojciechowski, M. F., Gasson, P., Hughes, C. E. and Wheeler, E. 2003. Phylogeny of robinioid legumes (Fabaceae) revisited: Coursetia and Gliricidia recircumscribed, and a biogeographical appraisal of the Caribbean endemics. *Syst. Bot.* **28**: 387–409.

Lawton R. O., Nair U. S., Pielke, R. A. S., and Welch R. M. 2001. Climate impact of tropical lowland deforestation on nearby montane cloud forests. *Science* **294**, 584–587.

Lean, J. and Rowntree, P. R. 1993. A GCM simulation of the impact of Amazonian deforestation on climate using an improved canopy representation. *Q. J. R. Meteor. Soc.* **119**, 509–530.

Lean, J. and Warrilow, D. 1989. Climatic impact of Amazon deforestation. *Nature* **342**, 311–313.

Ledru, M.-P., Cordeiro, R. C., Dominguez, J. M. L., Martin, L., Mourguiart, P., Sifeddine, A. *et al.* 2001. Late-glacial cooling in Amazonia inferred from pollen al Lagoa do Caçó, northern Brazil. *Quatern. Res.* **55**, 47–56.

Leigh, E. G. 1999. *Tropical Forest Ecology.* Oxford University Press, Oxford, UK, p. 245.

Leighton, M. 1983. *The El Nino—Southern Oscillation Event in Southeast Asia: Effects of Drought and Fire in Tropical Forest in Eastern Borneo.* Unpublished report, Department of Anthropology, Harvard University.

Leighton, M. and Wirawan, N. 1986. Catastrophic drought and fire in Borneo tropical rain forest associated with the 1982–1983 El Nino Southern Oscillation event. In *Tropical Forests and the World Atmosphere, AAAS Selected Symposium 101* (ed. G. T. Prance), pp. 75–102. Washington, DC: American Association for the Advancement of Science.

Lescure J.-P., Puig H., Riéra B., Leclerc D., Beekman A., and Bénéteau A. 1983. La phytomasse épigée d'une forêt dense en Guyane française. *Acta Oecol.* **4**, 237–251.

Levis, S., Foley, J. A., and Pollard, D. 1999. CO_2, climate and vegetation feedbacks at the Last Glacial Maximum. *J. Geophys. Res. Atmos.* **104**, 31191–31198.

Levis, S., Foley, J. A., and Pollard, D. 2000. Large-scale vegetation feedbacks on a doubled CO_2 climate. *J. Clim.* **13**, 1313–1325.

Lewis, S. 1998. Treefall gaps and regeneration: a comparison of continuous forest and fragmented forest in central Amazonia. Ph.D. thesis. Cambridge, UK: University of Cambridge.

Lewis, S. L., Malhi, Y., and Phillips, O. L. 2004a. Fingerprinting the impacts of global change on tropical forests. *Phil. Trans. R. Soc. Lond. B* **359**, 437–462.

Lewis, S. L. *et al.* 2004b. Concerted changes in tropical forest structure and dynamics: evidence from 50 South American long-term plots. *Phil. Trans. R. Soc. Lond. B* **359**, 421–436.

Lewis, S. L., Phillips, O. L., Sheil, D. Vinceti, B., Baker, T., Brown, S. *et al.* 2004c. Tropical forest tree mortality, recruitment and turnover rates: calculation, interpretation, and comparison when census intervals vary. *J. Ecol.* **92**(6), 929–944.

Lidgard, S. and Crane, P. R. 1990. Angiosperm diversification and Cretaceous floristic trends. *Palaeobiology* **16**, 77–93.

Liu, K.-B. and Colinvaux, P. A. 1985. Forest changes in the Amazon during the Last Glacial Maximum. *Nature* **318**, 556–557.

Lloyd, J. and Farquhar, G. D. 1996. The CO_2 dependence of photosynthesis, plant growth responses to elevated atmospheric CO_2 concentrations and their interaction with plant nutrient status. *Funct. Ecol.* **10**, 4–32.

Lloyd, J., Bird, M. I., Veenendaal, E. M., and Kruijt, B. 2001. Should photsphorus availability be constraining moist tropical forest responses to increasing CO_2 concentrations? In *Global Biogeochemical Cycles in the Climate System* (ed. E. D. Schulze), pp. 95–114. San Diego, CA: Academic Press.

Lobo, J. M. and Martin-Piera, F. 2002. Searching for a predictive model for species richness of Iberian dung beetle based on spatial and environmental variables. *Conserv. Biol.* **16**, 158–173.

Lovejoy, T. E. *et al.* 1986. Edge and other effects of isolation on Amazon forest fragments. In *Conservation Biology: The Science of Scarcity and Diversity* (ed. Soulé, M. E.), pp. 257–285 Sunderland, MA: Sinauer.

Lovelock, C. E., Kyllo, D. and Winter, K. 1996. Growth responses to vesicular-arbuscular mycorrhizae and elevated CO_2 in seedlings of tropical tree, *Beilschmiedia pendula. Funct. Ecol.* **10**, 662–667.

Lovelock, C. E., Virgo, A., Popp, M., and Winter, K. 1999. Effects of elevated CO_2 concentrations on photosynthesis, growth and reproduction of branches of the tropical canopy tree species, *Luehea seemannii* Tr. and Planch. *Plant, Cell Environ.* **22**, 49–59.

Lucht, W., Prentice, I. C., Myneni, R. B., Sitch, S., Friedlingstein, P., Cramer, W. *et al.* 2002. Climatic control of the high-latitude vegetation greening trend and Pinatubo effect. *Science* **296**, 1687–1689.

Lugo, A. E., Applefield, M., Pool, D., and McDonald, R. 1983. The impact of Hurricane David on the forests of Dominica. *Can. J. For. Res.* **132**, 201–211.

Lui, W. T. H., Massambani, O., and Nobre, C. A. 1994. Satellite recorded vegetation response to drought in Brazil. *Int. J. Clim.* **14**, 343–354.

Magallón, S., Crane, P. R., and Herendeen P. S. 1999. Phylogenetic pattern, diversity and diversification of eudicots. *Ann. M. Bot. Gard.* **86**, 297–372.

Mahlman, J. D. 1997. Uncertainties in projections of human-caused climate warming. *Science* **278**, 1416–1417.

Malcolm, J. R. 1994. Edge effects in central Amazonian forest fragments. *Ecology* **75**, 2438–2445.

Malcolm, J. R. 1998. A model of conductive heat flow in forest edges and fragmented landscapes. *Clim. Change* **39**, 487–502.

Maley, J. and P. Brenac 1998. Vegetation dynamics, palaeoenvironments and climatic changes in the forests of western Cameroon during the last 28,000 years B. P. *Rev. Palaeobot. Palynol.* **99**(2), 157–187.

Malhi, Y. 2002. Carbon in the atmosphere and terrestrial biosphere in the 21st century. *Phil. Trans. R. Soc. Lond.* **360**, 2925–2945.

Malhi, Y. and Grace, J. 2000. Tropical forests and atmospheric carbon dioxide. *Trends Ecol. Evol.* **15**, 332–337.

Malhi, Y. and Phillips, O. L. 2004. Tropical forests and global atmospheric change: a synthesis. *Phil. Trans. R. Soc. B* **359**, 549–555.

Malhi, Y. and Wright, J. 2004. Spatial patterns and recent trends in the climate of tropical forest regions. *Phil. Trans. R. Soc. Lond. B* **359**, 311–329.

Malhi, Y., Nobre, A. D., Grace, J., Kruijt, B., Pereira, M. G. P., Culf, A. *et al.* 1998. Carbon dioxide transfer over a Central Amazonian rain forest. *J. Geophys. Res. Atmos.* **103**(D24), 31593–31612.

Malhi Y. *et al.* 2002a. An international network to monitor the structure, composition and dynamics of Amazonian forests (RAINFOR). *J. Veg. Sci.* **13**, 439–450.

Malhi, Y., Phillips, O. L., Baker, T. R., Almeida, S., Frederiksen, T., Grace, J. *et al.* 2002b. An international network to understand the biomass and dynamics of Amazonian forests (RAINFOR). *J. Veg. Sci.* **13**, 439–450.

Malhi, Y. *et al.* 2004. The above-ground coarse wood productivity of 104 Neotropical forest plots. *Glob. Change Biol.* **10**, 563–591.

Marengo, J. A. 1995. Variations and change in South American streamflow. *Clim. Change* **31**, 99–117.

Marengo, J. A. 2004. Interdecadal variability and trends of rainfall across the Amazon basin, *Theor. Appl. Climatol.* **78**, 79–96.

Marengo, J. A., Druyan, L. M., and Hastenranth, S. 1993. Observational and modelling studies of Amazonia interannual climate variability. *Clim. Change* **23**, 267–286.

Marengo, J. A., Tomasella, J., and Uvo C. B. 1998. Trends in stream-flow and rainfall in tropical South America: Amazonia, Eastern Brazil and Northwest Peru. *JGR-Atmos* **103**, 1775–1783.

Martins, J. V., Artaxo, P., Liousse, C., Reid, J., Hobbs, P., and Kaufman, Y. 1998. Effects of black carbon content, particle size and mixing on light absorbtion by aerosol particles from biomass burning in Brazil. *J. Geophys. Res.* **103**, 32041–32050.

Maslin, M. A. and Burns, S. J. 2000. Reconstruction of the Amazon Basin effective moisture availability over the past 14,000 Years. *Science* **290**, 2285–2287.

Matson, P. A., Lohse, K. A., and Hall, S. J. 2002. The globalization of nitrogen deposition: consequences for terrestrial ecosystems. *Ambio.* **31**, 113–119.

Mayer, J. H. 1989. Socioeconomic aspects of the forest fire of 1982/83. and the relation of local communities towards forestry and forest management. FR Report No. 8, Samarinda.

Mayle, F. E. 2004. Assessment of the Neotropical dry forest refugia hypothesis in the light of palaeoecological data and vegetation model simulations. *J. Quaternary Sci.* **19**, 713–720.

Mayle, F. E., Burbidge, R., and Killeen, T. J. 2000. Millennial-scale dynamics of southern Amazonian rain forests. *Science* **290**, 2291–2294.

Mayle, F. E., Beerling, D. J., Gosling, W. D., and Bush, M. B. 2004. Responses of Amazonian ecosystems to climatic and atmospheric CO_2 changes since the last glacial maximum. *Phil. Trans. R. Soc. Lond. B* **359** (1443), 499–514.

McElwain, J. C. 1998. Do fossil plants signal palaeo-atmospheric CO_2 concentrations in the geological past. *Phil. Trans. R. Soc. Biol. Sci.* **353**, 1–15.

McWilliam, A.-L. C., Cabral, O. M. R., Gomes, B. M., Esteves, J. L., Roberts, J. M. 1996. Forest and pasture leaf gas-exchange in south-west Amazonia. In *Amazonia Deforestation and Climate* (eds. J. H. C. Gash *et al.*), pp. 265–286. John Wiley, Chichester, UK.

Medlyn, B. E., Barton, C. V. M., Broadmeadow, M. S. J., Ceulemans, R., De Angelis, P., Forstreuter, M. *et al.* 2001. Stomatal conductance of forest species after long-term exposure to elevated CO_2 concentration: a synthesis. *New Phytol.* **149**, 247–264.

Meggers, B. J. 1994. Archeological evidence for the impact of mega-Nino events on Amazonia during the past two millenia. *Clim. Change* **28**, 321–338.

Meinzer, F. C. 2003. Functional convergence in plant responses to the environment. *Oecologia*, **134**, 1–11.

Meinzer, F. C., Andrade, J. L., Goldstein, G., Holbrook, N. M., Cavelier, J., and Jackson, P. 1997. Control of transpiration from the upper canopy of a tropical forest: the role of stomatal, boundary layer and hydraulic architecture components. *Plant. Cell Environ.* **20**, 1242–1252.

Meinzer, F. C., Goldstein, G., and Andrade, J. L. 2001. Regulation of water flux through tropical forest canopy trees: do universal rules apply? *Tree Physial.* **21**, 19–26.

Meir, P. and Grace, J. 2002. Scaling relationships for woody tissue respiration in two tropical rain forests. *Plant. Cell Environ.* **25**, 963–973.

Melillo, J. M. *et al.* 2002. Soil warming and carbon-cycle feedbacks to the climate system. *Science* **298**, 2173–2176.

Mesquita, R., Delamonica, P., and Laurance, W. F. 1999. Effects of surrounding vegetation on edge-related tree mortality in Amazonian forest fragments. *Biol. Conserv.* **91**, 129–134.

Miller, D. R., Lin, J. D., and Lu, Z. 1991. Some effects of surrounding forest canopy architecture on the wind field in small clearings. *For. Ecol. Manage.* **45**, 79–91.

Minnich, R. A. 2001. An integrated model of two fire regimes. *Conserv. Biol.* **15**, 1549–1553.

Monnin, E., Indermühle, A., Dällenbach, A., Flückiger, J., Stauffer, B., Stocker, T. F. *et al.* 2001. Atmospheric CO_2 concentrations over the last glacial termination. *Science* **291**, 112–114.

Mooney, H. A., Bullock, S. H., and Medina, E. 1995. Introduction. In *Seasonally Dry Tropical Forests* (eds. S. H. Bullock, H. A. Mooney, and E. Medina), pp. 1–8. Cambridge: Cambridge University Press.

Mora, G. and Pratt, L. M. 2001. Isotopic evidence for cooler and drier conditions in the tropical Andes during the last glacial stage. *Geology* **29**, 519–522.

Moraes, J. L., Cerri, C. C., Mellilo, J., Kicklighter, D., Neill, C., Skole, D. *et al.* 1995. Soil carbon stocks of the Brazilian Amazon basin. *J. Am. Soil. Sci. Soc.* **59**, 244–247.

Morgan, J. A., Pataki, D. E., Körner, Ch., Clark, H., Del Grosso, S. J., Grünzweig, J. M. *et al.* 2004. Water relations in grassland and desert ecosystems exposed to elevated atmospheric CO_2. *Oecologia* (in press).

Mori, S. and Becker, P. 1991. Flooding affects survival of Lecythidaceae in terra-firme forest near Manaus, Brazil. *Biotropica* **23**, 87–90.

Moritz, C., Patton, J. L., Schneider, C. J., and Smith T. B. 2000. Diversification of rainforest faunas: an integrated molecular approach. *Annu. Rev. Ecol. Syst.* **31**, 553–563.

Morley, R. J. 2000. *Origin and Evolution of Tropical Rain Forests*, Chichester: John Wiley and Sons. p. 362, 378.

Mourguiart, P. and Ledru, M.-P. 2003. Last glacial maximum in an Andean cloud forest environment (eastern Cordillera, Bolivia). *Geology* **31** (3), 195–198.

Moy, C. M., Seltzer, G. O., Rodbell, D. T., and Anderson, D. M. 2002. Variability of El Niño/Southern Oscillation activity at millennial timescales during the Holocene epoch. *Nature* **420**, 162–165.

Muller, J. 1981. Fossil pollen records of extant angiosperms. *Bot. Rev.* **47**, 1–142.

Murphy, P. and Lugo, A. E. 1986. Ecology of tropical dry forest. *Ann. Rev. Ecol. Syst.* **17**, 67–88.

Myers, N. and Knoll, A. H. 2001. The biotic crisis and the future of evolution. *Proc. Natl. Acad. Sci. USA* **98**, 5389–5392.

Nakagawa, M. *et al.* 2000. Impact of severe drought associated with the 1997–1998 El Niño in a tropical forest in Sarawak. *J. Trop. Ecol.* **16**, 355–367.

Nascimento, F. E. M. and Laurance, W. F. 2002. Total aboveground biomass in central Amazonian rainforests: a landscape-scale study. *For. Ecol. Manage.* **168**, 311–321.

Nascimento, H. E. M. and Laurance, W. F. 2004. Biomass dynamics in Amazonian forest fragments. *Ecol. Appl.* **14**, S127–S138.

Nelson, B. W. 1994. Natural forest disturbance and change in the Brazilian Amazon. *Remote Sens. Rev.* **10**, 105–125.

Nelson, B. W. 2005. Commentary on: W. F. Laurance, *et al.* Pervasive alteration of tree communities in undisturbed Amazonian forests. *Biotropica*, 37.

Nelson, B. W., Ferreira, C. A. C., da Silva, M. F., and Kawasaki, M. L. 1990. Endemism centres, refugia and botanical collection density in the Brazilian Amazonia. *Nature* **345**, 714–716.

Nelson, B. W., Kapos, V., Adams, J., Oliveira, W., Braun, O., and do Amaral, I. 1994. Forest disturbance by large blowdowns in the Brazilian Amazon. *Ecology* **75**, 853–858.

Nemani, R. R., Keeling, C. D., Hashimoto, H., Jolly, W. M., Tucker, C. J., Myneni, R. B. *et al.* 2003. Climate driven increases in global terrestrial net primary production from 1982 to 1999. *Science* **300**, 1560–1563.

Nepstad, D. C., Carvalho, C., Davidson, E., Jipp, P., Lefebre, P., Negreiros, P. *et al.* 1994. The role of deep roots in the hydrological and carbon cycles of Amazonian forests and pastures. *Nature* **372**, 666–669.

Nepstad, D. C., Moreira, A. G., and Alencar, A. A. 1999a. *Flames in the Rain Forest: Origins, Impacts, and Alternatives to Amazonian fires*. Brasilia, Brazil: Pilot Program to Preserve the Brazilian Rain Forest.

Nepstad, D. C., Verissimo, A., Alencar, A., Nobre, C., Lima, E., Lefebvre, P. *et al.* 1999b. Large-scale impoverishment of Amazonian forests by logging and fire. *Nature* **398**, 505–508.

Nepstad, D., Carvalho, G., Barros, A. C., Alencar, A., Capobianco, J. P., Bishop, J. *et al.* 2001. Road paving, fire regime feedbacks, and the future of Amazon forests. *For. Ecol. Manage.* **154**, 395–407.

Nepstad, D., McGrath, D., Alencar, A., Barros, A. C., Carvalho, G., Santilli, M. *et al.* M. D. V. 2002a. Frontier governance in Amazonia. *Science* **295**, 629–630.

Nepstad, D., McGrath, D., Alencar, A., Barros, C., Carvalho, G., Santilli, M. *et al.* 2002b. Issues in Amazonian development—Response. *Science* **295**, 1643–1644.

Nepstad, D. C. *et al.* 2004. Amazon drought and its implications for forest flammability and tree growth: a basin-wide analysis. *Glob. Change Biol.* **10**, 704–717.

New, M., Hulme, M., and Jones, P. 1999. Representing twentieth-century space-time climate variability. Part I: development of a 1961–90 mean monthly terrestrial climatology. *J. Clim.* **12**(3), 829–856.

New, M., Hulme, M., and Jones, P. 2000. Representing twentieth-century space-time climate variability. Part II: development of 1901–96 monthly grids of terrestrial surface climate. *J. Clim.* **13**(13), 2217–2238.

New M., Todd, M., Hulme, M. and Jones, P. D. 2001. Precipitation measurements and trends in the twentieth century. *Int. J. Climatol.* **21**, 1899–1922.

Newbery, D. M., Kennedy, D. N., Petol, G. H., Madani, L., and Ridsdale, C. E. 1999. Primary forest dynamics in lowland dipterocarp forest at Danum Valley, Sabah, Malaysia, and the role of the understorey. *Phil. Trans. R. Soc. Lond. B* **354**, 1763–1782.

Nicholson S. E., Some B., and Kone B. 2000. An analysis of recent rainfall conditions in West Africa, including the rainy seasons of the 1997. El Niño and the 1998. La Nina years. *J. Clim.* **13**, 2628–2640.

Noble, I. R., Apps, M., Houghton, R. A., Lashof, D., Makundi, W., Murdiyarso, *et al.* 2000. *IPCC Special Report on Land Use, Land-Use Change and Forestry*, chapter 2: Implications of different definitions and generic issues. Cambridge University Press, Cambridge, UK.

Nobre, C. A., Sellers, P. J., and Shukla, J. 1991. Amazonian deforestation and regional climate change. *J. Clim.* **4**, 957–988.

Norby, R. J., Todd, D. E., Fults, J., and Johnson, D. W. 2001. Allometric determination of tree growth in a CO_2-enriched sweetgum stand. *New Phytol.* **150**, 477–487.

Norby, R. J. *et al.* 2002. Net primary productivity of a CO_2-enriched deciduous forest and the implications for carbon storage. *Ecol. Appl.* **12**, 1261–1266.

O'Brien, T. G., Kinnaird, M. F., Nurcahyo, A., Prasetyaningrum, M., and Iqbal, M. 2003. Fire, demography and the persistence of siamang (*Symphalangus syndactylus*: Hylobatidae) in a Sumatran rainforest. *Animal conserv.* **6**, 115–121.

Oliveira, de, A. A. and Mori, S. A. 1999. A central Amazonian terra firme forest. I. High tree species richness on poor soils. *Biodivers. Conserv.* **8**, 1219–1244.

Oliveira-Filho, A. T., de Mello, J., and Scolforo, J. 1997. Effects of past disturbance and edges on tree community structure and dynamics within a fragment of tropical semideciduous forest in south-eastern Brazil over a five-year period (1987–1992). *Plant Ecol.* **131**, 45–66.

Olson, J. S., Watts, J. A., and Allison, L. J. 1983. *Carbon in Live Vegetation of Major World Ecosystems*. Environ-mental Sciences Division publication # 1997. ORNL-5862, Oak Ridge National Laboratory, Oak Ridge.

Oren, R., Sperry, J. S., Katul, G. G., Pataki, D. E., Ewers, B. E., Phillips, N. *et al.* 1999. Survey and synthesis of intra- and interspecific variation in stomatal sensitivity to vapour pressure deficit. *Plant Cell Environ* **22**, 1515–1526.

Oren, R., Ellsworth, D. S., Johnsen, K. H., Phillips, N., Ewers, B. E., Maier, C. *et al.* 2001. Soil fertility limits carbon sequestration by forest ecosystems in a CO_2-enriched atmosphere. *Nature* **411**, 469–472.

Overman, J. P. M., Witte, H. J. L., and Saldarriaga, J. G. 1994. Evaluation of regression models for above-ground biomass determination in Amazon rainforest. *J. Trop. Ecol.* **10**, 207–218.

Paduano, G. M., Bush, M. B., Baker, P. A., Fritz, S. L., and Seltzer, G. O. 2003. The Late Quaternary vegetation history of Lake Titicaca, Peru/Bolivia. *Palaeogeogr. Palaeoclimatol. Palaeoecol.* **194**, 259–279.

Pagani, M., Freeman, K. H., and Arthur, M. A. 1999. A Late Miocene atmospheric CO_2 concentrations and expansion of C4 grasses. *Science* **285**, 876–879.

Palmer, J. R. and Totterdell, I. J. 2001. Production and export in a global ocean ecosystem model. *Deep-Sea Res.* **48**, 1169–1198.

Parmesan, C. and Yohe, G. 2003. A globally coherent fingerprint of climate change impacts across natural systems. *Nature* **421**, 37–42.

Parresol, B. R. 1999. Assessing tree and stand biomass: a review with examples and critical comparisons. *For. Sci.* **45**, 573–593

Paruelo, J. M., Garbulsky, M., Guerschman, J., and Jobbágy, E. 2004. Two decades of normalized difference vegetation index change in South America: identifying the imprint of global change. *Int. J. Remote Sens.* **20**, 2793–2806.

Passioura, J. B. 1988. Water transport in and to roots. *Annu. Rev. Plant. Physiol. Plant. Mol. Biol.* **39**, 245–265.

Pearson, P. N. and Palmer, M. R. 2000. Atmospheric carbon dioxide concentrations over the past 60 million years. *Nature* **406**, 695–699.

Pennington, R. T., Prado, D. E., and Pendry, C. A. 2000. Neotropical seasonally dry forests and Quaternary vegetation changes. *J. Biogeogr.* **27**, 261–273.

Pennington, R. T., Lavin, M., Prado, D. E., Pendry, C. A., Pell, S., and Butterworth, C. 2004. Historical climate change and speciation: neotropical seasonally dry forest plants show patterns of both Tertiary and Quaternary diversification. *Phil. Trans. R. Soc. Lond. B* **359**, 515–538.

Peres, C. A. 1999. Ground fires as agents of mortality in a Central Amazonian forest. *J. Trop. Ecol.* **15**, 535–541.

Peres, C. A. 2001. Paving the way to the future of Amazonia. *Trends in Ecol. Evol.* **16**, 217–219.

Peres, C. A., Barlow, J., and Haugaasen, T. 2003. Vertebrate responses to surface fires in a Central Amazonian forest. *Oryx* **37**, 97–109.

Pessenda, L. C. R., Gomes, B. M., Aravena, R., Ribeiro, A. S., Boulet, R., and Gouveia, S. E. M. 1998. The carbon isotope record in soils along a forest-cerrado ecosystem transect: implication for vegetation changes in Rondônia State, southwestern Brazilian Amazon region. *The Holocene* **8**, 631–635.

Pepin, S., Kürner, Ch. 2002. Web-FACE: a new canopy free-air CO_2 enrichment system for tall trees in mature forests. *Oecologia*, **133**, 1–9.

Petit, J. R., Jouzel, J., Raynaud, D., Barkov, N. I., Barnola, J. M., Basile, I. *et al.* 1999. Climate and atmospheric history of the past 420,000 years from the Vostok ice core, Antarctica. *Nature* **399**, 429–436.

Phillips, O. L. 1995. Evaluating turnover in tropical forests. *Science* **268**, 894–895.

Phillips, O. L. 1996. Long-term environmental change in tropical forests: increasing tree turnover. *Environ. Conserv.* **23**, 235–248.

Phillips O. L. and Baker T. R. 2002. Field manual for plot establishment and remeasurement. November 2002. version; see www.geog.leeds.ac.uk/projects/rainfor/.

Phillips, O. L. and Gentry, A. H. 1994. Increasing turnover through time in tropical forests. *Science* **263**, 954–958.

Phillips, O. L. and Sheil, D. 1997. Forest turnover, diversity and CO_2. *Trends Ecol. Evol.* **12**, 404.

Phillips, O. L., Hall, P., Gentry, A. H., Sawyer, S. A., and Vásquez, R. 1994. Dynamics and species richness of tropical forests. *Proc. Natl. Acad. Sci. U.S.A.* **91**, 2805–2809.

Phillips, O. L., Hall, P., Sawyer, S. A., and Vásquez, R. 1997. Species richness, tropical forest dynamics and sampling: response to Sheil. *Oikos* **79**, 183–187.

Phillips, O. L., Nuñez, V. P., and Timaná, M. 1998a. Tree mortality and collecting botanical voucher in tropical forests. *Biotropica* **30**, 298–305.

Phillips, O. L., Malhi, Y., Higuchi, N., Laurance, W. F., Nuñez, P. V., Vásquez, R. M. *et al.* 1998b. Changes in the carbon balance of tropical forest: evidence from long-term plots. *Science* **282**, 439–442.

Phillips, O. L., Malhi, Y., Vinceti, B., Baker, T., Lewis, S. L., Higuchi, N. *et al.* 2002a. Changes in the biomass of tropical forests: evaluating potential biases. *Ecol. Appl.* **12**, 576–587.

Phillips, O. L., Martinez, R. V., Arroyo, L., Baker, T. R., Killeen, T., Lewis, S. L. *et al.* 2002b. Increasing dominance of large lianas in Amazonian forests. *Nature* **418**, 770–774.

Phillips, O. L., Baker, T. R., Arroyo, L., Higuchi, N., Killeen, T., Laurance, W. F. *et al.* 2004. Pattern and process in Amazon tree turnover: 1976–2001. *Phil. Trans. R. Soc. Lond. B*, **359**, 381–408.

Pielke, R. A., Marland, G., Betts, R. A., Chase, T. N., Eastman, J. L., Niles, J. O. *et al.* 2002. The influence of land-use change and landscape dynamics on the climate system: relevance to climate-change policy beyond the radiative effect of greenhouse gasses. *Phil. Trans. R. Soc. Lond.* **360**, 1705–1719.

Pinard, M. A. and Huffman, J. 1997. Fire resistance and bark properties of trees in a seasonally dry forest in eastern Bolivia. *J. Trop. Ecol.* **13**, 727–740.

Pinard, M. A., Putz, F. E., and Licona, J. C. 1999. Tree mortality and vine proliferation following a wildfire in a subhumid tropical forest in eastern Bolivia. *For. Ecol. Manage.* **116**, 247–252.

Pinot, S., Ramstein, G., Harrison, S. P., Prentice, I. C., Guiot, J., Stute, M. *et al.* 1999. Tropical paleoclimates at the Last Glacial Maximum: comparison of Paleoclimate Modelling Intercomparison Project (PMIP) simulations and paleodata. *Clim. Dyn.* **15**, 857–874.

Piperno, D. R. and Becker, P. 1996. Vegetation history of a site in the central Amazon Basin derived from phytolith and charcoal records from natural soils. *Quatern. Res.* **45**, 202–209.

Poorter, H. and Perez-Soba, M. 2001. The growth response of plants to elevated CO_2 under non-optimal environmental conditions. *Oecologia* **129**, 1–20.

Potter, C., Klooster, S., de Carbalho, C. R., Genovese, V. B., Torregrosa, A., Dungan, J. *et al.* 2001. Modeling seasonal and interannual variability in ecosystem carbon cycling for the Brazilian Amazon region. *J. Geophys. Res. Atmos.* **106**, 10423–10446.

Potter, C., Klooster, S., Steinbach, M., Tan, P. N., Kumar, V., Shekhar, S. *et al.* 2004. Understanding global teleconnections of climate to regional model estimates of Amazon ecosystem carbon fluxes. *Glob. Change Biol.* **10**, 693–703.

Potts, M. D. 2003. Drought in a Bornean everwet rainforest, *J. Ecol.* **91**(3), 467–474.

Prado, D. E. and Gibbs, P. E. 1993. Patterns of species distributions in the dry seasonal forests of South America. *Ann. M. Bot. Gard.* **80**, 902–927.

Prentice, I. C., Cramer, W., Harrison, S. P., Leemans, R., Monserud, R., and Solomon, A. M. 1992. A global biome model based on plant physiology and dominance, soil properties and climate. *J. Biogeogr.* **19**, 117–134.

Prentice, I. C., Harrison, S. P., Jolly, D., and Guiot, J. 1998. The climate and biomes of Europe at 6000 yr BP:

comparison of model simulations and pollen-based reconstructions. *Quatern. Sci. Rev.* **17**, 659–668.

Prentice, I. C., Jolly, D., and BIOME6000 participants. 2000. Mid-Holocene and glacial-maximum vegetation geography of the northern continents. *J. Biogeogr.* **27**, 507–519.

Prentice, I. C., Farquhar, G. D., Fasham, M. J. R., Goulden, M. L., Heimann, M., Jaramillo, V. J. *et al.* 2001. The carbon cycle and atmospheric carbon dioxide, chapter 3. In *IPCC Climate Change 2001, Working Group I, The scientific basis*. Cambridge University Press, Cambridge, UK.

Procopio, A. S., Artaxo, P., Kaufman, Y., Remer, L., Schafer, J., and Holben, B. 2004. Multiyear analysis of Amazonian biomass burning smoke radiative forcing of climate. *Geophys. Res. Lett.* **31**, L03108, doi:10.1029/2003GL018646.

Pyke, C. R., Condit, R., Aguilar, S., and Lao, S. 2001. Floristic composition across a climatic gradient in a neotropical lowland forest. *J. Veg. Sci.* **12**, 553–566.

Ranta, P., Blom, T., Niemela, J., Joensuu, E., and Siitonen, M. 1998. The fragmented Atlantic rain forest of Brazil: size, shape and distribution of forest fragments. *Biodiv. Conserv.* **7**, 385–403.

Ratter, J. A., Ribeiro, J. F., and Bridgewater, S. 1997. The Brazilian cerrado vegetation and threats to its biodiversity. *Ann. Bot.* **80**, 223–230.

Reekie, E. G. and Bazzaz, F. A. 1989. Competition and patterns of resource use among seedlings of five tropical trees grown at ambient and elevated CO_2. *Oecologia* **79**, 212–222.

Retallack, G. J. 2001a. A 300-million-year record of atmospheric carbon dioxide from fossil plant cuticles. *Nature* **411**, 287–290.

Retallack, G. J. 2001b. Cenozoic expansion of grasslands and climatic cooling. *J. Geol.* **109**, 407–426.

Rice, A. H., Pyle, E. H., Saleska, S. R., Hutyra, L., de Camargo, P. B., Portilho, K. *et al.* 2004. Carbon balance and vegetation dynamics in an old-growth Amazonian forest. *Ecol. Applic.* **14**(4), S55–S71.

Richards, P. W. 1996. *The Tropical Rain Forest, 2nd Ed*, Cambridge University Press, Cambridge, p. 575.

Richardson, J. E., Pennington, R. T., Pennington, T. D., and Hollingsworth, P. M. 2001. Rapid Diversification of a Species-Rich Genus of neotropical rainforest trees. *Science*, **293**, 2242–2245.

Roberts, S. J. (2000). Tropical fire ecology. *Prog. Phys. Geogr.* **24**(2), 281–288.

Robinson, J. G. and Bennett, E. L. (eds.) 2000. *Hunting for Sustainability in Tropical Forests*. New York: Columbia University Press.

Roosevelt, A. C., Lima da Costa, M., Lopes Machado, C., Michab, M., Mercier, N., Valladas, H. *et al.* 1996. Paleoindian cave dwellers in the Amazon: the peopling of the Americas. *Science* **272**, 373–384.

Rosenfeld, D. 1999. TRMM observed first direct evidence of smoke from forest fires inhibiting rainfall. *Geophys. Res. Lett.* **26**, 3105–3108.

Sage, R. F. 1995. Was low atmospheric CO_2 during the Pleistocene a limiting factor for the origin of agriculture? *Glob. Change Biol.* **1**, 93–106.

Sage, R. F. and Coleman, J. R. 2001. Effects of low atmospheric CO_2 on plants: more than a thing of the past. *Trends Plant Sci.* **6**, 18–24.

Sage, R. F. and Cowling, S. A. 1999. Implications of stress in low CO_2 atmospheres of the past: are today's plants too conservative for a high CO_2 world. In *Carbon dioxide and Environmental Stress* (eds. Y. Luo and H. A. Mooney), pp. 289–304. San Diego, CA: Academic Press.

Saleh, C. 1997. Wildlife survey report from burned and unburned forest areas in central Kalimantan. Unpublished report, WWF Indonesia programme.

Saleska, S. R. *et al.* 2003. Carbon in Amazon forests: unexpected seasonal fluxes and disturbance-induced losses. *Science* **302**, 1554–1557.

Salo, J., Kalliola, R., Hakkinen, I., Makinen, Y., Niemela, P., Puhakka, M. *et al.* 1986. River dynamics and the diversity of Amazon lowland forest. *Nature* **322**, 254–258.

Sanderson, M. J. 1997. A nonparametric approach to estimating divergence times in the absence of rate constancy. *Mol. Biol. Evol.* **14**, 1218–1231.

Sanderson, M. J. 2001. *R8s, version 1*(beta), User's manual (June 2001) University of California, Davis: distributed by M. J. Sanderson. http://ginger.ucdavis.edu/r8s/.

Sanderson, M. J. 2002. Estimating absolute rates of molecular evolution and divergence times: a penalized likelihood approach. *Mol. Biol. Evol.* **19**, 101–109

Sanford, R. L., Saldarriaga, J., Clark, K. E., Uhl, C., and Herrera, R. 1985. Amazon rain-forest fires. *Science* **227**, 53–55.

Sanford, R. L., Jr., Braker, H. E., and Hartshorn, G. S. 1986. Canopy openings in a primary neotropical lowland forest. *J. Trop. Ecol.* **2**, 277–282.

Santos, J. 1996. Análise de modelos de regressao para estimar a fitomassa de floresta tropical umida de terra-firme da Amazonia Central. Minas Gerais: Universidade Federal de Vicosa.

Santos, G. M., Gomes, P., Anjos, R., Cordeiro, R., Turcq, B., Sifeddine, A. *et al.* 1996. ^{14}C AMS dating of fires in the central Amazon rain forest. *Nucl. Instr. Meth. Phys. Res. B* **172**, 761–766.

Saugier, B. 1983. Plant growth and its limitations in crops and natural communities. In *Disturbance and Ecosystems: Components of Response* (eds. H. A. Mooney and M. Godron), pp. 159–174. Berlin: Springer.

Savill, P. S. 1983. Silviculture in windy climates. *For. Abstr.* **44**, 473–488.

Saxe, H., Ellsworth, D. S., and Heath, J. 1998. Tree and forest functioning in an enriched CO_2 atmosphere. *New Phytol.* **139**, 395–436.

Scariot, A. 2001. Seedling mortality by litterfall in Amazonian forest fragments. *Biotropica* **32**, 662–669.

Schenk, H. J. and Jackson, R. B. 2003. The global biogeography of roots. *Ecol. Monogr.* **72**, 311–328.

Schneider, H., Schuettpelz, E., Pryer, K. M., Cranfill, R., Magallon, S., and Lupia, R. 2004. Ferns diversified in the shadow of angiosperms. *Nature* **428**, 553–557.

Schneider, R. R., Arima, E., Veríssimo, A., Barreto P., and Souza, C., Jr. 2000. *Amazônia sustentável: limitantes e oportunidades para o desenvolvimento rural*. Brasilia, Brazil: The World Bank.

Schnitzer, S. A. and Bongers, F. 2002. The ecology of lianas and their role in forests. *Trends Ecol. Evol.* **17**, 223–230.

Sellers, P. J., bounoua, L., Collatz, G. J. *et al.* 1996. Comparison of radiation and physiological effects of doubled atmospheric CO_2 on climate. *Science* **271**, 1402–1406.

Seltzer, G., Rodbell, D., and Burns, S. 2000. Isotopic evidence for late Quaternary climatic change in tropical South America. *Geology* **28**(1), 35–38.

Seltzer, G. O., Rodbell, D. T., Baker, P. A., Fritz, S. C., Tapia, P. M., Rowe, H. D. *et al.* 2002. Early warming of tropical South America at the last glacial–interglacial transition. *Science* **296**, 1685–1686.

Sheil, D. 1995a. Evaluating turnover in tropical forests. *Science* **268**, 894.

Sheil, D. 1995b. A critique of permanent plot methods and analysis with examples from Budongo forest, Uganda. *For. Ecol. Manage.* **77**, 11–34.

Sheil, D. 2003. Observations of long-term change in an African rain forest. In *Long-term Changes in Composition and Diversity as a Result of Natural and Man Made Disturbances: Case Studies from the Guyana Shield, Africa, Borneo and Melanesia*, Tropenbos series 22 (ed. H. ter Steege), pp. 37–59. Wageningen, The Netherlands: Tropenbos.

Sheil, D. and Burslem, D. F. R. P. 2003. Disturbing hypotheses in tropical forests. *Trends Ecol. Evol.* **18**, 18–26.

Sheil, D. and May, R. M. 1996. Mortality and recruitment rate evaluations in heterogeneous tropical forests. *J. Ecol.* **84**, 91–100.

Sheil, D., Burslem, D. F. R. P., and Alder, D. 1995. The interpretation and misinterpretation of mortality-rate measures. *J. Ecol.* **83**, 331–333.

Shukla, J., Nobre, C., and Sellers, P. 1990. Amazon deforestation and climate change. *Science* **247**, 1322–1326.

Shuttleworth, W. J. 1989. Micrometeorology of temperate and tropical forest. *Philosophical Transactions of the Royal Society of London Series, B-Biological Sciences*, **324**, 299–334.

Siegert, F., Ruecker, G., Hinrichs, A., and Hoffmann, A. A. 2001. Increased damage from fires in logged forests during droughts caused by El Nino. *Nature* **414**, 437–440.

Sifeddine, A., Martin, L., Turcq, B., Volkmer-Ribeiro, C., Soubies, F., Cordeiro, R. C. *et al.* 2001. Variations of the Amazon rainforest environment: a sedimentological record covering 30,000 years. *Palaeogeogr. Palaeoclimatol. Palaeoecol.* **168**, 221–235.

Sifeddine, A., Albuquerque, A. L. S., Ledru, M.-P., Turcq, B., Knoppers, B., Martin, L. *et al.* 2003. A 21,000 cal years paleoclimatic record from Caco Lake, northern Brazil: evidence from sedimentary and pollen analyses. *Palaeogeogr. Palaeoclimatol. Palaeoecol.* **189**, 25–34.

Silva Dias, M. A. F., Williams, E., Pereira, L., Pereira Filho, A., and Matsuo, P. 2002. Shallow convection response to land use and topography in SW Amazon in the wet season. *J. Geophys. Res.*

Silva Dias, P. L. and Regnier, P. 1996. Simulation of mesoscale circulations in a deforested area of Rondônia in the dry season. In *Amazonian Deforestation and Climate*. eds. J. Gash, C. Nobre, J. Roberts, and R. Victoria pp. 531–547. San Francisco, CA: John Wiley and Sons.

Silver W. L. 1994. Is nutrient availability related to plant nutrient use in humid tropical forests. *Oecologia*, **98**, 336–343.

Sitch, S., Smith, B., Prentice, I. C., Arneth, A., Bondeau, A., Cramer, W. *et al.* (2003). Evaluation of ecosystem dynamics, plant geography and terrestrial carbon cycling in the LPJ Dynamic Global Vegetation Model. *Glob. Change Biol.* **9**, 161–185.

Sizer, N. and Tanner, E. V. J., 1999. Responses of woody plant seedlings to edge formation in a lowland tropical rainforest, Amazonia. *Biol. Conserv.* **91**, 135–142.

Sizer, N. C., Tanner, E. V. J., and Kossman Ferraz, I. D. 2000. Edge effects on litterfall mass and nutrient concentrations in forest fragments in central Amazonia. *J. Trop. Ecol.* **16**, 853–863.

Skole, D. and Tucker, C. J. 1993. Tropical deforestation and habitat fragmentation in the Amazon: satellite data from 1978 to 1988. *Science* **260**, 1905–1910.

Slik, J. W. F. 2004. El Niño droughts and their effects on tree species composition and diversity in tropical rain forests. *Oecologia* **141**, 114–120.

Slik, J. W. F., Verburg, R. W., and Kessler, P. J. A. 2002. Effects of fire and selective logging on the tree species composition of lowland dipterocarp forest in East Kalimantan, Indonesia. *Biodiver. Conserv.* **11**, 85–98.

Soepadmo, E. 1993. Tropical rain forests as carbon sinks. *Chemosphere* **27**, 1025–1039.

Sombroek, W. G. 1984. Soils of the Amazon region. In *The Amazon. Limnology and Landscape Ecology of a Mighty Tropical River and its Basin.* (H. Sioli ed.) pp. 521–535. Dordrecht: W. Junk.

Somerville, A. 1980. Wind stability: forest layout and silviculture. *N. Z. J. For. Sci.* **10**, 476–501.

Sommer, R., Sa, T. D. D., Vielhauer, K., de Araujo, A. C., Folster, H., and Vlek, P. L. G. 2002. Transpiration and canopy conductance of secondary vegetation in the eastern Amazon. *Agri. For. Meteor.* **112**, 103–121.

Spall, S. A., Jones, C. D., and Cox, P. M. 2003. Simulation of the climate-carbon cycle system at the Last Glacial Maximum, (in preparation).

Sperry, J. S., Hacke, U. G., Oren, R., and Comstock, J. P. 2002. Water deficits and hydraulic limits to leaf water supply. *Plant Cell Environment* **25**, 251–263.

Stanhill, G. and Cohen, S. 2001. Global dimming: a review of the evidence for a widespread and significant reduction in global radiation with discussion of its probable causes and possible agricultural consequences. *Agric. For. Meteorol.* **107**, 255–278.

Stape, J. L., Binkley, D., and Ryan, M. G. 2004. Eucalyptus production and the supply, use and efficiency of use of water, light and nitrogen across a geographic gradient in Brazil. *For. Ecol. Manage.* **193**, 17–31.

Stark, N. M. and Jordan, C. F. 1978. Nutrient retention by the root mat of an Amazonian rain forest. *Ecology* **59**, 434–437.

Stebbins, G. L. 1974. *Flowering Plants: Evolution above the Species Level.* Cambridge, MA: Belknap press, Harvard University Press.

Stratton, L. C., Goldstein, G., and Meinzer, F. C. 2000. Temporal and spatial partitioning of water resources among eight woody species in a Hawaiian dry forest. *Oecologia* **124**, 309–317.

Stuiver, M., Reimer, P. J., Bard, E., Beck, J. W., Burr, G. S., Hughen, K. A. *et al.* 1998. INTCAL98 radiocarbon age calibration, 24,000–0 cal BP. *Radiocarbon* **40**(3), 1041–1083.

Sud, Y., Yang, R., and Walker, G. 1996. Impact of in situ deforestation in Amazonia on the regional climate: general circulation model simulation study. *J. Geophys. Res.* **101**, 7095–7109.

Suzuki, A. 1988. The socioecological study of orangutans and forest conditions after the big forest fire and drought, 1983. In Kutai National Park, Indonesia. In *Occasional Paper No. 14, Research Center for the South Pacific, Kagoshima University* (ed. H. Tagawa and N. Wirawan), University of Kagoshima, Japan.

Telles, E. de C. C., de Camargo, P. B., Martinelli, L. A., Trumbore, S. E., da Costa, E. S., Santos, J. *et al.* 2003. Influence of soil texture on carbon dynamics and storage potential in tropical forest soils in Amazonia. 2003. *Glob. Biogeochem. Cycles* **17**, 1040–1051.

ter Steege, H. and Hammond, D. S. 2001. Character convergence, diversity, and disturbance in tropical rain forest in Guyana. *Ecology* **82**, 3197–3212.

Ter-Mikaelian, M. T. and Korzukhin, M. D. 1997. Biomass equations for sixty-five North American tree species. *For. Ecol. Manage.* **97**, 1–24.

Thomas, S. C. 1996. Asymptotic height as a predictor of growth and allometric characteristics in Malaysian rain forest trees. *Am. J. Bot.* **83**, 556–566.

Thomas, W. W. 1999. Conservation and monographic research on the flora of Tropical America. *Biodivers. Conserv.* **8**, 1007–1015.

Thompson, L. G., Mosley-Thompson, E., Davis, M. E., Lin, P.-N., Henderson, K. A., Cole-Dai, J. *et al.* 1995. Late glacial stage and Holocene tropical ice core records from Huascaran, Peru. *Science* **269**, 46–50.

Thompson, L. G., Davis, M. E., Mosley-Thompson, E., Sowers, T. A., Henderson, K. A., Zagorodnov, V. S. *et al.* 1998. 25,000 year tropical climate history from Bolivian ice cores, *Science* **282**, 1858–1864.

Thorpe, R. B., Gregory, J. M., Johns, T. C., Wood, R. A., and Mitchell, J. F. B. 2001. Mechanisms determining the Atlantic thermohaline circulation response to greenhouse gas forcing in a non-flux adjusted coupled climate model. *J. Climatol.* **14**, 3102–3116.

Tian, H. Q., Melillo, J. M., Kicklighter, J. M., McGuire, A. D., Helfrich, J. V. K., Moore, B. *et al.* 1998. Effect of interannual climate variability on carbon storage in Amazaonian ecosystems. *Nature* **396**, 664–667.

Timmerman, A., Oberhuber, J., Bacher, A., Esch, M., Latif, M., and Roeckner, E. 1999. Increased El Niño frequency in a climate model forced by future greenhouse warming. *Nature* **398**, 694–697.

Tognetti, R., Cherubini, P., and Innes, J. L. 2000. Comparative stem-growth rates of Mediterranean trees under background and naturally enhanced ambient CO_2 concentrations. *New Phytol.* **146**, 59–74.

Tomasella, J. and Hodnett, M. G. 1997. Estimating unsaturated hydraulic conductivity of Brazilian soils using soil-water retention data. *Soil Sci.* **162**, 703–712.

Torn, M. S., Trumbore, S. E., Chadwick, O. A., Vitousek, P. M., and Hendricks, D. M. 1997. Mineral control of soil organic carbon storage and turnover. *Nature* **389**, 170–173.

Trumbore, S. 2000. Age of soil organic matter and soil respiration: radiocarbon constraints on belowground C dynamics. *Ecol. Appl.* **10**, 399–401.

Trumbore, S. E., Davidson, E. A., Barbosa de Carmago, P., Nepstad, D. C., and Martinelli, L. A. 1995. Belowground cycling of carbon in forests and pastures of Eastern Amazonia. *Glob. Biogeochem. Cycles* **9**, 515–528.

Tudhope, A. W., Chilcott, C. P., McCulloch, M. T., Cook, E. R., Chappell, J., Ellam, R. M. *et al.* 2001. Variability in the El Niño-Southern Oscillation through a glacial-interglacial cycle. *Science* **291**, 1511–1517.

Turcq, B., Sifeddine, A., Martin, L., Absy, M. L., Soubies, F., Suguio, K. *et al.* 1998. Amazonia rainforest fires: a lacustrine record of 7000 years. *Ambio* **27**(2), 139–142.

Turton, S. M. 1992. Understorey light environments in a north-east Australian rain forest before and after a tropical cyclone. *J. Trop. Ecol.* **8**, 241–252.

Turton, S. M. and Freiberger, H. J. 1997. Edge and aspect effects on the microclimate of a small tropical forest remnant on the Atherton Tableland, northeastern Australia. In *Tropical Forest Remnants: Ecology, Management, and Conservation of Fragmented Communities*, (eds. W. F. Laurance, and R. O. Bierregaard), pp. 45–54. Chicago, IL: University of Chicago Press.

Uehara, G. and Gillman, G. P. 1981. *The mineralogy, Chemistry, and Physics of Tropical Soils with Variable Charge Clays*. Westview tropical agriculture series; no. 4. Boulder, Co: Westview Press.

Uhl, C. and Kauffman, J. B. 1990. Deforestation effects on fire susceptibility and the potential response of the tree species to fire in the rainforest of the eastern Amazon. *Ecology* **71**, 437–449.

Uhl, C., Kauffman, J. B., and Cummings, D. L. 1988. Fire in the Venezuelan Amazon 2: environmental conditions necessary for forest fires in the evergreen rainforest of Venezuela. *Oikos*, **53**, 176–184.

Uhl, C., Verissimo, A., Mattos, M. M., Brandino, Z., and Vieira, I. C. G. 1991. Social, economic, and ecological consequences of selective logging in an Amazon frontier—the case of Tailandia. *For. Ecol. Manage.* **46**, 243–273.

Upchurch, G. R. and Wolfe, J. A. 1987. Mid-Cretaceous to Early Tertiary vegetation and climate: evidence from fossil leaves and woods. In *The Origins of Angiosperms and Their Biological Consequences*. E. M. Friis, W. G. Chaloner, and P. H. Crane, pp. 75–105. Cambridge University Press.

Van der Hammen. T. and Absy, M. L. 1994. Amazonia during the last glacial. *Palaeogeogr. Palaeoclimatol. Palaeoecol.* **109**, 247–261.

Vasconcelos, H. L. and Luizão, F. J. 2004. Litter production and litter nutrient concentrations in a fragmented Amazonian landscape. *Ecol. Appl.* **14**, 884–892.

Vellinga, M. and Wood, R. 2001. Global climatic impacts of a collapse of the Atlantic thermohaline circulation. *Clim. Change* (submitted).

Vellinga, M., Wood, R., and Gregory, J. 2001. Processes governing the recovery of a perturbed thermohaline circulation in HadCM3. *J. Clim.* (in press).

Verissimo, A., Cochrane, M. A., and Souza, C. 2002. National forests in the Amazon. *Science* **297**, 1478–1478.

Vertessy, R. A., Benyon, R. G., Osullivan, S. K., and Gribben, P. R. 1995. Relationships between stem diameter, sapwood area, leaf area and transpiration in a young mountain ash forest. *Tree Physiol.* **15**, 559–567.

Viana, V. M., Tabanez, A. A. and Batista, J. 1997. Dynamics and restoration of forest fragments in the Brazilian Atlantic moist forest. In *Tropical Forest Remnants: Ecology, Management, and Conservation of Fragmented Communities* (eds. W. F. Laurance and R. O. Bierregaard), pp. 351–365. Chicago, IL: University of Chicago Press.

Victoria R. L., Martinelli L. A., Moraes J. M., Ballester M. V., Krusche A. V., Pellegrino G. *et al.* 1998. Surface air temperature variations in the Amazon region and its borders during this century. *J. Clim.* **11**, 1105–1110.

Vitousek, P. M. 1984. Litterfall, nutrient cycling and nutrient limitation in tropical forests. *Ecology* **65**, 285–298.

Volk, M., Niklaus, P. A., and Körner, C. 2000. Soil moisture effects determine CO_2 responses of grassland species. *Oecologia* **125**, 380–388.

Walker, G. K., Sud, Y. C., and Atlas, R. 1995. Impact of ongoing Amazonian deforestation on local precipitation: a GCM simulation study. *Bull. Am. Meteor. Soc.* **76**, 346–361.

Wardlaw, I. F. 1990. The control of carbon partitioning in plants. Tansley Review No. 27. *New Phytol.* **116**, 341–381.

Watanabe, N., Evans, J. R., and Chow, W. S. 1994. Changes in the photosynthetic properties of Australian wheat cultivars over the last century. *Aust. J. Plant Physiol.* **21**, 169–183.

Weathers, K. C., Cadenasso, M. L., and Pickett, S. T. A. 2001. Forest edges as nutrient and pollution concentrators: potential synergisms between fragmentation, forest canopies, and the atmosphere. *Conserv. Biol.* **15**, 1506–1514.

Webb, L. J. 1958. Cyclones as an ecological factor in tropical lowland rain forest, North Queensland. *Aust. J. Bot.* **6**, 220–228.

Weishampel, J., Godin, J. R., and Henebry, G. M. 2001. Pantropical dynamics of 'intact' rain forest canopy texture. *Glob. Ecol. Biogeogr.* **10**, 389–397.

Weng, C., Bush, M. B., and Athens, J. S. 2002. Holocene climate change and hydrarch succession in lowland Amazonian Ecuador. *Rev. Palaeobot. Palynot.* **120**, 73–90.

Went, F. W. and Stark, N. M. 1968. Mycorrhiza. *BioScience* **18**, 1035.

Werth, D. and Avissar, R. 2002. The local and global effects of Amazon deforestation. *J. Geophys. Res.* **107**, (D20), 8087, doi: 10.1029/2001JD000717.

Whitmore T. and Burslem D. F. R. P. 1998. Major disturbances in tropical rainforests. In *Dynamics of Tropical Communities* (ed. D. M. Newbery, H. H. T. Prins. and N. D. Brown), pp. 549–565. Oxford: Blackwell Science.

Wielicki, B. A. *et al.* 2002. Evidence for large decadal variability in tropical mean radiative energy budget. *Science* **295**, 841–844.

Williams, M., Malhi, Y., Nobre, A. D., Rastetter, E. B., Grace, J., and Pereira, M. G. 1998. Seasonal variation in net carbon exchange and evapotranspiration in a Brazilian rain forest: a modelling analysis. *Plant Cell Environ.* **21**, 953–968.

Williamson, G. B. and Ickes, K. 2002. Mast fruiting and ENSO cycles—does the cue betray a cause? *Oikos* **97**, 459–461.

Williamson, G. B., Laurance, W. F. Oliveira, A., Delamonica, P., Gascon, C., Lovejoy, T. *et al.* 2000. Amazonian wet forest resistance to the 1997–98 El Niño drought. *Conserv. Biol.* **14**, 1538–1542.

Willis, K. J. and McElwain, J. C. 2002. *The Evolution of Plants*, p. 378. Oxford: Oxford University Press.

Wilson, M. F. and Henderson-Sellers, A. 1985. A global archive of land cover and soils data for use in general circulation climate models. *J. Clim.* **5**, 119–143.

Wing, S. L., Harrington, G. J., Bowen, G. J., and Koch, P. L. 2003. Floral change during the Initial Eocene Thermal Maximum in the Powder River Basin, Wyoming, Geological Society of America Special Paper 369, pp. 425–440.

Winter, K. and Lovelock, C. E. 1999. Growth responses of seedlings of early and late successional tropical forest trees to elevated atmospheric CO_2. *Flora* **194**, 221–227.

Winter, K., Garcia, M., Lovelock, C. E., Gottsberger, R., and Popp, M. 2000. Responses of model communities of two tropical tree species to elevated atmospheric CO_2: growth on unfertilized soil. *Flora* **195**, 289–302.

Winter, K., Aranda, J., Garcia, M., Virgo, A., and Paton, S. R. 2001a. Effect of elevated CO_2 and soil fertilization on whole-plant growth and water use in seedlings of a tropical pioneer tree, Ficus insipida Willd. *Flora* **196**, 458–464.

Winter, K., Garcia, M., Gottsberger, R., and Popp, M. 2001b. Marked growth response of communities of two

tropical tree species to elevated CO_2 when soil nutrient limitation is removed. *Flora* **196**, 47–58.

Wirawan, N. 1985. *Kutai National Park Management Plan 1985–1990*. WWF/IUCN, Bogor, Java, Indonesia.

Wolfe, J. A. 1990. Palaeobotanical evidence for a marked temperature increase following the Cretaceous/ Tertiary boundary. *Nature* **343**, 153–156.

Woodruff, D. S. 2001. Declines of biomes and biotas and the future of evolution. *Proc. Natl. Acad. Sci. USA* **98**, 5471–5476.

Woods, P. 1989. Effects of logging, drought, and fire on structure and composition of tropical forests in Sabah, Malaysia. *Biotropica* **21**, 290–298.

Wright, I. R., Gash, J. H. C., da Rocha, H. R., and Roberts, J. M. 1996. Modelling surface conductance for Amazonian pasture and forest. In *Amazonian Deforestation and Climate* (ed. J. H. C. Gash, C. A. Nobre, J. M. Roberts, and R. L. Victoria), San Francisco, CA: John Wiley and Sons.

Wright, S. J., Carrasco, C., Calderón, O., and Paton, S. 1999. The El Niño Southern Oscillation, variable fruit production, and famine in a tropical forest. *Ecology* **80**, 1632–1642.

Wright, S. J., Calderon, O., Hernandez, A., and Paton, S. 2004. Are lianas increasing in importance in tropical forests? A 17-year record from Panama. *Ecology* **85**, 484–489.

Würth, M. K. R., Winter, K., and Korner, C. 1998. In situ responses to elevated CO_2 in tropical forest understorey plants. *Funct. Ecol.* **12**, 886–895.

Würth, M. K. R., Winter, K., and Körner, C. 1998b. Leaf carbohydrate responses to CO_2 enrichment at the top of a tropical forest. *Oecologia* **116**, 18–25.

Yamakura, T., Hagihara, A., Sukardjo, S., and Ogawa H. 1986. Aboveground biomass of tropical rain forest stands in Indonesian Borneo. *Vegetation* **68**, 71–82.

Zachos, J. C., Pagani, M. N., Sloan, L., Thomas, E., and Billups, K. 2001. Trends, rhythms and aberrations in global climate 65 Ma to present. *Science* **292**, 686–693.

Zachos, J. C., Wara, M. W., Bohaty, S., Delaney, M. L., Petrizzo, M. R., Brill, A. *et al.* 2003. A transient rise in tropical sea surface temperature during the Paleocene-Eocene thermal maximum, *Science* **302**, 1551–1554, (10.1126/Science. 1090110).

Zhang, H., Henderson-Sellers, A., and McGuffie, K. 2001. The compounding effects of tropical deforestation and greenhouse warming on climate. *Clim. Change* **49**, 309–338.

Zobler, L. 1986. A world soil file for global climate modelling. NASA Technical Memorandum 87802, NASA.

Index